Results and Problems in Cell Differentiation

Series Editors:
W. Hennig, L. Nover, U. Scheer

30

Springer
Berlin
Heidelberg
New York
Barcelona
Hong Kong
London
Milan
Paris
Singapore
Tokyo

André M. Goffinet · Pasko Rakic (Eds.)

Mouse Brain
Development

With 69 Figures

 Springer

ANDRÉ M. GOFFINET
Neurobiology Unit
University of Namur Medical School
61, rue de Bruxelles
B-5000 Namur
Belgium

PASKO RAKIC
Section of Neurobiology
Yale University School of Medicine
333, Cedar Street
New Haven, CT 06510

ISSN 0080-1844
ISBN 978-3-642-53684-7

Library of Congress Cataloging-in-Pulication Data

Mouse brain development / André M. Goffinet, Pasko Rakic (eds.)
    p. cm. -- (Results and Problems in Cell Differentiation ; 30)
    Includes bibliographical references.
    ISBN 978-3-642-53684-7   ISBN 978-3-540-48002-0 (eBook)
    DOI 10.1007/978-3-540-48002-0

    1. Developmental neurobiology. 2. Mice as laboratory animals. I. Goffinet, A. II.
Rakic, Pasko, 1933- II. Series.

QP363.5 .M68 2000
573.8'619353--dc21

Springer-Verlag is a company in the BertelsmannSpringer publishing group.
© Springer-Verlag Berlin Heidelberg 2000
Softcover reprint of the hardcover 1st edition 2000

Cover design: Meta Design, Berlin
Typesetting: Scientific Publishing Services (P) Ltd., Madras
SPIN: 10681947   39/3136 - 5 4 3 2 1 0 - Printed on acid-free paper

Preface

Our understanding of the molecular mechanisms involved in mammalian brain development remains limited. However, the last few years have witnessed a quantum leap in our knowledge, due to technological improvements, particularly in molecular genetics. Despite this progress, the available body of data remains mostly phenomenological and reveals very little about the grammar that organizes the molecular dictionary to articulate a phenotype. Nevertheless, the recent progress in genetics will allow us to contemplate, for the first time, the integration of observation into a coherent view of brain development. Clearly, this may be a major challenge for the next century, and arguably is the most important task of contemporary developmental biology.

The purpose of the present book is to provide an overview that synthesizes up-to-date information on selected aspects of mouse brain development. Given the format, it was not possible to cover all aspects of brain development, and many important subjects are missing. The selected themes are, to a certain extent, subjective and reflect the interests of the contributing authors. Examples of major themes that are not covered are peripheral nervous system development, including myelination, the development of the hippocampus and several other CNS structures, as well as the developmental function of some important morphoregulatory molecules.

Although the use of the mouse as an animal model for studies of brain development is not new and was pioneered in the 1960s, mice did not become the favored animal model for studying brain development until the introduction of transgenic techniques (see Chap. 1 for historical perspectives). During the last few years, a number of new mutations have been generated. Initially, mutations were induced more or less randomly by transgenic inactivation. More recently, homologous recombination has been used to selectively inactivate specific genes. No doubt many mutants are available that have not yet been published, and it can be safely predicted that thousands will be generated during the next decade. In parallel, a large number of human genetic diseases that affect brain development are being characterized in molecular terms, and the pathophysiological understanding of each of them will require development studies in transgenic mice. Moreover, the validation of the Cre-loxP and Tet-on/off systems will allow modification of the temporal and cellular specificity of expression of any given form of a selected gene: the task ahead seems almost infinite! Thus, although basic developmental questions will remain to be tackled in comparatively simple models such as invertebrates (*Drosophila*, *C. elegans*) or in vertebrates such as *Xenopus* or zebrafish, the mouse will take central stage in the elucidation of brain development. However, the work in other mammals will also remain essential at some stage, particularly for developmental studies on the highly evolved structures like the cerebral cortex. The larger mammals such as carnivores and primates will probably be reserved to address specific developmental questions downstream of developmental genetic work carried out in simpler models and mice.

For all these reasons, the field of mouse developmental neurobiology will probably expand even further the next decade. For example, at many research institutions, the population of mice has quadrupled in the last 10 years, largely due to an increased demand for maintaining colonies of transgenic mice. The recently created International Behavioral and Neural Genetics Society (IBANGS) as well as the proliferation of international symposia, discussion forums on the Internet (e.g. http://www.bres-forum), and special issues in journals (e.g., *Brain Research*, *Cerebral Cortex*, *Molecular and Cellular Biology*) reflect this increased interest and phenomenal growth. Although various publications and web sites keep researchers informed, the problem of maintaining as well as cataloging the complex and uncoordinated nomenclature is formidable.

Given the gaps in the present book, which are mentioned above, 10 years from now the idea of writing a single volume on mouse brain development may sound unrealistic. So, this book is among the first and perhaps the last of such enterprises, and therefore may become a collector's item. However, before it is auctioned at Sotheby's, we hope it will find some use in the growing scientific community of dedicated colleagues who are working in this field. We wish to thank all of the contributors for taking a significant part of their time to help make this book possible. We also wish to thank the staff of Springer Verlag, particularly Ursula Gramm, for their understanding with some inevitable practical problems and for their most competent assistance.

André M. Goffinet and Pasko Rakic

Contents

Genetic Interactions During Hindbrain Segmentation
in the Mouse Embryo

Paul A. Trainor, Miguel Manzanares, and Robb Krumlauf 51

The Role of the p35/cdk5 Kinase in Cortical Development
Yong T. Kwon and Li-Huei Tsai   241

The Reelin-Signaling Pathway and Mouse Cortical Development
Isabelle Bar, Catherine Lambert de Rouvroit,
and André M. Goffinet   255

The Subpial Granular Layer in the Developing Cerebral Cortex
of Rodents
Gundela Meyer, Rafael Castro, José Miguel Soria, and Alfonso Fairén   277

From Spontaneous to Induced Neurological Mutations: A Personal Witness of the Ascent of the Mouse Model

Pasko Rakic[1]

1
Introduction

The mouse is arguably the most versatile animal model for studying the mechanisms of normal development and the pathogenesis of congenital malformations of the human nervous system. It does not offer the enormous potentials for manipulation by molecular genetics of small invertebrates, nor does it have the advantage of a large brain size and the similarity to the human that has been exploited by neuroanatomists and neurophysiologists in the nonhuman primates. However, as a small, relatively fast-reproducing mammal, with several well-defined inbred strains, it has became an unexcelled tool in modern developmental neurobiology. In addition, with the impeding completion of its genome sequence, the enlargement of the repertoire of spontaneous and induced mutations, combined with the creation of mosaic animals, the mouse has become an essential model system in almost every area of neurobiology and experimental neuropathology.

The individual chapters in this book, *Mouse Brain Development*, provide examples of the rapid growth of this field in the past three decades and highlight the ascent of the mouse as an increasingly favored experimental model system for the study of mammalian brain development. I will take the liberty of reviewing the history of this subject in a somewhat personal way. During the late 1960s, when I began to study brain development, the term "neuroscience" had just been coined. This magical concept of an interdisciplinary approach to these complex problems has inspired the transformation of research on the nervous system from a description of histological findings from postmortem human brain tissue or interpolation of its function from changes in an EEG pattern, to a highly sophisticated experimental discipline. I was introduced to developmental neurobiology by studying the cytoarchitectonic formation of the various brain structures using Yakovlev's collection of normal and pathological human fetal brain specimens (e.g. Rakic and Yakovlev 1968; Rakic and Sidman 1969, 1970). Although this

[1] Section of Neurobiology, Yale University School of Medicine, 333 Cedar Street, SHM, C303, New Haven, Connecticut 06510, USA Tel.: 203-785-5288, Fax: 203-785-5263

Results and Problems in Cell Differentiation, Vol. 30
Goffinet and Rakic (Eds.): Mouse Brain Development
© Springer-Verlag Berlin Heidelberg 2000

classical approach provided essential insight into the timing and general principles of human brain development, it also had serious limitations. Methodological advances in neuroanatomy, cell biology, molecular biology, and molecular genetics created a means to manipulate cellular events genetically, to study the basic molecular and cellular mechanisms involved in the formation of the nervous system.

I was lucky to benefit from participation as a scribe in some of the discussions about the advantages and disadvantages of various animal model systems at the early meetings of the Neuroscience Research Program (NRP), a multidisciplinary organization created by Francis O. Schmitt at MIT in the late 1960s. I remember eloquent and prophetic advocacy for the use of *C. elegans* and *Drosphila* by Sidney Brenner and Seymour Benzer, but also equally convincing arguments for some other candidate model systems, such as the *Daphnia magna*, made by Cyrus Levinthal, and the leech (*Hirudo medicinals*) by Gunther Stent, that in spite of specific advantages have eventually proved to be less used. Among the vertebrates, zebrafish (*Dario rerio*), presently a highly valued model system, was not, at the time, even contemplated. However, the mouse (*Mus musculus*) was from the start an attractive choice for the analysis of mammalian brain development. Since about one-third of all known mutants in mice had behavioral abnormalities due to disturbances of brain development, it was likely that a large portion of the genome in this species was devoted to building the nervous system. Furthermore, it was the only animal model that had all the major divisions of the human brain, including the cerebral neocortex. Therefore, for those researchers interested in normal and pathological development of the human brain, the selection of the mouse was an obvious, although not easy, choice, since the other model systems have some great advantages. In fact, as evident below, the research in mouse would not have advanced without input from the data and principles generated using other model systems.

In collaboration with Richard Sidman in the Department of Neuropathology at Harvard Medical School, I began analyses of selected neurological mutant mice that were available from the Jackson Laboratories at Bar Harbor. For me, the most fascinating mutants were the *reeler* and *weaver*, because the resulting abnormality in the homozygous mice appeared to involve a defect in neuronal migration, the subject of my main research interest which was inspired by autoradiographic and electronmicroscopic findings in human and non-human embryonic brains (Rakic and Sidman 1968, 1970; Rakic 1971, 1972). The initial results of our analyses were promising, and in fact the first description of changes in synaptic organization due to single gene deficits was not obtained from a simple model organism but, rather, was based on electronmicroscopic examination of these neurological mutant mice (Rakic and Sidman 1972, 1973a–c; reviewed in Rakic 1976; Caviness and Rakic 1978). It is not the purpose of this chapter to review this rapidly moving and changing field. Rather, I will provide selected examples from the early studies on spontaneous and induced mutations to provide a contrast to the present approaches used in induced gene manipulation and to illustrate the advantage of the mouse as a model system for studying mechanisms of normal and pathological development in the human brain.

2
Beginnings: The Values and Limits of Spontaneous Mutations

Among about 100 spontaneous mutations in mice that were known in the 1970s to affect the central nervous system (Sidman et al. 1965), two – *weaver* and *reeler* – were particularly suitable for the analysis of neuronal cell migration and the formation of synaptic connections. Their chromosomal locations were known: *weaver* mapped to chromosome 16 and *reeler* to chromosome 5. It was already established that both were autosomal recessive mutations carried on the same, C57BL/6J, genetic background (Falconer 1951; Hamburgh 1963; Lane 1964), and, thus, their phenotypes and timing and sequences of cellular events could be effectively compared. These two mutations were discovered because of visible tremor and locomotion abnormalities in mice, presumably due to the cerebellar malformation that was evident upon dissection. This is why they were initially attractive to me, although the changes in the cerebral cortex in *reeler* were equally intriguing. The power of electron microscopy, which had just become available for brain research (e.g. Peters et al. 1970) was particularly effective in establishing the three-dimensional cellular lattice of the neuronal organization and synaptic architecture of the cerebellar cortex (Eccles et al. 1967; Palay and Chan-Palay 1974). Finally, ^{3}H-thymidine autoradigraphy allowed the time and place of the origins of cells to be permanently marked and permitted cellular migration to be followed from the point of origin to the final destination (Sidman 1970). Thus, to examine the phenotypes of these mutants, the methods of choice were ^{3}H-thymidine labeling for the timing of cellular genesis, followed by Nissl stain for the description of cytoarchitectonic features, Golgi silver impregnation for the study of dendritic morphology, and electron microscopy for the analysis of cellular ultrastructure (for a description of the value of this strategy see Sotelo and Mariani 1999). I will here review the early studies of cerebellar development in *weaver* and *reeler* as examples of this approach and for comparison with the contemporary strategies used in induced mutations.

An examination of the cerebellar tissue of mice homozygous for the *weaver* mutant allele (*wv/wv*) showed that the majority of the granule cells degenerate at the inner border of the external granular layer prior to their incorporation into synaptic circuitry (Rakic and Sidman 1973a–c). The ultrastructural abnormalities found in the Bergmann glia fibers led us to propose that the primary effect in this mutation might be in preventing the

normal differentiation of the glial cell population, which then curtails neuronal migration and, thus, through a cascade of cellular events, leads to the
abnormality of the cerebellar cortex (Rakic and Sidman 1973a). Although our
initial interpretation that the death of granule cells could be a secondary
event due to the failure of their migration over defective Bregmann glial
fibers was overzealous (e.g. see Sotelo and Changeux 1974; Goldowitz and
Mullen 1982; Patil et al. 1995), it has established the validity of studying the
role of cell–cell interactions in pathogenesis of brain malformations. Furthermore, the elimination of cerebellar granule cells shortly after their last
division, before they have had a chance to establish connections with the
mossy fibers, provided an opportunity to study the role of cell–cell interactions in the formation of the synaptic architecture and seemd to justify the
popular term "genetic dissection". The selective absence of granule cells from
the cerebellar circuitry made it possible to test whether mossy afferents might
take a "shortcut" and form synapses directly upon the Purkinje cells when
the normal synaptic link, the granule cell, was absent. A third, somewhat
different issue that could be resolved by studying the *weaver* cerebellum, was
whether the intact synapses that normally form contacts with other segments
of Purkinje cells, such as axons of stellate and basket cells or climbing fiber
butons, expand their synaptic fields to fill the surfaces of dendritic spines left
vacant by the absence of granule cell axons.

We found that, in spite of the absence of granule cells, Purkinje cells in
wv/wv mice develop a dendritic arbor which, although somewhat reduced in
size, is nevertheless richly covered with spines (Fig. 1; Rakic and Sidman
1973a,b). In this mutant, during development the "molecular layer" in the
vermis region consists of the dendritic branches of Purkinje cells that are
studded with spines lacking presynaptic elements. These spines develop a
typical, bud-shaped profile as well as other normal ultrastructural characteristics. In addition, we and other investigators observed a normal-looking
"postsynaptic thickening", an electron-dense, filamentous material that accumulates at the inner membrane surface of the Purkinje cell dendritic
spines (Fig. 1C–E; see also Hirano and Dembitzer 1973; Rakic and Sidman
1973b; Sotelo 1973, 1975). Freeze-fracture studies demonstrated the normal
aggregates of particles on the inner surface of the dendritic spines in weaver
animals (Landis and Reese 1977). Although biochemical and immunological
characterizations of this undercoating were lacking, we interpreted its formation as an expression of the ability of the postsynaptic cell to form the
characteristic synaptic thickening in the absence of the presynaptic elements
(Fig. 1C–E). To study the long-term effect on of the absence of granule cells
on Purknje cell dendritic spines, we performed analyses of the cerebellar
cortex in 1- to 2-year-old, outcrossed, homozygous *weaver* animals, which
display more vigor than inbred animals and often have a normal life span
(Caviness et al. 1972; Rakic and Sidman 1973c). Electron microscopic
analysis of the cerebellar cortex in these adult *weaver* mice showed that the
postsynaptic densities, which appear to develop autonomously, are also

maintained indefinitely despite the absence of synaptic input. These "naked" spines remain asynaptic even in the immediate vicinity of mossy fiber terminals or of stellate and basket cell axons that have established normal synaptic junctions with adjacent smooth segments of dendritic shafts or cell somata. Thus, the *weaver* mutation provided the first direct evidence that, with regard to the Purkinje cell, development of a specialized postsynaptic site can proceed autonomously and can be maintained, independent of direct contact with specific presynaptic partners (Rakic and Sidman 1973c; Rakic 1976).

The *reeler* (*rl/rl*) mutation was also known to affect the cerebellum (Sidman et al. 1965) and thus offered an additional opportunity to test the general validity of the conclusions about synaptic specificity inferred from the analysis of *weaver*. The *reeler* mouse also exhibited pronounced cortical abnormalities in forebrain structures (Caviness and Sidman 1973), and eventually became the most thoroughly analyzed neurological mutant (reviewed in Rakic and Caviness 1995; Lambert de Rouvroit and Goffinet 1998; Bar et al., this volume). I will focus here only on the abnormality of the cerebellar cortex to compare some of the findings with those in *weaver*. Most Purkinje cells in homozygous *reeler* are malpositioned and dispersed throughout the cerebellar white matter; in addition the majority of granule cells do not attain their normal positions (Fig. 2A,B). Therefore, although all the cerebellar neuronal elements are present, many Purkinje cells do not come into direct contact with granule cell axons, due to their abnormal distributions.

Electron microscopic analysis reveals that in areas where parallel fibers are absent or are severely reduced in number, the displaced somata of Purkinje cells exhibit normal basket cell axon contacts, but their misaligned and disoriented dendritic trees are heavily invested by spines that are devoid of contacts with parallel fibers (Fig. 2C and Rakic and Sidman 1972; Rakic 1976). In the course of development most of the spines on aberrant dendrites become enveloped in astroglial sheaths and, as in the weaver, they remain asynaptic even though numerous synaptic terminals of various origin may lie in the immediate vicinity. The data from *reeler* were less decisive than the *weaver* for testing the autonomy of the dendritic spine development, because a certain number of granule cell axons are present even in the most malformed areas. Since the severe alignment-orientation disorder in the *reeler* cerebellum has surprisingly little effect on the specificity of the basic synaptic relationships, the pronounced cerebellar symptoms seen in this mutant may be the result of derangements in the numerical balance and in the sequential order of the connectivity between different elements within the synaptic circuitry (Caviness and Rakic 1978).

This research approach applied to the *reeler* and *weaver* mutants in the 1970s made it possible for the first time to study the effects of genetically induced cell misalignment on cyto- and synaptoarchitecture in the cerebellar cortex. In fact, these initial results have generated an air of optimism that

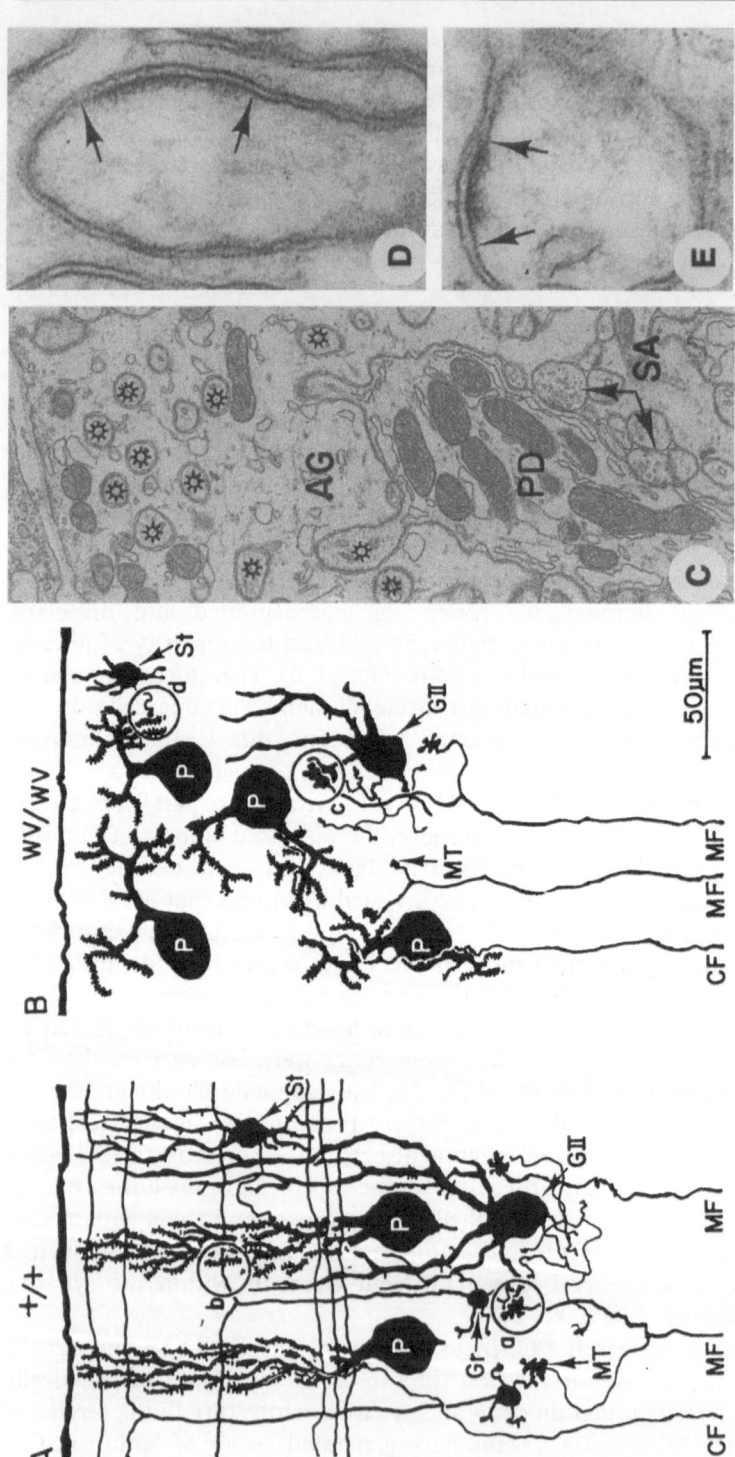

such mutations, occurring at a single genetic locus, might allow the uncovering of the molecular mechanisms underlying the regulation of the major developmental processes of the mammalian brain (Caviness and Rakic 1978; Goffinet 1979). This optimism was due to the finding that the systematic changes in the neuronal positions of well-defined neuronal populations – the abnormal patterns of their death and migration – alter synaptic connectivity in a similar way, despite variations in genetic background and despite variations in the mutation itself (Caviness et al. 1972; Rakic and Sidman 1973a; Wilson et al. 1981). Furthermore, the complex brain abnormalities appeared to reflect an error in specific molecular mechanisms caused by a mutation in a single gene that was required for cellular interactions involved in pattern formation during the course of brain differentiation.

After the initial description of the phenotype in these mutants, it was difficult, at least for me, to design the next experiment that would elucidate the pedigree of causes in the chain of molecular events or that would help in the identification of the up and downstream acting genes involved in the pathogenesis of the observed defects. Even though the number of spontaneous mutants has doubled since than (Lyon and Searle 1989), it seemed unrealistic to expect to find relevant combinations in our life span. In fact, the quest to elucidate the nature of and the molecular mechanisms underlying developmental phenomena and their complex cellular interactions through spontaneous mutants appeared unacceptably slow. After the peak in the mid 1970s, the use of neurological mutations in mice declined in the otherwise rapidly growing field of developmental neurobiology. I, myself, began to focus on the experimental manipulation of brain development in the larger and more slowly developing primates to address basic cellular issues relevant to human brain development (e.g. Rakic 1976, 1981; Rakic and Riley 1983).

Fig. 1A–E. Composite drawings of mouse cerebellar cortex impregnated according to the Golgi method in a 3-week-old wild-type (C57BL/6J +/+) mouse (A) and homozygous *weaver* (C57BL/6J wv/wv) littermate (B). Both cerebella were cut longitudinal to the folium. In the wild-type Purkinje cell (P) dendritic spheres are oriented 90° to parallel fibers or axons of granule cells (*Gr*). In the *weaver* littermate, which at the midline region lacks granule cells, Purkinje cell dendritic arbors, although reduced in size and misoriented, are nevertheless studded with spines. CF Climbing fibers; GII Golgi type II neurons; MF mossy fibers; PA Purkinje cell axon; StL stellate cells. Normal and abnormal cell connections are enclosed in *circles* (a–d). C Cerebellar tissue from an outcrossed, 13-month-old, homozygous *weaver* (wv/wv) mouse. At this age, most dendritic spines (*asterisks*) are completely enveloped in large plazas of (*AG*) astroglial cytoplasm. Postsynaptic thickenings are present in several spines in this field, but they do not show up well at this magnification. Note that stellate cell axons (*SA*) form symmetrical synapses with a smooth segment of Purkinje cell dendrite. Magnification 16,200x. D,E Asynaptic (unattached) dendritic spines in a 10-day-old C57BL/6J wv/wv mouse. Spines were cut longitudinally in D and transversely in E to display approximate size and location of "postsynaptic" densities (between *arrows*). Magnification 50,000x. (Assembled from Rakic 1976)

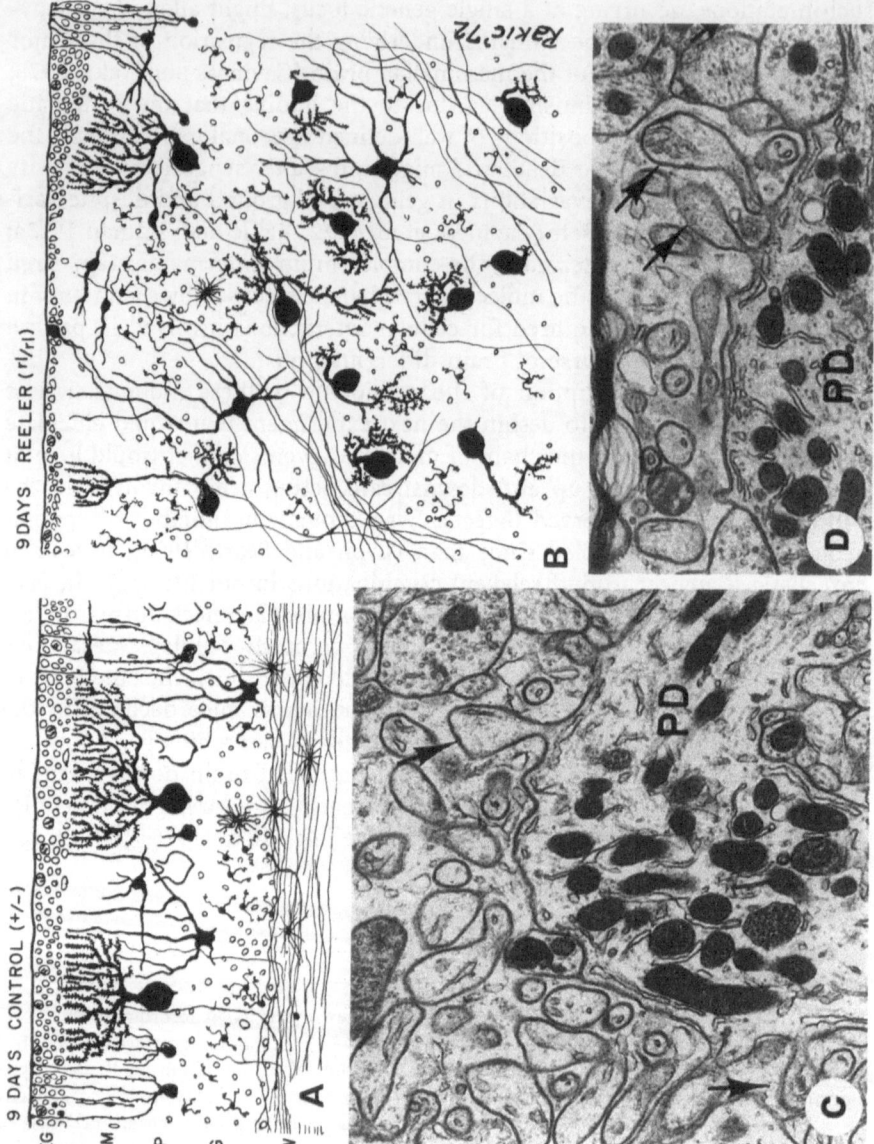

◄──

Fig. 2A–D. Composite drawings from Golgi impregnated preparations of the cerebellar cortex of a 9-day-old normal (+/+) mouse (**A**) and its homozygous *reeler* (*rl/rl*) littermate (**B**). C,D Tissue samples from the cerebellar cortex of a 14-day-old C57BL/6J *rl/rl* mouse. A displaced Purkinje cell dendrite (**PD**) in the territories lacking parallel fibers has numerous asynaptic spines (*arrows*) enveloped by astroglial cytoplasm. *G* Granular layer; *EG* external granular layer; *M* molecular layer; *P* Purkinje cell layer; *W* white matter. Magnification 16 000

3
Renaissance: New Opportunities and Induced Mutations

The progress from neurological mutant mice in the 1980s was relatively slow (Rakic and Caviness 1995). One reason for this seems to be the low probability of obtaining new mutations with the relevant functional outcome, compared with the opportunities in the fast-propagating invertebrates where it was possible to introduce genetically engineered mutations (Benzer 1973; Metzstein et al. 1998). The other, equally important reason was that we did not know the defective gene responsible for the mutant phenotypes. Without knowing the identity of the defective genes, after describing the phenotype at the light and electron microsopic levels, relatively little could be done to understand the underlying molecular mechanisms.

In the mid 1990s the use of neurological mutant mice began again to generate optimism and excitement. Because of the progress in positional cloning, it was now more feasible to identify the mutated genes. For example, the illusive *reeler* and *weaver* genes were finally identified, opening the possibility of studying their primary effect (e.g. D'Archangelo et al. 1995; Bar et al. 1995; Hirotsune et al. 1995; Ogawa et al. 1995; Patil et al. 1995). Unexpectedly, it turned out that the *weaver* mutation causes a gycine-to-serine substitution in the pore region of G-protein-gated inwardly rectifying potassium channel subunit GIRK2, a member of a greater GIRK family of heterotrimeric channels with distinct regional localization in the brain. The locus in the mouse has been mapped to chromosome 16 to a region pertaining to the Down's syndrome region in human chromosome 21. Although this discovery, by itself, did not explain the mechanism of cerebellar defect, it pointed out to the relationship with ion channels that may be involved in granule cell migration (Herrup 1996), in particular the connection with the NMDA receptor complex (Komuro and Rakic 1993; Liesi et al. 1999). The *reeler* gene, which was isolated independently by groups using different strategies, turned out to be a protein of nearly 3500 amino acids that may have evolved to mediate cell–cell interaction during neural development (reviewed in Bar et al., this volume). The possibility of understanding the pathogenesis of this abnormality has increased with the characterization of several spontaneous mutations that have similar effects on brain development, such as *scrambler* and *yatori* (e.g. Goldowitz et al. 1997; Ware et al. 1997; Pearlman et al. 1998; reviewed in Hatten 1999; Sotelo and Mariani

1999). Advances in positional cloning have also helped in the search for genes involved in the control of visible phenotypic differences, such as the size of the various brain structures, as exemplified by the recent work of Robert Williams (Williams et al. 1998; Williams, this volume). In addition, the mosaic and aggregation chimeras, which in the 1970s were produced only in a few laboratories, became more common, increasing the opportunities for creating brains or in vitro systems that contained a combination of cells of different genetic composition (e.g. Herrup and Mullen 1979; Kuan et al. 1997; Mullen et al. 1997; Tan et al. 1998). Finally, the use of genetically altered cell lines has opened the possibility of studying mechanisms of identified gene action in vitro.

The number of available neurological mutants in mice has also been expanded by the increased use of various mutagens that have been used successfully in other species. However, the application of large-scale mutagenesis to mice started only recently, and is presently underway in the Jackson Laboratories as well as in several other centers. Germline mutations, which are passed onto subsequent generations, can be induced in mice as in other species, either by manipulation of the DNA in the embryo or by chemical manipulation of the DNA in the sperm (Brown and Peters 1996; Schimenti and Bucan 1998). The most efficient mutagen in the mice, the alkylating agent ethylnitrosourea (ENU), which predominantly produces point mutations in sperm, has already generated some celebrated breakthroughs such as the discovery of the circadian rhythm gene clock (Takahashi et al. 1994). Improvements in screening strategies for induced genetic mutations have been instrumental in identifying genes involved in specific biological functions. Nevertheless, in the nervous system the major problem remains in the screening, which favors the discovery of more obvious behavioral defects such as those involved in locomotion or sleep patterns, but leaves undetected even severe deficits in the visual or association systems.

Perhaps the most important reason for the renewed interest in the mouse model system for studying brain development has been the ability to generate targeted mutations (Capecchi 1989). Transgenic mice have a new or an extra copy of a desired gene inserted into the genome. The advances made in this technology permit creation of new mutations and open up the possibility of investigating specific developmentally and medically defined questions. The transgenes are transmitted in Mendelian fashion and the homozygous animals with a null mutation are generated in the F2 and subsequent generations by a well-established breeding routine, or, when necessary, by ovarian transplantation (Sundberg and Boggess 2000). The use of this strategy has produced an impressive array of mouse models for various neurodegenerative diseases in which mice expressing disease-causing genes recapitulate at least some features of these disorders (e.g. Price et al. 1998; Walsh 1999). Thus, although most inherited neuropsychiatric disorders are not monogenic, analysis of brain phenotypes in properly selected transgenic mouse mutants can elucidate specific molecular pathways involved in their symptomatology. The increased knowledge of the affected DNA sequences in humans will further enhance the use of the mouse as a primary experimental model system for the study of the pathogenesis of specific congenital abnormalities in the human.

Another strategy that is increasingly gaining support is the search for the genes and molecules that regulate neurogenesis that were initially identified through the use of hybridoma technology (e.g. lamp: Levitt 1994; PC34.1: Arimatsu et al. 1992; astrotactin: Hatten 1999; NJPA1: Anton et al. 1996) which then can be used to create transgenics. The repertoire of molecules that need to be examined using this strategy is rapidly increasing (e.g. see Rubenstien and Rakic 1999) and extends to the homeotic series of "master genes" as determinants of compartmental boundaries or lineage assignment (e.g. Acampora et al. 1999; Hatanaka and Jones 1999; Rubinstein et al. 1999). Transgenic manipulations can also be used to uncover genes of specific and limited regional jurisdiction (Gitton et al. 1999). Some transgenes that are useful for the study of specific neurodevelopmental issues have been discovered as a byproduct of research into other systems (e.g. Cohen-Tannoudji et al. 1994; Soriano et al. 1995). Others have been identified initially in human congenital disorders, but the precise mechanism of their action in the developing brain can now be tested in mouse. For example, the defective gene LI on chromosome 17, which is responsible for the Miller-Dieker form of lissencephaly in humans, has as its mouse homologue Lis1, which appears to be associated with the disruption of neuronal migration in the cerebrum (Reiner et al. 1993; Des Portes et al. 1998; Gleeson et al. 1999).

The current technology is also capable of producing a complete loss of function of a particular gene (knockout). Either this loss can be expressed throughout the mouse life span or, as in the case of the so-called conditional and inducible knockouts, the generation of a null-mutation can be focused on a selected neuronal structure or phase in neural development (e.g. Chen et al. 1998). The selection of genes to target can come from the large set of mammalian equivalents of well-defined genes identified initially in invertebrates. Knockout mice have a targeted gene deleted so that no product of this gene can be synthesized in the null mutant. These approaches seems to be the most powerful, though they are not without drawbacks (see Crabbe et al. 1999).

Suddenly, one does not have to rely on luck or wait until a desired spontaneous or induced mutation occurs; rather, scientists can decide which, and even at what time during development, well-defined, preselected genes will be activated or shut off. In collaboration with Richard Flavell, we began a series of studies using "knockout" technology to delete mammalian homologues of nematode and fruit flie genes that can in principle explain some aspects of brain development or that have relevance to the pathogenesis of certain congenital brain malformations in humans. For example, we found that the selected genes regulating programmed cell death in the nematode play essentially the same role in mice and presumably in other mammals including primates (Kuida et al. 1996, 1998; see review in Kuan et al., this volume). With the forthcoming completion of the sequences for both the mouse and human genome, this approach is going to be even more effective. However, it is important to recognize that many knockouts and induced mutations are embryonic lethals, and that many do not produce noticeable changes in brain phenotype. Those in the latter group may hold a hidden benefit for the explanation of many genetic and acquired neurological disorders that occur after a long incubation and in response to environmental factors. These mutants may appear normal in their general health, neurological reflexes, and sensory and motor tasks, but show an abnormal response to external stimuli such as physical and chemical challenges. For example, although knockout mice lacking the neural isoform of the stress-activated protein kinase SAPK/JNK (JNK3), develop and reproduce normally, without visible behavioral deficits, they exhibit a remarkable resistance to kainic acid-induced seizures and neuronal apoptosis in the hippocampal region (Fig. 3; Yang et al. 1997). This resistance is dose dependent and highly reproducible (compare Behrens et al. 1999 with Yang et al. 1997). Another instructive example is a knockout mouse lacking the β_2 subunit of the nicotinic Ach receptor. This mouse has a normal phenotype and behavior, but when exposed to nicotine does not respond normally (Piccioto and Wickman 1998). This type of "hidden phenotype" may be particularly useful as a model for studying the pathogenesis of human conditions in which the effect of mutations becomes expressed only upon environmental stress.

The use of spontaneous and induced mutations in studying the mechanisms of neurogenesis has produced some additional conceptually important benefits. One of the most obvious is that these studies have revealed the extent of the preservation of some basic gene functions throughout evolution. For example, the apoptotic genes, such as the Ced and Bcl families from nematode, as well as their mammalian equivalents CPP32 and Blx, play remarkably similar positive and negative roles in both orders (Kuida et al. 1996, 1998; see review in Kuan et al., this volume). Another side benefit is the realization that a basic gene function can be utilized for a different overall strategy in ontogenetic and phylogenetic development. For example, the *reeler* gene, which at the early stages of neurogenesis in the mouse plays a role in the positioning of neurons in their target structures (Bar et al., this

-->

Fig. 3. Characterization of c-Jun N-terminal Kinase (JNK) 3 deficient mice as an example of choosing the right experimental paradigm to uncover the otherwise subtle phenotypes of knockout animals. JNK3 null-mutant mice had normal development and showed no obvious phenotypes compared to wild-type mice. However, when challenged with excitotoxin kainic acid (30 mg/kg), wild-type mice developed seizures (A), whereas JNK3 (−/−) mice had remarkable resistance to kainic-acid induced seizures (B). In addition, wild-type mice exhibited robust neuronal apoptosis in the CA1 subfield of hippocampus following the kainic acid challenge (C). In contrast, JNK3 (−/−) mice showed no sign of CA1 hippocampal neuron deaths under either regular dosage (30 mg/kg) or seizure-inducing dosage of kainic acid (45 mg/kg) (Yang et al. 1997)

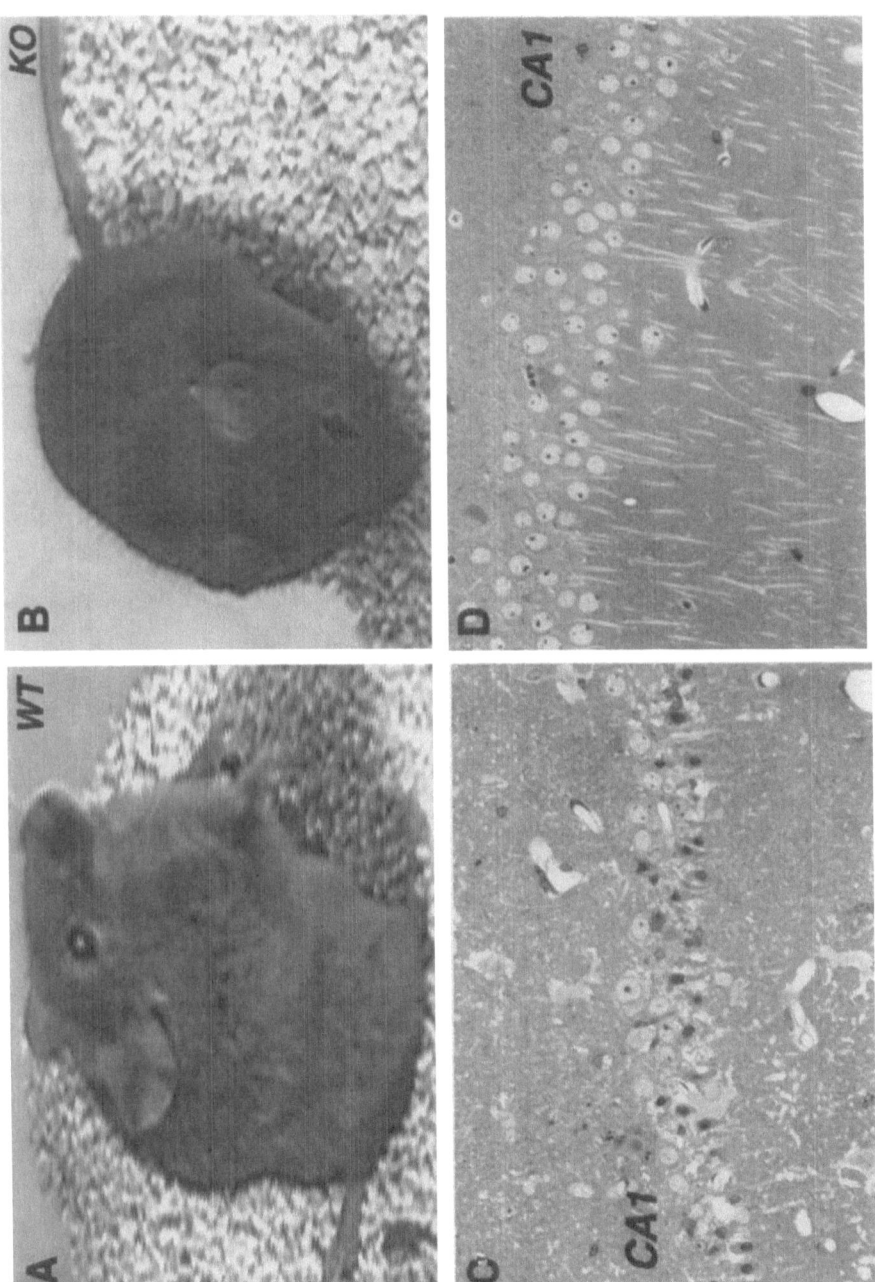

volume), has been implicated in the pathogenesis of schizophrenia, a disease expressed during postdevelopmental ages in humans (Impagnatiello et al. 1998). Some of the genes identified in invertebrates have several homologues in mammals, each of which may be expressed in different subregions of the brain (e.g. Kuan et al., this volume). Thus, it appears that the evolutionary process takes advantage of a basic gene function that, in a different context, can perform a more specialized and specific developmental role or a highly modified function in the maintenance of the adult brain.

With over a hundred spontaneous neurological mutants already available, as well as the number of transgenic and knockout mice increasing at a rapid pace, the priority in selecting a research subject is a paradoxical problem facing contemporary developmental neurobiologists. With the completion of the mouse genome project the choices will be even greater, but the selection process may be somewhat simplified and more rational. Researchers should first select a specific developmental problem, then search for the genes that could provide an answer to their question, or they should create a trans-genenic mouse with which they could test their problem experimentally. These are choices that were not available two decades ago.

4
Epilogue

In view of the developments in the field of developmental neurobiology, the return to the mouse model was for me as natural as returning home. However, I must emphasize that the enthusiasm for the mouse model was not sufficient to cause me to abandon higher mammalian models, which still provide a unique advantage for studying specific questions. For example, we continue to study cortical specification or mechanisms of ocular dominance column segregation in nonhuman primates (e.g. Donoghue and Rakic 1999; Kornack and Rakic 1997; Meissirel et al. 1997), while the basic molecular mechanisms involved in neuronal differentiation are being studied more effectively in simple organisms (Huilin et al. 1999; Sestan et al. 1999). It is also now possible to introduce foreign genes via retrovirus vectors into monkey embryonic brain (Kornack and Rakic 1995) and thus begin to explore the consequences on neural development. However, in the overall scheme, the mouse occupies a pivotal position in our investigations, as it allows the testing of the developmental roles of regulatory genes whose functions have been identified in invertebrates, as well as enabling us to search for the functions of genes that have been isolated in the human. The lessons from this field during the past two decades have been that one cannot predict where the next breakthrough will come from. It is clear, however, that an understanding of the roles of various genes involved in normal and abnormal histogenesis of the human brain can benefit enormously from the use of the mouse model and that gene manipulation will figure pre-eminently in this strategy.

Acknowledgement. I gratefully acknowledge the contribution of members of my laboratory and numerous colleagues who over the years have inspired changes in approaches and stimulated discussion about the basic conceptual issues in brain development. This research was continuously supported by the US Public Health Service.

References

Acampora D, Barone P, Simeone A (1999) Otx genes in corticogenesis and brain development. Cerebral Cortex 9:533–542

Anton SA, Cameron RS, Rakic P (1996) Role of neuron–glial junctional proteins in the maintenance and termination of neuronal migration across the embryonic cerebral wall. J Neurosci 16:2283–2293

Arimatsu Y, Miyamoto M, Nihonmatsu I, Hirata K, Urataini Y, Hatanka Y, Takiguchi-Hoyash Y (1992) Early regional specification for a molecular neuronal phenotype in the rat neocortex. Proc Natl Acad Sci USA 89:8879–8883

Bar I, Lambert de Rouvroit C, Royaux I, Kritzman DB, Dernoncourt C, Rulelle D, Beckers MC, Goffinett A (1995) YAC containing the reeler locus with preliminary characterization of candidate gene fragments. Genomics 26:543–546

Behrens A, Sibilia M, Wagner EF (1999) Amino-terminal phosphorylation of c-Jun regulates stress-induced apoptosis and cellular proliferation. Nat Genet 21:326–329

Benzer S (1973) Genetic dissection of behavior. Sci Am 229:24–37

Brown SDM, Peters J (1996) Combining mutagenesis and genomics in the mouse – closing the phenotype gap. Trends Genet 12:443–445

Capecchi MR (1989) Altering the genome by homologous recombination. Science 244:1288–1292

Caviness VS Jr (1982) Neocortical histogenesis in normal and reeler mice: a developmental study based upon 3[H]-thymidine autoradiography. Dev Brain Res 4:293–302

Caviness VS Jr, Crandall JE, Edwards A (1988) The reeler malformation. In: Peters A, Jones EG (eds) Cerebral Cortex. Plenum, New York, pp 59–89

Caviness VS Jr, Rakic P (1978) Mechanisms of cortical development: a view from mutations in mice. Annu Rev Neurosci 1:297–326

Caviness VS Jr, Sidman RL (1973) Time of origin of corresponding cell classes in the cerebral cortex of normal and reeler mutant mice: an autoradiographic analysis. J Comp Neurol 148:141–152

Caviness VS Jr, So DK, Sidman RL (1972) The hybrid reeler mouse. J Hered 63:241–246

Chen JS, Kelz MB, Zeng GQ, Sakai N, Steffen C, Shockett PE, Picciotto MR, Duma RS, Nestler EJ (1998) Transgenic animals with inducible, targeted gene expression in brain. Mol Pharmacol 54:495–503

Cohen-Tanoudji M, Babinet C, Wassef M (1994) Early intrinsic regional specification of the mouse somatosensory cortex. Nature 368:460–463

Crabbe JC, Wahlsten D, Dudek BC (1999) Genetics of mouse behavior: interactions with laboratory environment. Science 284:1670–1672

D'Arcangelo G, Miao GG, Chen S-C, Soares HD, Morgan JI, Curren T (1995) A protein related to extracellular matrix proteins deleted in the mouse mutant reeler. Nature 374:719–723 des Portes V, Pinard JM, Billuart P, Vinet MC, Koulakoff A, Carrie A, Gelot A, Dupuis E, Motte J, Berwald-Netter Y, Catala M, Kahn A, Beldjord C, Chelly JA (1998) Novel CNS gene required for neuronal migration and involved in X-linked subcortical laminar heterotopia and lissencephaly syndrome. Cell 92:51-61

Donoghue MJ, Rakic P (1999) Molecular gradients and compartments in the embryonic primate cerebral cortex. Cerebral Cortex 9:586-600

dctlparEccles JC, Ito M, Szentagothai J (1967) The Cerebellum as a Neuronal Machine. Springer, Berlin

Falconer DS (1951) Two new mutants "trembler" and "reeler" white neurological actions in the house mouse (*Mus musculus*). J Genet 50:192-201

Gitton Y, Cohen-Tannoudji M, Wassef M (1999) Role of thalamic axons in the expression of H-2Z1, a mouse somatosensory cortex specific marker. Cerebral Cortex 9:611-620

Gleeson JG, Lin PT, Flanagan LA, Walsh CA (1999) Doublecortin is a microtubule-associated protein and is expressed widely by migrating neurons. Neuron 23:257-271

Goffinet AM (1979) An early developmental defect in cerebral cortex of the reeler mouse. A morphological study leading to a hypothesis concerning the action of the mutant gene. Anat Embryol 157:205-216

Godowitz D, Cushing R, Lowell E, D'Arcangelo G, Sheldon M, Sweet H, Davisson M, Staindler D, Curren T (1997) Çerebellar disorganization characteristic of reeler in scrambler mutant mice despite the presence of Reelin. J Neurosci 17:8767-8777

Goldowitz D, Mullen RJ (1982) Granule cell as a site of gene action in the weaver mouse cerebellum: evidence for heterozygous mutant chimeras. J Neurosci 2:1474-1485

Hamburgh M (1963) Analysis of the postnatal developmental effects of "reeler", a neurological mutation in mice. A study in developmental genetics. Dev Biol 8:165-185

Hatanaka Y, Jones EG (1999) Novel genes expressed in the developing medial cortex. Cerebral Cortex 9:577-585

Hatten ME (1999) Central nervous system neuronal migration. Annu Rev Neurosci 22:511-539

Herrup K (1996) The weaver mouse: a most cantankerous rodent. Proc Natl Acad Sci USA 93:10541-10542

Herrup K, Mullen RJ (1979) Staggerer chimeras: intrinsic nature of Purkinje cell defects and implications for neuronal cerebellar development. Brain Res 178:443-457

Hirano A, Dembitzer HM (1973) Cerebellar alteration in the weaver mouse. J Comp Biol 56:478-486

Hirotsune S, Takahara T, Sasaki N, Hirose K, Yoshiki A, Ohashi T, Kusakabe M, Murakami Y, Muramatsu M, Watanabe S, Nakao K, Katsuki M, Hayashizaki Y (1995) The reeler gene encodes a protein with an EGF-like motif expressed by pioneer neurons. Nat Genet 10:77-83

Impagnatiello F, Guidotti AR, Pesold C, Dwivedi Y, Caruncho H, Pisu MG, Uzunov PD, Smalheiser NR, Davis JM, Pandey NG, Pappas GD, Tueting P, Sharma RP, Costa E (1998) A decrease of reelin expression as putative vulnerability factor in schizophrenia. Proc Natl Acad Sci USA 95:15718-15723

Komuro H, Rakic P (1993) Modulation of neuronal migration by NMDA receptors. Science 260:95-97

Kornack DR, Rakic P (1995) Radial and horizontal deployment of clonally related cells in the primate neocortex: relationship to distinct mitotic lineages. Neuron 15:311-321

Kornack DR, Rakic P (1998) Changes in cell cycle kinetics during the development and evolution of primate neocortex. Proc Natl Acad Sci USA 95:1242-1246

Kuan C, Elliot E, Flavell RA, Rakic P (1997) Restrictive clonal allocation in the chimeric mouse brain. Proc Natl Acad Sci USA 94:3374-3379

Kuan C, Yang DD, Semanta Roy DR, Davis RJ, Rakic P, Flavell RA (1999) The Jnk1 and Jnk2 protein kinases are required for regional-specific apoptosis during early brain development. Neuron 22:667-676

Kuida K, Zheng TS, Na S, Kuang C, Yang D, Karasuyama H, Rakic P, Flavell RA (1996) Decreased apoptosis in the brain and premature lethality in CPP32-deficient mice. Nature 384:368-372

Kuida K, Haydar T, Kuan C, Yong G, Taya C, Karasuyama A, Su S-H, Rakic P, Flavell RA (1998) Reduced apoptosis and cytochrome c-mediated caspase activation in mice lacking Caspase-9. Cell 94:325-333

Lambert de Rouvroit C, Goffinet AM (1998) The reeler mouse as a model of brain development. Adv Anat Embryol Cell Biol 150:1-108

Landis DMD, Reese TS (1977) Structure of the Purkinje cell membrane in staggerer and weaver mutant mouse. J Comp Neurol 171:247-260

Lane PW (1964) Personal communication. Mouse Lett 30:32

Leisi P, Stewart RR, Akinshuola E, Wright J (1999) Weaver cerebellar granule neurons show altered expression of NMDA receptor subunit both in vivo and in vitro. J Neurobiol 38:441-454

Levitt P (1994) Experimental approaches that reveal principles of cerebral cortical development. In: The Cognitive Neurosciences. MIT Press, Cambrige, pp 147-163

Lyon MF, Searle G (1989) Genetic variants and strains of the laboratory mouse. 2nd Edn. Oxford University Press, Oxford

Meissirel C, Wikler KC, Chalupa LM, Rakic P (1997) Early divergence of M and P visual subsystems in the embryonic primate brain. Proc Natl Acad Sci USA 94:5900-5905

Metzstein MM, Stanfield GM, Horvitz HR (1998) Genetics of programmed cell death in C. elegans: past, present and future. Trends Genet 14 410-14 416

Miyashita-Lin EM, Hevner R, Montzka Wassarman K, Martinez S, Rubenstein JLR (1999) Early neocortical regionalization in the absence of thalamic innervation. Science 285:906-909

Mullen RJ, Herrup K (1979) Chimeric analysis of mouse cerebellar mutants. In: Breakefield XO (ed) Genetic Approacehes to the Nervous System. Elsevier, Amsterdam, pp 173-196

Mullen RJ, Hamre KM, Goldowitz D (1997) Cerebellar mutant mice and chimeras revisited. Perspec Dev Neurobiol 5:43-55

Ogawa M, Miyata T, Nakajima K, Yagyu K, Selke M, Ikenaka K, Yamamoto H, Mikoshiba K (1995) The reeler gene-associaterd antigen on Cajal-Retzius neurons is a crucial molecule for laminar organization of cortical neurons. Neuron 14:1-20

Palay SL, Chan-Palay V (1974) Cerebellar cortex: cytology and organization. Springer, Berlin, Heidelberg, New York

Patil N, Cox DR, Bhat D, Faham RT, Spencer C, Davidson MT (1995) A potassium channel muation in weaver mice implicates membrane excitability in granule cell differentiation. Nat Genet 11:126-129

Pearlman AL, Faust PL, Hatten ME, Brunstorm JE (1998) New directions for neuronal migration. Curr Opin Neurobiol 8:45-54

Peters A, Palay S, Webster HF (1970) The fine structure of the nervous system: the cells and their processes. Harper and Row, New York

Picciotto MR, Wickman K (1998) Using knockout and transgenic mice to study neurophysiology and behavior. Physiol Rev 78:1131-1163

Pinto-Lord CM, Evrard E, Caviness VS Jr (1982) Obstructed neuronal migration along radial glial fibers in the neocortex of the reeler mouse: a Golgi-EM analysis. Dev Brain Res 4:379-339

Price DL, Sisodia SS, Borchelt DR (1998) Genetic neurodegenerative dieseaes: the human illness and transgenic models. Science 282:1079-1983

Qu H, Rand MD, Wu X, Sestan N, Wang W, Rakic P, Xu T, Artavanis-Tsakonas S (1999) Processing of the Notch ligand Delta by metalloprotease Kuzbanian. Science 283:94-98

Rakic P (1971) Neuron-glia relationship during granule cell migration in developing cerebellar cortex. A Golgi and electronmicroscopic study in *Macacus rhesus*. J Comp Neurol 141:283-312

Rakic P (1972) Mode of cell migration to the superficial layers of fetal monkey neocortex. J Comp Neurol 145:61-84

Rakic P (1976) Synaptic specificity in the cerebellar cortex: study of anomalous circuits induced by a single gene mutation in mice. In: The Synapse. Cold Spring Harbor Symp Quant Biol 40:333-346

Rakic P (1981) Development of visual centers in the primate brain depends on binocular competition before birth. Science 214:928-931   the

Rakic P (1988) Specification of cerebral cortical areas. Science 241:170-176

Rakic P (1995) A small step for the cell - a giant leap for mankind: a hypothesis of neocortical expansion during evolution. Trends Neurosci 18:383-388

Rakic P, Caviness VS Jr (1995) Cortical development: view from neurological mutants two decades later. Neuron 14:1101-1104

Rakic P, Riley KP (1983) Regulation of axon numbers in the primate optic nerve by prenatal binocular competition. Nature 305:135-137

Rakic P, Sidman RL (1968) Supravital DNA synthesis in the developing human and mouse brain. J Neuropath Exp Neurol 27:246-276

Rakic P, Sidman R (1969) Telencephalic origin of pulvinar neurons in the fetal human brain. Z Anat Entwickl-Gesch 129:53-82

Rakic P, Sidman RL (1970) Histogenesis of cortical layers in human cerebellum, particularly the lamina dissecans. J Comp Neurol 139:473-500

Rakic P, Sidman RL (1972) Synaptic organization of displaced and disoriented cerebellar cortical neurons in reeler mice. J Neuropath Exp Neurol 31:192

Rakic P, Sidman RL (1973a) Weaver mutant mouse cerebellum; defective neuronal migration secondary to specific abnormality of Bergmann glia. Proc Natl Acad Sci USA 70:240-244

Rakic P, Sidman RL (1973b) Sequence of developmental abnormalities leading to granule cell deficit in cerebellar cortex of weaver mutant mice. J Comp Neurol 152:103-132

Rakic P, Sidman RL (1973c) Organization of cerebellar cortex secondary to deficit of granular cells in weaver mutant mice. J Comp Neurol 152:133-162

Rakic P, Yakovlev PI (1968) Development of the corpus callosum and cavum septi in man. J Comp Neurol 132:45-72

Reiner O, Carrozzo R, Shen Y, Wehnert M, Faustinella F, Dobyns WB, Caskey CT, Ledbetter DH (1993) Isolation of a Miller-Dieker lissencephaly gene containing G protein b-subunit-like repeats. Nature 364:717-721

Rubenstein JLR, Rakic P (1999) Genetic control of cortical development. Cerebral Cortex 9:521-523

Rubenstein JLR, Anderson S, Shi L, Miyashita-Lin E, Bulfone A, Hevner R (1999) Genetic control of cortical regionalization and connectivity. Cerebral Cortex 9:524-532

Schimenti J, Bucan M (1998) Functional genomics in the mouse: phenotype-based mutagenesis screens. Genome Resarch 8:698-710

Sestan N, Artavanis-Tsakonas S, Rakic P (1999) Contact-dependent inhibition of cortical neurite growth mediated by Notch signaling. Science 286:741-746

Sidman RL (1970) Autoradiographic methods and principles for study of the nervous system with thymidine-H3. In: Nauta WJH, Ebbeson SOE (eds) Contemporary Research Methods in Neuroanatomy. Springer, Berlin, Heildelberg, New York, pp 252-274

Sidman RL, Green MC, Appel SH (1965) Catalog of the Neurological Mutants of the Mouse. Cambridge Harvard University Press, Cambridge

Soriano E, Dumesnil N, Auladell C, Cohen-Tannoudji M, Sotelo C (1995) Molecular heterogeneity of progenitors and radial migration in the developing cerebral cortex revealed by transgenic expression. Proc Natl Acad Sci USA 92:11 676-11 680

Sotelo C (1973) Permanence and fate of paramembranous synaptic specializations in mutants and experimental animals. Brain Res 62:345-351

Sotelo C (1975) Dendritic abnormalities of Purkinje cells in cerebellum of neurological mutant mice (Weaver and Staggerer). Adv Neurol 12:335-351

Sotelo C, Changeux J-P (1974) Trans-synaptic degeneration "en cascade" in the cerebellar cortex of staggerer mutant mice. Brain Res 67:519–526

Sotelo C, Mariani J (1999) Resarch strategies for the analysis of neurological mutants of the mouse. In: Crusio WE, Gerlai RT (eds) Molecular genetics. Techniques for behavioral neuroscience. Elsevier, Amsterdam

Sundberg JP, Boggess D (2000) Systematic approach to evaluation of mouse mutations. CRC Press, Boca Raton

Takahashi JS, Pinto LH, Holtz Vitaterna M (1994) Forward and reverse genetic approaches to behavior in the mouse. Science 264:1724–1733

Tan S-S, Kalloniatis M, Sturm K, Tam PPL, Reese BE, Faulkner-Jones B (1998) Separate progenitors for radial and tangential cell dispersion during development of the cerebral neocortex. Neuron 21:295–304

Ware ML, Fox JW, Gonzalez JL, Lambert de Rouvroit C, Russo CJ, Chua SC, Goffinet AM, Walch CA (1997) Aberrant splicing of a mouse disabled homolog, mdab1, in the scrambler mouse. Neuron 19:239–246

Walsh CA (1999) Genetic malformations of the human cerebral cortex. Neuron 23:19–29

Williams RW, Storm RC, Godowitz D (1998) Natural variation in neuron number in mice is linked to a major quantitative trait locus on Chr 11. J Neurosci 18:138–146

Wilson L, Sotelo C, Caviness VS Jr (1981) Heterologous synapses upon Purkinje-cells in the cerebellum of the reeler mutant mouse – an experimental light and electron-microscopic study. Brain Res 213:63–82

Yang D, Kuan C, Whitmarsh AJ, Rincon M, Zheng TS, Davis RJ, Rakic P, Flavell RA (1997) Absence of excitotoxicity-induced apoptosis in the hippocampus of mice lacking the Jnk3 gene. Nature 389:865–870

Mapping Genes that Modulate Mouse Brain Development: A Quantitative Genetic Approach

Robert W. Williams[1]

1
Introduction

> In my opinion there are only quantitative differences, not qualitative differences, between the brain of a man and that of a mouse. (Ramón y Cajal 1890)

> The difference in behavioral capacity between man and chimpanzee may be no more than the addition of one cell generation in the segmentation of the neuroblasts which form the cerebral network. (Lashley 1949)

The complexity of CNS development is staggering. In mice a total of approximately 75 million neurons and 25 million glial cells are generated, moved, connected, and integrated into hundreds of different circuits over a period of 1 month. The process is coordinated by the expression of a large fraction of the genome – as many as 40 000 genes are involved (Sutcliffe 1988; Adams et al. 1993). These same genes coordinate the development of the human brain, but a thousand times more neurons are generated (Williams and Herrup 1988) and their integration and training take more than a decade. While 5000 of these genes have common roles in cellular metabolism, this still leaves a huge complement that have selective, transient, and partially redundant roles in the development of different parts of the brain (Usui et al. 1994; Gautvik et al. 1996). Reductionist approaches that focus on isolated processes and molecules may seem hopelessly inadequate, but they are an absolute necessity at this early stage of analysis and understanding.

This chapter introduces a comparatively new reductionist approach called complex trait analysis that my research group is using to explore the genetic basis of CNS development. Complex trait analysis is a field that developed rapidly in the 1990s as a result of the hybridization of quantitative and molecular genetics. The suite of techniques associated with complex trait analysis greatly extends the variety of CNS phenotypes that can be subjected to systematic molecular analysis. It is in essence a forward

[1] Center for Neuroscience and Department of Anatomy and Neurobiology, University of Tennessee, 855 Monroe Avenue, Memphis, Tennessee, 38163 USA <http://nervenet.org>

Results and Problems in Cell Differentiation, Vol. 30
Goffinet and Rakic (Eds.): Mouse Brain Development
© Springer-Verlag Berlin Heidelberg 2000

genetic approach that proceeds from phenotypic variation to single genes. This approach has been embraced by behavioral geneticists and neuro-pharmacologists (Plomin et al. 1991; Johnson et al. 1992; Crabbe et al. 1994; Takahashi et al. 1994; Kanes et al. 1996), and these techniques can now be applied with equal vigor to explore genetic sources of variation in brain structure and development. This chapter begins with a genetic analysis of sources of variation in brain weight and illustrates how we have mapped quantitative trait loci (QTLs) that control brain weight and neuron number in mice.

2
Why Brain Weight and Neuron Number Matter

2.1
Metabolic Constraints

There are several reasons why differences in brain size and neuron number are interesting and biologically significant. First, relative to its size, the brain with its large population of neurons consumes a disproportionate amount of energy (Clark et al. 1994). The high cost of making, training, and main-taining this metabolically demanding organ has wide-ranging effects on an animal's development and behavior (Sacher and Staffeldt 1974; Eisenberg and Wilson 1978; Martin 1981; Armstrong 1983; Hofman 1983; Pagel and Harvey 1990; Allman et al. 1993). Humans are an extreme example, with a brain that is ten times heavier than expected on the basis of body weight. We afford this luxury by developing slowly and by having an efficient diet (Aiello and Wheeler 1995). Given the fact that we are such a large-brained species, it may be a surprise to learn that mice have brains that are pro-portionally just as large as those of humans. A 22-g adult mouse typically has a 450-mg brain, whereas a 66-kg human typically has a 1350-g brain – 2% in both cases.

2.2
Functional Correlates

A second and almost self-evident reason to be interested in brain weight and neuron number is that variation in these simple parameters is associated with variation in behavior (Lashley 1949; Rensch 1956; Wimer and Prater 1966; Fuller and Herman 1972; Fuller 1979; Roderick et al. 1979; Crusio et al. 1989; Jacobs et al. 1990; Belknap et al. 1992; Aboitiz 1996; Keverne et al. 1996). This is most clear cut when specific regions of the brains of different species or individuals are compared. For example, in songbirds the volume of song system nuclei and numbers of neurons tend to be positively correlated with different features of song production (e.g. DeVoogd et al. 1993; Ward et al. 1998). Another fine example, although strongly negative in this case, is the correlation in mice between avoidance learning and the size of the infrapyramidal projection from dentate gyrus to CA3 (Schwegler and Lipp 1983; Lipp et al. 1989).

2.3
Insights into CNS Development

My colleagues and I are interested in brain weight and neuron number for a third reason: as a means to map, clone, and characterize genes that control the proliferation, differentiation, and death of cells in the CNS (Williams and Herrup 1988; Williams et al. 1998a). These genes are entry points into molecular networks that control brain development. Differences in brain weight are proportional to total brain DNA content and consequently to total CNS cell numbers (Zamenhof and von Marthens 1976). This is true even in neonatal mice, before appreciable glial cell production (Zamenhof et al. 1971; Zamenhof and von Marthens 1976). For this reason, brain weight is a surprisingly good surrogate measure for total cell number in mice, as in humans (Pakkenberg and Gundersen 1997).

The initial tactical and technical problem is how to go about identifying genes that modulate cell proliferation and death either in specific nuclei or in the brain as a whole. Mutants may be useful in some instances, but we need more generic methods that can target any and all CNS regions and cell types. Instead of depending on rare mutations and knockouts, we need methods that provide information about common gene variants – the normal gene polymorphisms that are responsible for the far more pervasive and important natural variation found within typical populations of animals.

Natural variation can be impressive. Numbers of neurons in the human neocortex vary from 15 to 32 billion (Pakkenberg and Gundersen 1997). The volume of human primary visual cortex varies threefold (Stensaas et al. 1974; Gilissen and Zilles 1996). Numbers of ocular dominance columns within the primary visual cortex of rhesus monkeys vary more than 50% (Horton and Hocking 1996). These robust differences are not caused by mutations but are caused by the cumulative action of many normally variable genes and by the action of numerous developmental and environmental factors. In the long run, normal genetic polymorphisms are the most critical source of variance: they are the substrate for evolutionary and developmental modification of brain size and cellular architecture (Williams and Herrup 1988; Lipp 1989; Williams et al. 1993).

3
Biometric Analysis of the Size and Structure of the Mouse CNS

3.1
Precedents

In the late 1960s, Thomas Roderick, John Fuller, Douglas Wahlsten, and Richard and Cynthia Wimer began an ambitious program to manipulate neuroanatomical traits in mice by selective breeding (Roderick 1979). Their aim was to explore correlated changes in behavior. They gave the rapidly expanding field of behavioral neurogenetics a rigorous foundation in quantitative and statistical neuroanatomy (Wimer et al. 1969; Fuller and Geils 1972; Wahlsten 1975; Roderick et al. 1976; Fuller 1979; Wimer 1979; Wimer and Wimer 1985). Rather than relying on mutants, they exploited the substantial variation among standard inbred strains of mice. This work led to some important breakthroughs and some brick walls. One of the breakthroughs was successfully selecting for substantial differences in brain weight over less than 20 generations (Fuller 1979). An obvious limitation, highlighted by Roderick (1976), was that it was not possible to map gene loci responsible for the remarkable quantitative variation in CNS size, regional architecture, or behavior.

3.2
A New Opportunity

The situation has changed radically in the past decade (Lander and Botstein 1989; Plomin et al. 1991; Belknap et al. 1992; Johnson et al. 1992; Tanksley 1993; Frankel 1995). Computational methods and molecular reagents – particularly the polymerase chain reaction method – have become so powerful and economical that it is now practical to systematically dissect complex polygenic traits such as brain weight into sets of single well-defined QTLs. Virtually any heritable trait in mice, whether structural, physiological, pharmacological, or behavioral, can be targeted for analysis. Recent examples in mice include epilepsy (Rise et al. 1991), effects of ethanol and haloperidol (Belknap et al. 1993; Plomin and McClearn 1993; Hitzemann et al. 1994; Kanes et al. 1996; Buck et al. 1997); patterns of sleep and activity (Toth and Williams 1998, 1999), and the mouse equivalent of anxiety (Flint et al. 1995). As illustrated in the work of Belknap and colleagues (1992), it is now feasible to continue the systematic genetic dissection of the mouse CNS begun in the late 1960s and to start identifying genes that underlie heritable variation in CNS size and structure.

Variation in brain weight is a classic polygenic trait, one that is influenced during development by the activity of hundreds, if not thousands, of genes.

Brain weight is also affected by maternal factors and myriad environmental factors (e.g. Collins 1970; Eleftheriou et al. 1975; Katz and Davies 1983; Wahlsten 1983). Finally, many factors that target body size have important pleiotropic or correlated effects on brain size, making the selectivity of action a critical problem (Lande 1979). From the point of view of genetic complexity, it is hard to imagine a morphometric trait that would be more difficult to resolve into individual QTLs.

We began this biometric analysis by weighing brains of numerous different types of mice. Table 1 is taken from a database that has been assembled over a 5-year period with contributions from Drs. Dan Goldowitz, Richelle Strom, Lu Lu, David Airey, and Guomin Zhou. For the great majority of animals, we have information on sex, body weight, age, and type and quality of fixation. For animals born at the University of Tennessee, we also generally know the size of the litter and the mother's parity. Most cases that we have studied were fixed by perfusion with mixed aldehydes (Williams et al. 1996a). This process leads to a reduction in brain weight of 3–4%, for which these data have been corrected. Weights include the olfactory bulbs, the paraflocculi, and the entire brainstem but exclude the dura, the pineal, and the pituitary.

3.3
Brain Weight Is Highly Variable

Brain weight is highly variable among strains reared in a common environment. For example, both A/J and DBA/2J have average brain weights close to 410 mg, whereas C57BL/6J and BALB/cJ have weights close to 510 mg. The variation within each strain is considerable even after compensating for differences in age, body weight, and sex by multiple regression (Williams et al. 1997). Two animals of the same sex and body weight taken from the same litter often have brain weights that differ by 10–20 mg. The coefficient of variation within isogenic groups shown in Table 1 averages about 5.5%, but when technical errors associated with fixation and dissection are taken into account, true non-genetic variation is close to 4%. In comparison, the retinal ganglion population of isogenic mice has a coefficient of variation that averages 3.6% (Williams et al. 1996a). We have explored the possibility that some of these differences in brain weight are due to variation in water content and the volume of the ventricles, and the short answer is that neither factor is important in mice older than 30 days. Wet and dry brain weights are very tightly correlated.

3.4
Sex and Age Effects on Brain Weight

Both sexes and a wide range of ages were studied. Surprisingly, in mice sex has no detectable effect on adult brain weight (Williams et al. 1997) and this

Table 1. Brain weights of 28 common inbred strains of laboratory mice, with a comparison to two previous studies

Inbred strains	Brain			Liters	Roderick et al. 1973	Fuller and Wimer 1966
	SA[a]	SD	CV%			
129/J	423	15	3.1	4	454	444
129/SvJ	430	17	3.9	4		
A/J	408	21	5.0	11	455	437
AKR/J	464	29	4.9	5	530	
BALB/cByJ	448	26	5.1	6		
BALB/cJ	524	28	5.2	12	540	502
C3H/HeJ	416	21	4.8	2		
C3H/HeSnJ	504	20	4.2	6		
C57BL/6J	499	21	4.4	23	489	449
C57BL/10J	459	20	4.2	3		
C57BLKS/J	463	19	4.0	8		
C57L/J	411	18	4.1	2	448	
C58/J	429	19	4.2	2	451	
CBA/J	462	7	3.0	1	508	
CBA/CaJ	437	21	4.8	3		
CE/J	472	23	5.0	7	476	
DBA/1J	403	23	5.9	4		409
DBA/2J	417	27	6.4	10	432	413
FVB/NJ	481	11	2.2	5		
LG/J	488	25	5.2	4	552	
LP/J	397	29	7.1	5	466	
NOD/LtJ	524	47	8.4	3		
NZB/BinJ	515	40	7.7	5		
NZW/LacJ	479	38	7.9	3		
PL/J	452	27	5.9	3	516	
SJL/J	419	26	6.3	7	450	413
SM/J	469	24	4.6	11	496	436
SWR/J	396	15	3.7	2	469	
Averages[b]	453	23	5.0	4.6	483/446[c]	438/446[c]

[a] Brain weights sex and age (SA) corrected. All values normalized to those of 75-day-old females without fixation. SD, standard deviation computed using individual values; CV, corresponding coefficient of variation (SD/mean).

[b] Litter average is geometric mean.

[c] Paired averages (483/446): first value from the original study; second value is average for the same set from our current database. Note the fair agreement with Fuller and Wimer (1966, $r = 0.83$). Values from Roderick et al. (1973) are consistently higher and the correlation is somewhat lower ($r = 0.78$). This difference may be due to their use of retired breeders killed by CO_2 asphyxiation.

otherwise important trait can be neglected for most purposes. In some strains, there is a significant age-related increase in brain weight even after sexual maturity is reached. There is also a significant correlation between body weight and brain weight. The correlation across strains listed in Table 1 is merely 0.2, but in some crosses, such as that between CAST/Ei and BALB/cJ, the correlation can rise to 0.8. Information on over 7500 mice and over 100 strains is available online at <http://nervenet.org>.

3.5
Large Differences Between Substrains

Perhaps the most remarkable aspect of the data summarized in Table 1 is the large differences in brain weight between several substrains of mice. For instance, brain weights of BALB/cByJ and BALB/cJ differ by 76 mg; C57L/J and C57BL/6J differ by 88 mg; C3H/HeJ and C3H/HeSnJ also differ by 88 mg. The closely matched and highly significant differences in these three pairs are intriguing. These differences were presumably generated by the recent fixation of variant alleles in a very small number of genes – probably one or two.

4
Mapping Brain Weight QTLs

4.1
QTLs Versus Mendelian Loci

QTLs are conventional genes that have two or more alleles that contribute to quantitative variation of specific traits (Roff 1997; Lynch and Walsh 1998). A trait may be a concentration or number, a size, weight or density, an activity or behavior, a severity index or an age-of-onset. QTLs are often contrasted with Mendelian loci that have discontinuous effects on phenotypes and predictable segregation patterns. In contrast, individual QTLs usually have more modest effects on a particular phenotype and are associated with phenotypes in a probabilistic way. A QTL might account for as little as 2% or as much as 50% of the total phenotypic variance. QTLs come in sets that collectively define a polygene. For example, at least three QTLs are currently known to control part of the twofold variation in numbers of retinal ganglion cells (Williams et al. 1998a; Strom 1999), and at least 30 QTLs appear to modulate body size (Cheverud et al. 1996; Brockmann et al. 1998). In the next several pages I explain the process of mapping a QTL, in this case one of the first QTLs demonstrated to modulate brain weight in the mouse. There are four key steps in mapping QTLs.

4.2
Step 1: Assessing Trait Variation

The first step is to identify significant variation in phenotypes among individuals, or, in the case of laboratory mice, among inbred strains. Variation is an absolute necessity. It is the signal we are trying to pinpoint on a map of the genome. The greater the heritable variation, the better the prospects of success. Figure 1 illustrates the wide variation in brain weight among two inbred strains (BALB/cJ and CAST/Ei) and among their inter-cross and backcross progeny. This is a cross that I will use throughout this section as a specific example of mapping a brain weight QTL. Note that brain weight in the F1 generation overlaps that of the BALB/cJ parental strain. Brain weight may be inherited as a dominant trait, but since all of these F1 progeny were born to BALB/cJ mothers, maternal nongenetic factors are also likely to be an important factor. The spread of points among

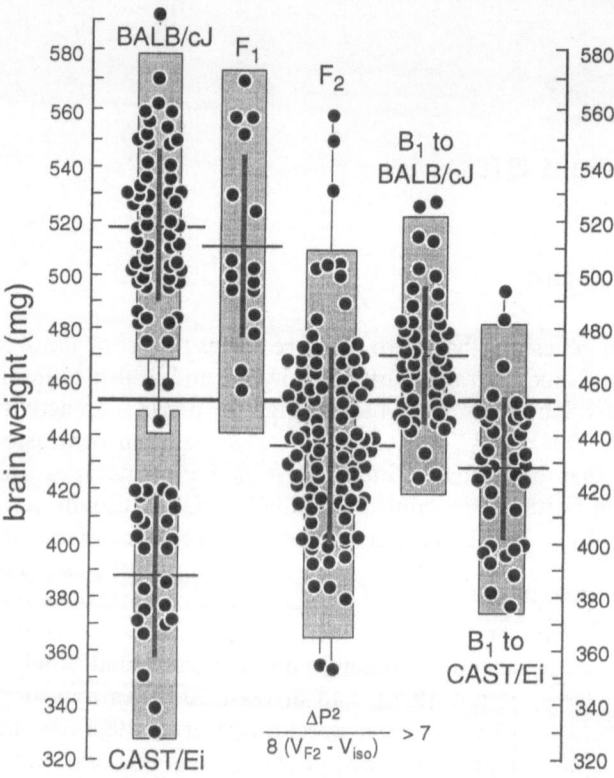

Fig. 1. Variation in brain weight between two inbred strains and their test cross progeny. The two parental strains BALB/cJ and CAST/Ei are shown far *left*. Each *dot* represents brain weight of an individual mouse; *short horizontal lines* through each *box* indicate group averages; *vertical bars* within each *box* mark indicate standard deviations; *horizontal line* at 454 mg is the mid-parental value (average of BALB/cJ and CAST/Ei). *Box heights* are generally ±2 SD. F1 animals were crossed back to both parental strains, giving rise to the two sets of B1 progeny shown *right*. *Equation* at bottom of figure is the Wright-Castle equation (Wright 1978) for estimating the minimum number of effective factors (single or linked QTLs) that contribute to the genetic variance of a trait. *Delta P*, difference between parental strain means; *VF2* and *Viso*, variances of F2 and isogenic strains, respectively. For these data, we estimate that at least eight polymorphic genes account for the increased variance of the F2 relative to that of the isogenic groups. Data are not corrected for variation in age, sex, or body weight

F2 individuals is somewhat greater than that of either parental strain. This increase in variance is due to the segregation and assortment of QTLs that affect brain weight. No fewer than seven QTLs are needed to account for the differences seen among members of this cross (Wright 1978), but using our small sample of F2 animals ($n = 98$), we have only succeeded in mapping one of these QTLs.

4.3
Step 2: Estimating Heritability

The second step in QTL mapping is to verify that a substantial fraction of the variability of the trait is heritable (Crucio 1992; Wahlsten 1992; Williams et al. 1996a). In a standard mouse colony, variation in brain weight has a heritability that ranges from 0.35 to 0.7 (Roderick et al. 1973, 1976; Seyfried and Daniel 1977; Fuller 1979; Henderson 1979; Atchley et al. 1984; Williams et al. 1996b; Strom and Williams 1997; Strom 1999). Heritability estimates can admittedly be problematic (Lewontin 1957; Eleftheriou et al. 1975), and in the context of the heritability of human intelligence, Wahlsten (1994) comments, "I would feel more secure riding a three legged moose over thin ice than relying on a heritability coefficient to help me understand the origins of individual differences or predict future levels of intelligence." However, it can still be useful to go through the process of computing heritability. The reason is that we need to have some idea of the approximate fraction of variance in our sample population that is due to heritable genetic factors before we attempt to map QTLs. The heritable variance is what we are trying to assign to a set of QTLs. While heritability estimates may be labile, the QTLs that we map are anchored in the genome itself.

Heritability is the fraction of the total variance in a trait that is generated by the segregation and assortment of allelic variants at the many gene loci that influence a trait. (New mutations contribute very little to heritability under all but extreme environmental conditions.) A simple way to measure heritability is to compare traits between parents and their offspring. Figure 2 compares the average brain weight of parents with that of their first litters. The correlation between values is a direct estimate of heritability, in this case what is called the narrow-sense, or additive, heritability (Lynch and Walsh 1998). The correlation for this particular dataset is 0.38. Broad-sense heritability which includes variance due to dominance effects and non-linear interactions between different genes is likely to be as high as 0.5. In comparison to these estimates, variation in neuron number has a broad-sense heritability of approximately 0.8 for granule cells in the dentate gyrus (Wimer and Wimer 1989) and between 0.7 and 0.9 for retinal ganglion cells (Williams et al. 1996a, 1998a; Strom 1999). These values are certainly sufficiently high to motivate a QTL analysis.

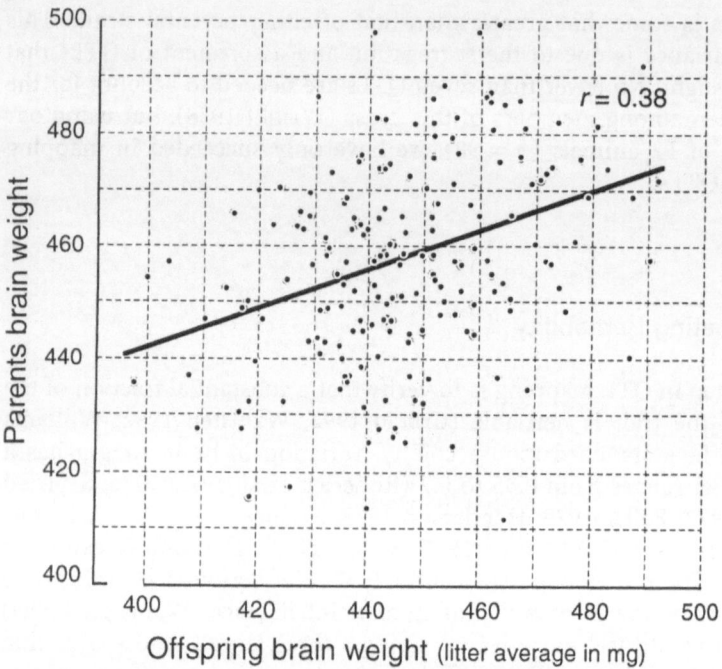

Fig. 2. Correlation between brain weights of parents and their offspring estimates heritability. Animals are from a multigenerational cross between C57BL/6J and DBA/2J inbred strains (G. Zhou and R. W. Williams, unpublished). Parental values are the average unfixed weights of mothers and fathers without correction for variation in age or body weight. Offspring data are the average per litter. Brains weights are unfixed and without correction for variation in body weight, sex, or age. Offspring weights tend to be slightly less because they are on average about 50 days younger. Correlation between pairs of values is 0.38 and is a direct estimate of narrow-sense heritability of brain weight in this cross and environment. Correlations between mothers and offspring and fathers and offspring do not differ significantly. Thus, this estimate of heritability is not inflated by maternal effect

4.4
Step 3: Phenotyping and Genotyping Members of an Experimental Cross

The third step is to gather phenotype and genotype data from a set of animals appropriate for QTL mapping. Several different types of crosses can be used to map QTLs (Taylor 1978; Groot et al. 1992; Frankel 1995; Darvasi 1998; Vadasz et al. 1998; Williams 1998b, <http://nervenet.org/papers/shortcourse98.html>). Figure 1 already introduced one of the most common, the F2 intercross. The central idea behind the intercross is to allow high and low alleles of QTLs inherited from the two inbred strains to segregate and assort independently from unlinked marker loci. The only marker loci that will consistently be associated with high, intermediate, and low trait values in the set of F2 progeny are those marker loci that are closely linked to QTLs (Tanksley 1993; Williams 1998b).

4.4.1
Phenotyping and Regression Analysis

Obtaining brain weights is quick and easy, but before we can use these weights to map QTLs, we need to deal with the issue of specificity of gene action. The brain weight data we have considered so far have not been corrected for significant differences in the mean body weight among mice. The heritability that we blithely assigned to brain weight may actually be a consequence of heritable variation in body size. Unless we adjust our brain weight phenotype appropriately, we risk mapping body weight QTLs (Hahn and Haber 1978; Lande 1979). To ensure that we are mapping what we want to map, we need to factor out variation in brain weight that is predictable from variation in body weight, sex, age, and other variables for which we have data.

A crude way of factoring out body size is to use the ratio of brain weight to body weight as a phenotype. A computationally and conceptually far more powerful approach, however, is to use multiple regression analysis to remove predictable variance associated with body size and any other important but perhaps extraneous variables (Williams et al. 1997). The same logic applies when the aim is to map QTLs that modulate the size of particular CNS cell populations (Williams et al. 1998a,b). We do not want to map generic brain weight QTLs and inadvertently ascribe the effects of these QTLs to single cell populations. In this case, we can exploit multiple regression to remove variance in cell number that is associated with variation in total brain weight, although this can be a tricky procedure because QTLs will often have graded or differential effects on different parts of the body and brain. For example, allelic variants of a hormone receptor may generate 5% of the variance in total brain weight, 10% of the total variance in hippocampal weight, and 50% of the variance in CA3 pyramidal cell numbers. The actions that we attribute to this hormone receptor QTL will therefore depend on what sort of data we collect and how we statistically process these data. In this example, a single QTL could easily end up with three different names – *brain weight modulator 1*, *hippocampal volume modulator 1*, and *CA3Py1*. A careful analysis of a QTL with pleiotropic effects may reveal that the intensity of its effects varies widely across different parts of the CNS. Until much more is known about the spatial and temporal distribution of QTL effects, we will have to accept the likelihood that QTLs will often have differential effects on multiple structures.

Figure 3 provides a graphic explanation of a simple regression analysis run on the set of F2 intercross animals previously illustrated in Fig. 1. For every 1-g increase in body weight there is approximately a 7.9-mg increase in brain weight. Sixty-six percent of the variance in brain weight can be predicted by body weight alone. Sex in this case is also a significant predictor, and at a

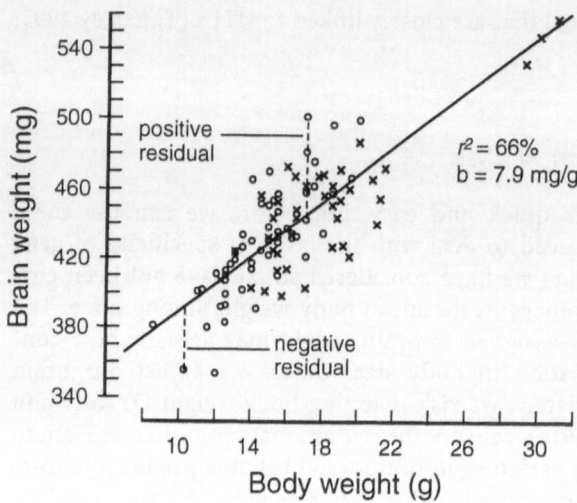

Fig. 3. Regression analysis of body and brain weight, used to minimize effects of variance in brain weight due to differences in body weight. *Crosses*, males; *open circles*, females. Rather than using each animal's actual brain weight as a phenotype, we compute a residual brain weight based on body weight and sex. Examples of positive and negative residuals are marked. In this dataset, the correlation is 0.81 and r^2 (the coefficient of determination) is 0.66; b is the coefficient (slope) of the regression equation

given body weight, females have brain weights that are on average 9.4 mg heavier than those of males. However, in this particular sample, neither age nor the logarithm of age were useful predictors ($P \sim 0.6$). Table 2 is a statistical synopsis of a multiple regression analysis that takes both body weight and sex into account. When we use the regression equation and coefficients in Table 2 to compensate for differences in body weight and sex, we absorb 67.4% of the variance in brain weight. The residual 32.6% of the variance is generated by technical error, by other non-controlled environmental effects, and most importantly to us by the QTLs that we are trying to locate on the map of the mouse genome. Rather than using the original brain weight data to map, we use the residual deviations illustrated in Figure 3. For each animal, we compute a derived phenotype that is the difference in milligrams between the predicted weight of that animal given its weight and sex and its actual brain weight. By mapping residuals we improve our ability to detect QTLs that are likely to have selective effects on CNS development.

Table 2. Regression analysis of brain weight in an F2 intercross. $r^2 = 67.4\%$

Variable	Coefficient	SE	P
Body (g)	8.5	0.64	≤0.0001
Sex (1 = F)	9.4	4.7	0.047

4.4.2
Genotyping

In a typical analysis of F2 progeny, three to five marker loci spaced about 15–25 centimorgans (cM) apart are genotyped on each of the 20 chromosome pairs. These marker loci are usually repetitive microsatellite DNA sequences that consist of variable numbers of cytosine–adenine (CA) dinucleotide repeats. One strain of mouse may have a microsatellite with 30 CA repeats, whereas another strain may have a microsatellite with 40 CA repeats. The 5′ and 3′ flanking sequences of each microsatellite are unique to that part of the genome, but they are also highly conserved among strains of mice. This makes it possible to design PCR primers that selectively amplify a microsatellite located at a precisely defined chromosomal position (Dietrich et al. 1994).

To map QTLs responsible for a part of the variation illustrated among the F2 progeny, genomic DNA from each animal is extracted and genotyped using the polymerase chain reaction. The procedures are simple and are described in detail at http://nervenet.org/papers/PCR.html. There are three possible genotypes at each microsatellite locus: *BB*, *BC*, and *CC*. Approximately 110 microsatellite loci that effectively sample the entire genome of

Table 3. Quantitative comparison of phenotypes and genotypes

Identifiers		Phenotypes (mg)		Geno-type	Models			Sorted		
Case	PCR plate order	Brain	Brain residuals	D6Mit327	Add	B Dom	C Dom	Add = 1	Add = 0	Add = −1
090894F	1	469	38	B	1	1	1	38		
041195K	2	496	26	H	0	1	−1		26	
071095A	3	502	8	H	0	1	−1		8	
040695M	4	489	21	H	0	1	−1		21	
051295G	5	475	−1	H	0	1	−1		−1	
041195I	6	489	18	C	−1	−1	−1			18
090894I	7	436	−23	H	0	1	−1		−23	
081595V	8	550	52	B	1	1	1	52		
041195M	9	463	−27	H	0	1	−1		−27	
...	10–89	...	...	...	...	...	...	...	...	...
081595M	90	496	−7	B	1	1	1	−7		
101295I	91	501	−1	H	0	1	−1		−1	
040695S	92	477	8	H	0	1	−1		8	
091895L	93	468	−11	H	0	1	−1		−11	
071095D	94	443	−31	C	−1	−1	−1			−31
080394Z	95	481	−3	C	−1	−1	−1			−3
072195G	96	496	3	C	−1	−1	−1			3
					0.39	**0.23**	**0.40**	*17*	*−2*	*−8*

Add, additive effects; Dom, dominance deviations.
Bold values at bottom are correlations between residuals and model values; italic values are means.

each animal were genotyped. Table 3 illustrates the organization of phenotype and genotype data for 96 animals as entered into a spreadsheet. The first two columns are case identifiers. The third and fourth columns list phenotypes in milligrams, while the fifth column lists the genotype of each animal at a particular microsatellite locus on chromosome (Chr) 6 called *D6Mit327*. The three genotypes are listed as *C* (corresponding to *CC*), *H* (the heterozygote *CB*), and *B* (corresponding to *BB*). As shown in the sixth column, these three genotypes can be converted into values of −1, 0, and +1. The sixth and seventh columns are values assigned to each genotype assuming that either the *B* allele or the *C* allele is dominant. For example, if the *C* allele is dominant then all of the heterozygous animals are assigned the low trait value of the CAST/Ei parent, −1 in this case.

4.5
Step 4: The Statistics of Mapping QTLs

We now have all the necessary data and are poised to assess whether QTLs have been discovered, and if so, with what precision and confidence (Churchill and Doerge 1994; Lander and Schork 1994). Mapping QTLs involves finding marker loci for which the three genotypes match up well with variation in the phenotype. BALB/cJ has a much larger brain than CAST/Ei. If a QTL modulating brain weight is located near one of the microsatellites then F2 animals that are homozygous for *B* alleles at that marker should have heavier brains than those homozygous for *C* alleles. Referring to Table 3, we test whether or not there is a significant correlation (or regression coefficient) between the numerical values (−1, 0, and +1) in the sixth through eighth columns and brain weight residuals in the fourth column. These correlations are listed at the bottom of Table 3.

A complementary way to explore these data is to determine whether brain weight residuals of animals with the *BB* genotype are greater than those of groups of animals with the other two genotypes. This type of categorization is shown on the right side of Table 3. The average residual of individuals with the *BB* genotype is 17.3 ± 4.8 mg (bottom right), whereas that of *CC* individuals is −7.8 ± 3.4 mg. Half of the difference between these means is an estimate of the additive effect of substituting a low *C* allele with a high *B* allele, a value of 12.6 mg in this case. The *BC* heterozygotes in this sample have an average phenotype that is 6.4 mg lower than the midpoint between the parental *BB* and *CC* genotypes. This deviation estimates the dominance of the *C* allele.

In this analysis we have tested whether a QTL that influences brain weight is located close to the single microsatellite marker *D6Mit327*. But we would like to scan the entire genome in the same way. Is the correlation of 0.39 between phenotypes and genotypes at *D6Mit327* the highest that there is across the entire set of 21 chromosomes in the mouse? If we do this analysis at each of the 110 marker loci, we discover that genotypes at *D6Mit327* match variation in brain weight much better than any other marker (Table 4). In fact, the probability of getting such a good match by chance alone if one performed only a single test is about 1 in 10 000. This value is referred to as the point-wise, or nominal, probability of linkage. In addition to listing nominal probabilities, Table 4 lists several other interesting statistics and coefficients. One of these is the likelihood ratio statistic (LRS), a value that, like the logarithm of the odds ratio (the LOD score), is used to assess whether or not a QTL is likely to be present close to the marker locus (Haley and Knott 1992). The next two columns list the additive effects of allele substitutions and the predicted dominance deviation described in the preceding paragraph. The last column lists the fraction of the variance that can be accounted for by genotypes at the marker locus. This latter value is just the square of the correlation coefficient that we already computed in Table 3. For example, at *D6Mit327* the estimate of explained variance is 16%.

To refine the analysis of this QTL near *D6Mit327*, we could genotype neighboring markers to determine whether they have even stronger association with variation in brain weight. This additional genotyping is usually not necessary because we can infer the genotypes that are likely to be present between neighboring marker loci. For example, if a mouse has a *BB* genotype at one marker and a *CC* genotype at a flanking marker, then halfway between these markers the genotype will most probably split the difference and be *BC*. Comparing predicted genotypes with actual phenotypes in the interval between marker loci is referred to as interval mapping (Lander and Botstein 1989). This refinement can significantly improve the statistical power of a QTL search and makes it possible to distinguish between a weak QTL that is

Table 4. Results of a genome-wide screen for a brain weight QTL

Locus	Chr	P^a	LRS^b	Add^c	Dom^c	$\%^d$
D2Mit295	2	0.02326	7.5	7.02	−3.29	5
D3Mit23	3	0.02901	7.1	6.56	−5.71	5
D4Mit172	4	0.01964	7.9	5.73	7.60	6
D4Mit151	4	0.02791	7.2	6.95	5.57	5
D6Mit273	6	0.04654	6.1	−2.84	−9.19	4
D6Mit327	6	**0.00011**	**18.3**	**12.59**	**−6.42**	**16**
D7Mit193	7	0.03257	6.8	8.68	3.24	5
D7Mit120	7	0.00747	9.8	12.13	0.78	9
D7Mit31	7	0.02386	7.5	7.85	1.92	5
D12Mit158	12	0.01966	7.9	6.68	6.99	6
D16Mit65	16	0.00891	9.4	7.05	−5.61	7
DXMit54	X	0.03059	7.0	5.73	4.05	5

[a] Point-wise or nominal probability of achieving an LRS value by chance.
[b] Likelihood ratio statistic (4.61 times the LOD score).
[c] Additive effects and dominance deviations in milligrams.
[d] Percentage of variance that can be explained by a QTL tightly linked to the marker locus.

near to a marker and a strong QTL that is located between markers. In other words, interval mapping improves the ability to locate a QTL and to estimate the effects that it is likely to have on the phenotype.

The results of the more fine-grained interval mapping analysis of Chr 6 are illustrated in Fig. 4. The horizontal line at the top represents most of Chr 6 (from 19 cM to 70 cM). We needed to type only four marker loci – *D6Mit273*, *D6Mit71*, *D6Mit327*, and *D6Mit113* – to effectively scan the entire chromosome. Using the genotype data and the program Map Manager QT (Manly and Olson 1999; <http://mcbio.med.buffalo.edu/mmQT.html>), the LRS was computed at 1-cM intervals. These values were then used to generate the shaded likelihood profile map. As we suspected on the basis of our initial point-wise statistical analysis, there appears to be a QTL influencing brain weight very near *D6Mit327*.

4.5.1
Permutation Analysis

The process of mapping QTLs involves computing hundreds of linkage statistics across the entire set of chromosomes. Given the large number of statistical tests there is a strong probability of getting a "significant" asso-

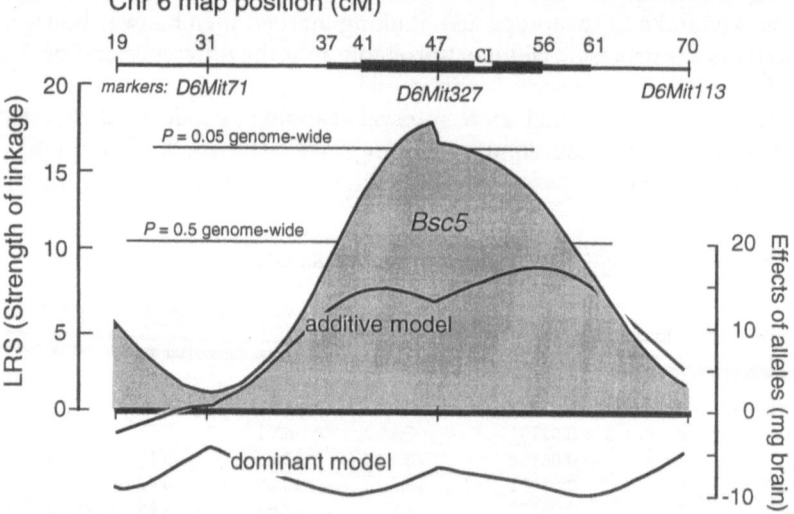

Fig. 4. Linkage of the QTL *Bsc5* to chromosome 6 in a cross between BALB/cJ and CAST/Ei. *X-axis* represents position along Chr 6. The most proximal marker that we typed (*D6Mit273*) maps at 19 centiMorgans (cM), whereas the most distal maps at 70 cM. The *Bsc5* locus is most likely to map about 1 cM proximal to microsatellite marker *D6Mit32*. The confidence interval (CI) of this estimate (*bold black lines*) is wide: from 37 to 61 cM for a two-LOD CI (95% probability), and from 41 to 56 cM for a one-LOD CI. Genome-wide probability thresholds (Fig. 5) are marked by *fine horizontal lines*. The *right scale* and *two lower curves* indicate approximate additive effect and dominance deviations generated by *Bsc5*. Substitution of a single BALB/cJ allele for a CAST/Ei allele at *Bsc5* may be responsible for a 15-mg gain in brain weight

ciation by chance alone. The nominal probabilities listed in Table 4 tell us little about the genome-wide probability that we have discovered a QTL (Lander and Kruglyak 1997). We need to compensate for these multiple tests. The appropriate correction factor depends on the particular distribution of trait values and the quality and quantity of genotype data.

A conceptually simple but computationally tedious permutation test can be used to estimate the distribution of best LRS scores that one might expect to get by chance with a given dataset (Churchill and Doerge 1994). This procedure randomly reassigns phenotype values listed in Table 3, and then remaps the jumbled dataset to get a new version of Table 4. For each per-mutation the program keeps track of the single highest LRS score. The process is carried out another 9999 times. Figure 5 shows a histogram of the best LRS scores for 10 000 permutations of the data in Table 3. The average of these best LRS scores was close to 10. The distribution of peak scores can now be used to gauge the probability of obtaining an LRS of 18.3 by chance alone. Only 2% of permutations do this well or better. We can therefore be rea-sonably confident that we have mapped a QTL modulating brain weight to Chr 6. This is the fifth QTL that Richelle Strom and I have mapped (Strom 1999; R. C. Strom and R. W. Williams, in preparation), and we have named it *brain size control 5* (*Bsc5*). *Bsc5* maps on Chr 6, approximately 1 cM proximal to *D6Mit327*. *Bsc5* has not been mapped with much precision: the 95% confidence interval is defined by the width of the map profile 2 LOD units (or 9.2 LRS units) to either side of the peak, in this case between 37 and 61 cM. This 24-cM interval contains approximately 1200 genes, and perusing a list of

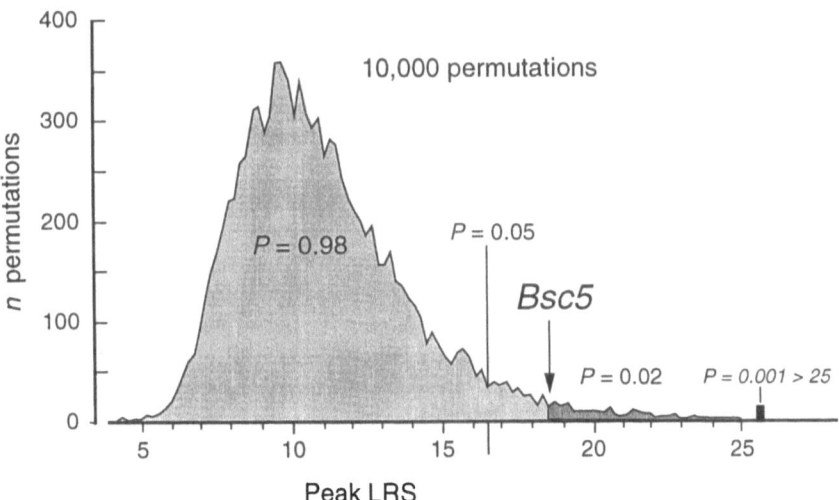

Fig. 5. Permutation analysis of the *Bsc5* locus. Genome-wide thresholds for estimating strength of linkage are computed by randomly permuting data such as those listed in Table 3. This histogram tallies single best LRS scores for each of 10000 permutations. The two-tailed probability of a random dataset having a peak LRS score better than 18.4 is 0.0215 ± 0.0015

candidates at this point is little more than an entertaining exercise in optimism. A quick scan of this region using the Mouse Genome Database
<http://www.informatics.jax.org/locus.html> reveals one interesting candi-
date, the thyrotropin releasing hormone gene that maps at 43 cM.

4.6
Cloning QTLs

Mapping QTLs is the initial reconnaissance stage in a systematic effort to
explore mechanisms that modulate the development of the CNS. The next step
is to match each QTL with a single gene and its alternative alleles. QTLs will
generally need to be mapped with a precision of 1 to 2 cM, a chromosomal
interval that will typically harbor 50–100 genes. Achieving this level of accuracy
is not impractical, although it will often require an analysis of 1000 or more
animals (Darvasi 1997, 1998). A small subset of positional candidate genes can
then be chosen for further analysis on the basis of expression patterns, known
function, and differences in DNA sequence among strains. The efficiency of the
candidate gene approach will improve greatly in the next decade. The genome
of C57BL/6J will have been sequenced within several years, and it is also likely
that the utility of this code will be enhanced with sequence data from other
major inbred strains such as 129, A, BALB/c, C3H, DBA/2, CAST/Ei, and
SPRET/Ei. Once sequence data have been combined with expression maps
for different parts of the mouse brain, it will be possible to winnow a set of
candidate genes to a very short list. If in the example that we have been using
the thyrotropin releasing hormone gene survives this filtration, then we may
be justified in comparing its sequence among strains with different phenotypes. The conversion of quantitative phenotypes (e.g. low to high) by
substituting alleles of one strain with that of another strain will provide the
final and most compelling support that the identity between a QTL and a
particular sequence variant has been made correctly (Frankel 1995).

It is important to realize that QTLs are not invariant across different
populations of mice. A QTL can be identified because it is polymorphic in a
particular population or cross. The same gene may not necessarily be polymorphic in another cross. If a gene is not polymorphic, it cannot generate
phenotypic variance; it is effectively a virtual QTL. While we succeeded in
mapping a QTL on Chr 6 that influences brain weight in a cross between
BALB/cJ and CAST/Ei, we will not necessarily find this QTL in a cross
between DBA/2J and C57BL/6J because the latter two strains may possess
the same allele at that particular locus.

4.7
Probability of Success

What is the probability of successfully mapping one or more QTLs? For CNS
traits that have heritabilities above 50%, it will usually be possible to map

several QTLs in even a small cross consisting of 200 to 500 individuals. Behavioral traits such as open field activity tend to have relatively low heritabilities (<30%), yet they have been successfully dissected into sets of QTLs (Flint et al. 1995). For example, Le Roy, Roubertoux, and colleagues (Le Roy et al. 1999.) have successfully mapped over 12 QTLs that modulate the development of several behavioral traits in preweanling mice. Morphometric CNS traits tend to have higher heritabilities (Roff 1997) and over the past few years we have identifying close to 20 QTLs that modulate the size and cellular composition of the brain and eye (Strom and Williams 1997; Williams et al. 1998a; Gilissen and Williams 1997; Airey et al. 1998; Strom 1999; Zhou and Williams 1999b; Williams and Zhou 1999). The main constraint is the number of animals that can be phenotyped. In most cases, approximately 150 animals were phenotyped per QTL. Note that if multiple polymorphic loci are clustered near one another, then a small intercross may only detect a single poorly localized QTL, in essence a polygenic QTL. A large high-resolution cross, especially an advanced intercross of the type described by Darvasi (1998), may adequately distinguish individual loci (Williams 1998).

5
Neuron and Glial Cell Numbers in Adult Mice

While it is satisfying to decompose variation in brain weight into sets of individual QTLs, we still need to determine what parts of the CNS and what cell populations are most and least affected. For example, is there a subset of QTLs that specifically modulates the size of the cerebellum or the proliferation of granule cells? We are taking two approaches to begin to answer these types of questions. The first approach is literally to disassemble each brain and map QTLs that modulate the size of the parts of the CNS. For example, after controlling for variation in brain weight, we have succeeded in mapping QTLs that have relatively intense effects on the weight of the cerebellum (Gilissen and Williams 1997; Airey et al. 1998; Airey et al. 1999). We have also mapped at least one QTL that has a selective effect on the weight of the hippocampus (Lu et al. 1999). This more fine-grained analysis puts us in a better position to get back to mechanistic explanations of QTL effects (Strom and Williams 1999).

A second approach is to carry out systematic stereological studies of individual nuclei and cell types. The main impediment is the stamina that is needed to section, stain, and count particular nuclei or regions in hundreds of cases. But if a single cross could be could be shared among many neurogeneticists and used to analyze many different CNS structures, then the effort required to map each QTL could be reduced substantially. A library of mouse brain sections for particular strains and particular crosses would be especially useful because investigators could map multiple QTLs without processing any tissue or genotyping any animals. As described in the next section, my colleagues and I are assembling resources to make such collaborative QTL mapping feasible.

5.1
The Mouse Brain Library at http://nervenet.org/mbl/mbl.html

A key requirement for collaborative QTL mapping is an extensive library of sectioned material suitable for quantitative analysis. Dr. Glenn Rosen (Beth Israel Deaconess Medical Center) and I have begun to assemble such a collection. The Mouse Brain Library (MBL) at nervenet.org consists of digital images taken from approximately 600 brains. The initial set of cases have all been processed in celloidin, cut at 30 μm in coronal or horizontal planes, and stained with cresyl violet. Figure 6, a slide taken from the MBL, shows a 1-in-10 series of horizontal sections cut through a 476-mg brain from a C57BL/6J

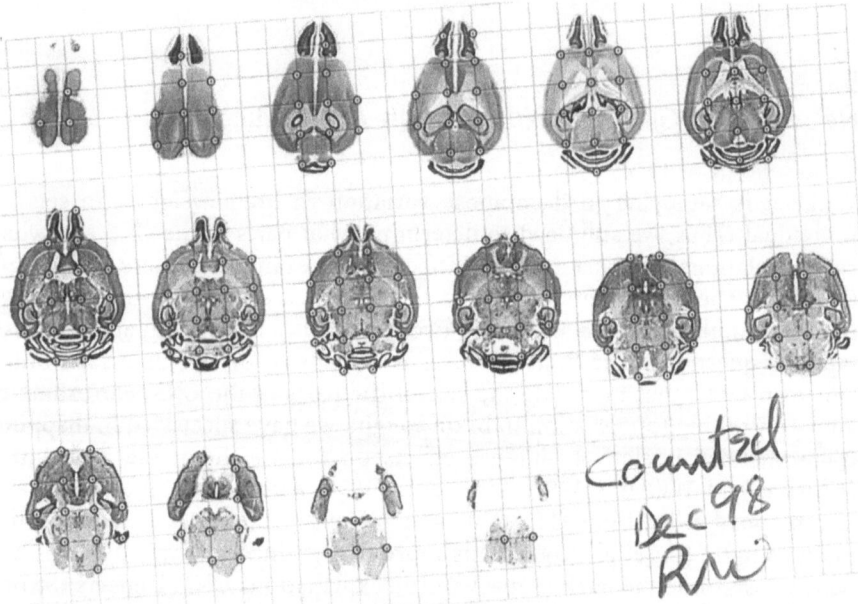

Fig. 6. Nissl-stained sections of an adult male mouse brain (C57BL/6J, case 232, slide A). This slide from the Mouse Brain Library contains every tenth horizontal section (30 μm thick; celloidin embedded). Dorsal-most section is in the *upper left corner*; caudal-most is in the *lower right quadrant*. *Small dots* over the tissue represent sample sites used for high-magnification cell counts and to estimate brain volume after processing. To minimize sample periodicity and ensure a more representative set of sample sites, the grid has been intentionally oriented at a randomly selected angle between 5° and 40° (in this case ~5°). Four such slides were prepared for each case

mouse. The minimum resolution of MBL images is 25 μm/pixel in the xy plane. Every fifth section was photographed, giving a resolution of 150 μm in the z axis. Images can be downloaded from the MBL and then analyzed using a program such as NIH Image. The quality is sufficient to segment the brain into well over 500 regions. A large subset of cases has been imaged at a resolution of 4.5 μm/pixel, a magnification at which single large cells can often be resolved. For purposes of QTL mapping, the collection contains an average of eight cases from each of the 35 BXD/Ty recombinant inbred (RI) strains, the set of AXB/Pg and BXA/Pg RI strains, and over 20 of the most common inbred strains listed in Table 1.

5.2
Numbers of Neurons and Glial Cells in the Brain of a Mouse

Slides such as the one shown in Fig. 6 can be used for a variety of detailed morphometric studies. For example, the tissue is ideally suited for analysis using either the disector or its close relative, direct three-dimensional counting (Williams and Rakic 1988). Our immediate objective is to uncover QTLs that control total CNS cell number. Estimating the total neuronal and glial population is simple, although in practice it takes about 10 hours. The slide reproduced in Fig. 6 was sampled at fixed 2000-μm steps, and fields were counted using differential interference contrast optics and a 100× oil-immersion objective (Williams and Rakic 1988; see <http://nervenet.org/papers/3DCounting.html> for details on methods and implementation). Each sample site, or counting box, had dimensions of 32 × 33 × 20 μm xyz. The final estimate of cell number was based on 175 counts obtained at grid points that intersected tissue. A total of 1908 cells were counted and categorized.

It is simple to compute the total cell number of this brain by multiplying the mean cell density by the estimated brain volume obtained by the point count. Each grid point represents a 1.2-mm^3 volume of tissue (section interval × section thickness × grid dimensions, or 10 × 30 × 2000 × 2000 μm^3). Since 175 sites intersected tissue, the brain volume after processing is ~210 mm^3 (24% linear shrinkage based on the fixed brain weight of 476 mg). The brain of this individual contained approximately 75 million neurons, 23 million glial cells, 7 million blood vessel-related endothelial cells, and 3-4 million miscellaneous pial, ependymal, and choroidal cells. Granule cells of the cerebellum made up just under half of the total neuronal population (34 million), a value in close agreement with a previous estimate (Wetts and Herrup 1983). The reliability of the method can be determined easily by an independent count of another slide. It will not be difficult to perform counts like these for all of the BXD strains, and this will make it possible to map QTLs that have specific effects on neuronal and glial cell population size.

6
Mapping QTLs that Modulate Neuron Number

6.1
Mapping Cell-Specific QTLs

It is practical to map QTLs that affect individual neuronal populations. We have done this type of fine-grained analysis for one of the more accessible population of neurons in the CNS: the projection neurons of the retina, also known as retinal ganglion cells (Williams et al. 1998a). One reason why we chose this population is that it is possible to count these cells easily and precisely. Each cell has one and only one axon in the optic nerve, and a quantitative electron microscopic census of axons in a single cross-section of the nerve provides a reliable and unbiased estimate of total neuron number (Rice et al. 1995; Williams et al. 1996a).

We began by estimating the size of this cell population in 5–10 individuals from each of 5–10 different inbred strains, but we have now extended the analysis to 20 common inbred strains (Williams et al. 1996a; Strom 1999; Zhou and Williams 1999a). The range of variation among strains was substantial and motivated us to count ganglion cells in ~8 individuals from each of the 26 BXD strains and 12 BXH strains. Strain averages tended to fall into well-defined modes that corresponded to the parental strain averages of 55 000 (C57BL/6J) and 63 000 (DBA/2J). This striking non-normal distribution of phenotypes across the set of BXD RI strains suggested that two or three QTLs were modulating neuron number.

6.2
The *Nnc1* Locus

Using 26 strains of BXD mice, we were able to map the major QTL that is primarily responsible for the bimodality of strain averages (Williams et al. 1998a). There is an excellent correspondence between variation in retinal ganglion cell numbers among these inbred strains and their genotypes on distal chromosome 11. At the *Tstap91* locus the correlation reached a peak of 0.69. The genome-wide probability of getting a correlation this high by chance is less than 0.01. We named this QTL *neuron number control 1* (*Nnc1*). Given that only a single neuronal population was studied, this name may seem too broad. Our justification is that we know little about possible pleiotropic effects of this QTL on other neuronal populations. We do know that the QTL is not associated with brain weight in adult mice (the parental strain with the heavier brain has lower ganglion cell number) and it is unlikely that *Nnc1* modulates cell populations throughout the CNS. However, preliminary analysis suggests that *Nnc1* may have reciprocal effects on cell populations in the inner nuclear layer; the high ganglion cell allele inherited from DBA/2J may actually be associated with low cell numbers in the inner nuclear layer (Williams et al. 1998a,b).

6.3
Mechanisms of QTL Action

A QTL does not need to be cloned before it can be used to study mechanisms of brain development or function. For example, we wanted to determine whether *Nnc1* modulates neurogenesis or cell death (Strom and Williams 1998). To answer this question we counted the cell population after neurogenesis, but before the onset of cell death. We were able to show convincingly that the bimodality is produced by a fundamental difference in the total production of retinal ganglion cells. *Nnc1* must influence the generation of retinal ganglion cells. There are several ways in which *Nnc1* could control this process: through variation in the progenitor pool size, pathways of cell differentiation, or cell cycle parameters. With such a robust effect and so many strains with alternate phenotypes and alleles, it should now be possible to explore the relative importance of these processes and define more precisely how *Nnc1* modulates neurogenesis.

6.4
Candidate Gene Analysis

Nnc1 illustrates the power of candidate gene approaches to cloning QTLs. *Nnc1* has been mapped to a 3-cM interval between *Hoxb* and *Krt1*. Remarkably, this short interval which makes up ~0.2% of the mouse genome includes three superb candidate genes whose products are expressed in the developing retina and are known to be involved in the differentiation or proliferation of retinal cells. These candidates are the retinoic acid alpha receptor (*Rara*), the thyroid hormone alpha receptor (*Thra*), and the ErbB2 receptor (Williams et al. 1998a). The thyroid hormone receptor is the most tantalizing of the three because its natural ligand, thyroxine, is known to modulate the rate and timing of ganglion cell differentiation in *Xenopus laevis* during metamorphosis (Hoskins 1985). Thyroid hormone receptors dimerize with other members of the steroid hormone receptor family (Piedrafita and Pfahl 1995), and a unique combination of isoforms could explain cellular and regional specificity of action. The most powerful way to test whether or not *Nnc1* and *Thra* are one and the same would be to reciprocally convert high and low strains by swapping the two alleles. This is a tough experiment, and an expedient intermediary step is to test effects of targeted deletions of candidate genes. For example, loss of one candidate gene, *Rara*, by homologous recombination (Luo et al. 1996) does not perturb the size of the ganglion cell population (Zhou et al. 1998). In contrast, loss of the *Thra* gene (Wikström et al. 1998) has effects on the phenotype that may rival those of *Nnc1*.

The level of variation that we exploited to map and characterize the *Nnc1* locus is by no means unusual. Numbers of granule cell in the dentate gyrus of mice vary from 320 000 in C58/J to 525 000 in LG/J (Wimer et al. 1978; Wimer and Wimer 1989; 15 strains); Purkinje cell numbers vary from at least 130 000 to 200 000 (Wetts and Herrup 1983; Herrup et al. 1996; 5 strains); and the number of neurons in the caudate varies from 2.75 million to 4.75 million (G. Rosen, pers. comm.). A persistent investigator should in many cases be able to map and ultimately clone QTLs modulating each of these populations. These differences in neuron number may in some cases reflect the diversity of biochemical and behavioral traits (Seyfried et al. 1979; Sprott and Staats 1981; Plomin et al. 1991; Festing 1993; Dains et al. 1996; Vadasz et al. 1998).

7
Conclusion

The quotations taken from the work of Ramón y Cajal and Lashley that introduced this chapter emphasize that the most prominent differences between the brains of mice, chimpanzees, and humans are quantitative; and yet, despite the pride we take in our burgeoning understanding of neuronal development, we have not yet identified a single gene or allele responsible for any part of this astonishing quantitative variation. This is humbling. In our drive to understand the fundamentals of development we can initially be forgiven the blind eye that we have collectively turned to normal variation, but to get beyond descriptions of neuronal development – to get to a second stage of purposeful design, modification, and repair – we will need to understand the subtle, intertwined, and messy molecular genetics that makes a human brain different from that of a mouse and the brain of one human different from that of another (Bartley et al. 1997).

Acknowledgements. I thank Dr. Richelle Strom for sharing data on the *Bsc5* locus. I extend thanks to Drs. Guomin Zhou, Dan Goldowitz, David Airey, Lu Lu, Kenneth Manly, and Glenn Rosen for their contributions to this work. This research was supported by NINDS RO1 35485.

References

Aboitiz F (1996) Does bigger mean better? Evolutionary determinants of brain size and structure. Brain Behav Evol 47:225–245

Adams MD, Soares MB, Kerlavage AR, Fields C, Venter JC (1993) Rapid cDNA sequencing (expressed sequence tags) from a directionally cloned human infant brain cDNA library. Nat Genet 4:373–380

Aiello LC, Wheeler P (1995) The expensive-tissue hypothesis. The brain and the digestive system in human and primate evolution. Curr Anthropol 36:199–221

Airey DC, Lu L, Strom RC, Gilissen G, Williams RW (1999) Cerebellum-specific QTLs in the mouse brain. Int Mouse Genome Cont 13:E9

Airey DC, Strom RC, Williams RW (1998) Genetic architecture of normal variation in cerebellar size. Soc Neurosci Abstr 24:303

Allman JM, McLaughlin T, Hakeem A (1993) Brain weight and life-span in primate species. Proc Natl Acad Sci USA 90:118–122

Armstrong E (1993) Relative brain size and metabolism in mammals. Science 220:1302–1304

Atchley WR, Riska B, Kohn LAP, Plummer AA, Rutledge JJ (1984) A quantitative genetic analysis of brain and body size associations, their origin and ontogeny: data from mice. Evolution 38:1165–1179

Bartley AJ, Jones DW, Weinberger DR (1997) Genetic variability of human brain size and cortical gyral patterns. Brain 120:257–269

Belknap JK, Phillips TJ, O'Toole LA (1992) Quantitative trait loci associated with brain weight in the BXD/Ty recombinant inbred mouse strains. Brain Res Bull 29:337–344

Belknap JK, Metten P, Helms ML, O'Toole LA, Angeli-Gade S, Crabbe JC, Phillips TJ (1993) Quantitative trait loci (QTL) applications to substances of abuse: physical dependence studies with nitrous oxide and ethanol in BXD mice. Behav Genet 23:213–222

Brockmann GA, Haley CS, Renne U, Knott SA, Schwerin M (1998) Quantitative trait loci affecting body weight and fatness from a mouse line selected for extreme high growth. Genetics 150:369–381

Buck KJ, Metten P, Belknap JK, Crabbe JC (1997) Quantitative trait loci involved in genetic predisposition to acute alcohol withdrawal in mice. J Neurosci 17:3946–3955

Cheverud JM, Routman EJ, Duarte FAM, van Swinderen B, Cothran K, Perel C (1996) Quantitative trait loci for murine growth. Genetics 142:1305–1319

Churchill GA, Doerge RW (1994) Empirical threshold values for quantitative trait mapping. Genetics 138:963–971

Clark JB, Bates TE, Almeida A, Cullingford T, Warwick J (1994) Energy metabolism in the developing mammalian brain. Biochem Soc Trans 22:980–983

Collins RA (1970) Experimental modification of brain weight and behavior in mice: an enrichment study. Dev Psychobiol 3:145–155

Crabbe JC, Belknap JK, Buck KJ (1994) Genetic animal models of alcohol and drug abuse. Science 264:1715–1723

Crusio WE (1992) Quantitative genetics. In: Goldowitz D, Wahlsten D, Wimer RE (eds) Techniques for the genetic analysis of brain and behavior. Elsevier, Amsterdam, pp 231–250

Crusio WE, Schwegler H, van Abeelen JHF (1989) Behavioral responses to novelty and structural variation of hippocampus in mice. I. Quantitative-genetic analysis of behavior in the open field. Behav Brain Res 32:75–80

Dains K, Hitzemann B, Hitzemann R (1996) Genetics, neuroleptic-response and the organization of cholinergic neurons in the mouse striatum. J Pharmacol Exp Ther 279:1430–1438

Darvasi A (1997) Interval-specific congenic strains (ISCS): an experimental design for mapping a QTL into a 1-centiMorgan interval. Mamm Gen 8:163–167

Darvasi A (1998) Experimental strategies for the genetic dissection of complex traits in animal models. Nat Genet 18:19–24

DeVoogd TJ, Krebs JR, Healy SD, Purvis A (1993) Relations between song repertoire size and the volume of brain nuclei related to song: comparative evolutionary analyses amongst oscine birds. Proc R Soc Lond B 254:75–82

Dietrich WF, Miller JC, Steen RG, Merchant M, Damron D, Nahf R, Gross A, Joyce DC, Wessel M, Dredge RD, Marquis A, Stein LD, Goodman N, Page DC, Lander ES (1994) A genetic map of the 4006 simple sequence length polymorphisms. Nat Genet 7:220–245

Eisenberg JF, Wilson DE (1978) Relative brain size and feeding strategies in the chiroptera. Evolution 32:740–751

Eleftheriou BE, Elias MF, Castellano C, Oliverio A (1975) Cortex weight: a genetic analysis in the mouse. J Hered 66:207-212

Festing MFW (1993) Origins and characteristics of inbred strains of mice. Mouse Genome 91:393-509 <http://www.informatics.jax.org/strtools.html>

Flint J, Corley R, DeFries JC, Fulker DW, Gray JA, Miller S, Collins AC (1995) A simple genetic basis for a complex psychological trait in laboratory mice. Science 269:1432-1435

Frankel WN (1995) Taking stock of complex trait genetics in mice. Trends Gen 11:471-477

Fuller JL (1979) Fuller BWS lines: history and results. In: Hahn ME, Jensen C, Dudek BC (eds) Development and evolution of brain size. Academic Press, New York, pp 187-204

Fuller JL, Geils HD (1972) Brain growth in mice selected for high and low brain weight. Dev Psychobiol 5:307-318

Fuller JL, Herman BH (1972) Effect of genotype and practice on behavioral development in mice. Dev Psychobiol 7:21-30

Fuller JL, Wimer RE (1966) Neural, sensory, and motor functions. In: Green EL (ed) The biology of the laboratory mouse, 2nd edn. Dover, New York, pp 609-628

Gautvik KM, De Lecea L, Gautvik VT, Danielson PE, Tranque P, Dopazo A, Bloom FE, Sutcliffe JG (1996) Overview of the most prevalent hypothalamus-specific mRNAs, as identified by directional tag PCR subtraction. Proc Natl Acad Sci USA 93:8733-8738

Gilissen E, Williams RW (1997) Genetic dissection and QTL analysis of forebrain, hindbrain, olfactory bulb, and cerebellum. Soc Neurosci Abstr 23:864

Gilissen E, Zilles K (1996) The calcarine sulcus as an estimate of the total volume of the human striate cortex: a morphometric study of reliability and intersubject variability. J Brain Res 37:57-66

Groot PC, Moen CJA, Dietrich W, Stoye JP, Lander ES, Demant P (1992) The recombinant congenic strains for analysis of multigenic traits: genetic composition. FASEB J 6:2826-2835

Hahn ME, Haber SB (1978) A diallel analysis of brain and body weight in male inbred laboratory mice (Mus musculus). Behav Genet 8:251-260

Haley CS, Knott SA (1992) A simple regression method for mapping quantitative trait loci in line crosses using flanking markers. Heredity 69:315-324

Henderson ND (1979) Dominance for large brains in laboratory mice. Behav Genet 9:45-49

Herrup K, Shojaeian-Zanjani H, Panzini L, Sunter K, Mariani J (1996) The numerical matching of source and target populations in the CNS: the inferior olive to Purkinje cell projection. Dev Brain Res 96:28-35

Hitzemann B, Dains K, Kanes S, Hitzemann R (1994) Further studies on the relationship between dopamine cell density and haloperidol-induced catalepsy. J Pharmacol Exp Ther 271:969-976

Hofman MA (1983) Evolution of the brain in neonatal and adult placental mammals: a theoretical approach. J Theor Biol 105:317-332

Horton JC, Hocking DR (1996) Intrinsic variability of ocular dominance column periodicity in normal macaque monkeys. J Neurosci 16:7228-7339

Hoskins SG (1985) Induction of the ipsilateral retinothalamic projection in Xenopus laevis by thryoxine: results and speculation. J Neurobiol 17:203-229

Jacobs LF, Gaulin SC, Sherry DF, Hoffman GE (1990) Evolution of spatial cognition: sex-specific patterns of spatial behavior predict hippocampal size. Proc Natl Acad Sci USA 87:6349-6352

Johnson TE, DeFries JC, Markel PD (1992) Mapping quantitative trait loci for behavioral traits in the mouse. Behav Genet 22:635-653

Kanes S, Dains K, Cipp L, Gatley J, Hitzemann B, Rasmussen E, Sanderson S, Silverman M, Hitzemann R (1996) Mapping the genes for haloperidol-induced catalepsy. J Pharmacol Exp Ther 277:1016-1025

Katz HB, Davies CA (1983) The separate and combined effects of early undernutrition and environmental complexity at different ages on cerebral measures in rats. Dev Psychobiol 16:47-58

Keverne EB, Martel FL, Nevison CM (1996) Primate brain evolution: genetic and functional considerations. Proc R Soc Lond B 263:689–696

Lande R (1979) Quantitative genetic analysis of multivariate evolution, applied to brain:body size allometry. Evolution 33:234–251

Lander ES, Botstein D (1989) Mapping Mendelian factors underlying quantitative traits using RFLP linkage maps. Genetics 121:185–199

Lander E, Kruglyak L (1997) Genetic dissection of complex traits: guidelines for interpreting and reporting linkage results. Nat Genet 11:241–247

Lander ES, Schork NJ (1994) Genetic dissection of complex traits. Science 265:2037–2048

Lashley KS (1949) Persistent problems in the evolution of mind. Q Rev Biol 24:28–42

Le Roy I, Perex-Diaz F, Cherfouh A, Roubertoux PL (1999) Preweanling sensorial and motor development in laboratory mice: quantitative trait loci mapping. Dev Psychobiol 34:139–158.

Lewontin RC (1957) The adaptations of populations to varying environments. Cold Spring Harbor Symp Quant Biol 22:395–408

Lipp HP (1989) Non-mental aspects of encephalization: the forebrain as a playground of mammalian evolution. Hum Evol 4:45–53

Lipp HP, Schwegler H, Crusio WE, Wolfer DP, Leisinger-Trigona MC, Heimrich B, Driscoll P (1989) Using genetically-defined rodent strains for the identification of hippocampal traits relevant for two-way avoidance behavior: a non-invasive approach. Experientia 45:845–859

Lu L, Airey DC, Zhou G, Williams RW (1999) New murine hippocampus-specific QTL maps to distal Chr 1. Int Mouse Genome Conf 13:E44

Luo J, Sucove HM, Bader JA, Evans RM, Giguere V (1996) Compound mutants for retinoic acid receptor (RAR) beta and RAR alpha 1 reveal developmental functions of multiple RAR beta isoforms. Mech Dev 55:33–44

Lynch M, Walsh B (1998) Genetics and analysis of quantitative traits. Sinauer, San Francisco

Manly KF, Olson JM (1999) Overview of QTL mapping software and introduction to Map Manager QT. Mamm Genome 10:327–334

Martin RD (1981) Relative brain size and metabolic rate in terrestrial vertebrates. Nature 293:57–60

Pagel MD, Havey PH (1990) Diversity in the brain sizes of newborn mammals: allometry, energetics, or life history tactics? Bioscience 40:116–122

Pakkenberg B, Gundersen HJG (1997) Neocortical neuron number in humans: effect of sex and age. J Comp Neurol 384:312–320

Piedrafita FJ, Pfahl M (1995) Thyroid hormone receptors. In: Baeuerle PA (ed) Inducible gene expression, vol 2. Birkhäuser, Basel, pp 157–185

Plomin R, McClearn GE (1993) Quantitative trait loci (QTL) analyses and alcohol-related behaviors. Behav Genet 23:197–211

Plomin R, McClearn GE, Gora-Maslak G, Neiderhiser JM (1991) Use of recombinant inbred strains to detect quantitative trait loci associated with behavior. Behav Genet 21:99–116

Ramón y Cajal S (1890) Estudios sobre la corteza cerebral humana. III. Cortez motriz. Rev Trimestr Micrográf 5:1–11. [Transl by Jacobson M (1991) In: Developmental neurobiology, 3rd edn. Plenum, New York, p 401]

Rensch B (1956) Increase of learning capability with increase of brain-size. Am Nat 90:81–95

Rice DS, Williams RW, Goldowitz D (1995) Genetic control of retinal projections in inbred strains of albino mice. J Comp Neurol 354:459–469

Rise ML, Frankel WN, Coffing JM, Seyfried TN (1991) Genes for epilepsy mapping in the mouse. Science 253:669–673

Roderick TH (1979) Genetic techniques as tools of analysis of brain–behavior relationships. In: Hahn ME, Jensen C, Dudek BC (eds) Development and evolution of brain size. Academic Press, New York, pp 133–145

Roderick TH, Wimer RE, Wimer CC, Schwartzkroin PA (1973) Genetic and phenotypic variation in weight of brain and spinal cord between inbred strains of mice. Brain Res 64:345–353

Roderick TH, Wimer RE, Wimer CC (1976) Genetic manipulation of neuroanatomical traits. In: Petrinovich L, McGaugh L (eds) Knowing, thinking, and believing. Plenum, New York, pp 143–178

Roff DA (1997) Evolutionary quantitative genetics. Chapman & Hall, New York

Sacher GA, Staffeldt EF (1974) Relation of gestation time to brain weight for placental mammals: implication for the theory of vertebrate growth. Am Nat 108:593–615

Schwegler H, Lipp HP (1983) Hereditary covariations of neuronal circuitry and behavior: correlations between the proportions of hippocampal synaptic fields in regio inferior and two-way avoidance in mice and rats. Behav Brain Res 7:297–305

Seyfried TN, Daniel WL (1977) Inheritance of brain weight in two strains of mice. J Hered 68:337–338

Seyfried TM, Glaser GH, Yu RK (1979) Genetic variability for regional brain gangliosides in five strains of young mice. Biochem Genet 17:43–55

Sprott RL, Staats J (1981) Behavioral studies using genetically defined mice – a bibliography. Behav Genet 11:73–84

Stensaas SS, Donald MA, Eddington DK, Dobelle WH (1974) The topography and variability of the primary visual cortex in man. J Neurosurg 40:747–755

Storer JB (1967) Relation of lifespan to brain weight, body weight, and metabolic rate among inbred mouse strains. Exp Gerontol 2:173–182

Strom RC (1999) Genetic control of neuron number. Dissertation, University of Tennessee, Memphis <http://nervenet.org>

Strom RC, Williams RW (1997) Mapping genes that control variation in brain weight using F2 intercross progeny. Soc Neurosci Abstr 23:864

Strom RC, Williams RW (1998) Cell production and cell death in the generation of variation in neuron number. J Neurosci 18:9948–9953

Sutcliffe JG (1988) mRNA in the mammalian central nervous system. Annu Rev Neurosci 11:157–198

Takahashi JS, Pinto LH, Vitaterna MH (1994) Forward and reverse genetic approaches to behavior in the mouse. Science 1724:1724–1733

Tanksley SD (1993) Mapping polygenes. Annu Rev Genet 27:205–233

Taylor BA (1978) Recombinant inbred strains. Use in gene mapping. In: Morse H (ed) Origins of inbred mice. Academic Press, New York, pp 423–438

Toth LA, Williams RW (1988) Genetic analysis of complex quantitative traits using inbred mice. Sleep Res Soc Bull 4:50–56

Toth LA, Williams RW (1999) Strain-related differences in slow wave sleep and rapid-eye-movement sleep in C57BL/6J and BALB/c mice. Behav Genet 29: in press

Usui H, Falk JD, Dopazo A, de Lecea L, Erlander MG, Sutcliffe JG (1994) Isolation of clones of rat striatum-specific mRNAs by directional tag PCR subtraction. J Neurosci 14:4915–4926

Vadasz C, Sziraki I, Sasvari M, Kabai P, Murthy LR, Saito M, Laszlovszky I (1998) Analysis of the mesotelencephalic dopamine system by quantitative-trait locus introgression. Neurochem Res 23:1337–1354

Wahlsten D (1975) Genetic variation in the development of mouse brain and behavior: evidence from the middle postnatal period. Dev Psychobiol 8:371–380

Wahlsten D (1983) Maternal effects on mouse brain weight. Dev Brain Res 9:216–221

Wahlsten D (1992) The problem of test reliability in genetic studies of brain–behavior correlation. In: Goldowitz D, Wahlsten D, Wimer RE (eds) Techniques for the genetic analysis of brain and behavior: focus on the mouse. Elsevier, Amsterdam, pp 407–422

Wahlsten D (1994) The intelligence of heritability. Can Psychol 35:244–267

Ward BC, Nordeen EJ, Nordeen KW (1998) Individual variation in neuron number predicts differences in the propensity for avian vocal imitation. Proc Natl Acad Sci USA 95:1277–1282

Wetts R, Herrup K (1983) Direct correlation between Purkinje and granule cell number in the cerebella of lurcher chimeras and wild-type mice. Dev Brain Res 10:41–47

Wikström L, Johansson C, Saltó C, Barlow C, Campos Barros A, Baas F, Forrest D, Thorén P, Vennström B (1998) Abnormal heart rate and body temperature in mice lacking thyroid hormone receptor-1. EMBO J 17:455–461

Williams RW (1998) Neuroscience meets quantitative genetics: using morphometric data to map genes that modulate CNS architecture. In: Morrison J, Hof P (eds) Short course in quantitative neuroanatomy. Society of Neuroscience, Washington DC, pp 66–78. <nervenet.org/papers/ShortCourse98.html>

Williams RW, Herrup K (1988) The control of neuron number. Annu Rev Neurosci 11:423–453

Williams RW, Rakic P (1988) Three-dimensional counting: an accurate and direct method to estimate numbers of cells in sectioned material. J Comp Neurol 278:344–352 <nervenet.org>

Williams RW, Zhou G (1999) Genetic control of eye size: a novel quantitative genetic approach. Invest Ophthalmol Vis Sci Suppl 40:S964

Williams RW, Cavada C, Reinoso-Suárez F (1993) Rapid evolution of the visual system: a cellular assay of the retina and dorsal lateral geniculate nucleus of the Spanish wildcat and the domestic cat. J Neurosci 13:208–228 <nervenet.org>

Williams RW, Strom RC, Rice DS, Goldowitz D (1996a) Genetic and environmental control of variation in retinal ganglion cells number in mice. J Neurosci 16:7193–7205 <nervenet.org>

Williams RW, Strom RC, Goldowitz D (1996b) Mapping quantitative trait loci that control normal variation in brain weight in the mouse. Soc Neurosci Abstr 22:518

Williams RW, Goldowitz DG, Strom RC (1997) Brain weight in relation to body weight, age, and sex: a multiple regression analysis. Soc Neurosci Abstr 23:864

Williams RW, Strom RC, Goldowitz D (1998a) Natural variation in neuron number in mice is linked to a major quantitative trait locus on Chr 11. J Neurosci 18:138–146 <nervenet.org>

Williams RW, Strom RC, Zhou G, Yan Z (1998b) Genetic dissection of retinal development. Sem in Cell Dev Biol 9:249–255 <nervenet.org>

Wimer C (1979) Correlates of mouse brain weight: search for component morphological traits. In: Hahn ME, Jensen C, Dudek BC (eds) Development and evolution of brain size. Academic, New York, pp 147–162

Wimer C, Prater L (1966) Behavioral differences in mice genetically selected for high and low brain weight. Psychol Rep 19:675–681

Wimer RE, Wimer CC (1985) Animal behavior genetics: a search for the biological foundation of behavior. Annu Rev Psychol 36:171–218

Wimer RE, Wimer CC (1989) On the sources of strain and sex differences in granule cell number in the dentate area of house mice. Dev Brain Res 48:167–176

Wimer RE, Wimer CC, Roderick TH (1969) Genetic variability in forebrain structures between inbred strains of mice. Brain Res 16:257–264

Wimer RE, Wimer CC, Vaughn JE, Barber RP, Balvanz BZ, Chernow CR (1978) The genetic organization of neuron number in the area dentata of mouse mice. Brain Res 157:105–122

Wright S (1978) Evolution and the genetics of populations, Vol 4. Variability within and among natural populations. U Chicago, Chicago

Zamenhof S, van Marthens E (1976) Neonatal and adult brain parameters in mice selected for adult brain weight. Dev Psychobiol 9:587–593

Zamenhof S, van Martens E, Gauel L (1971) DNA (cell number) in neonatal brain: second generation (F2) alternation by maternal (F0) dietary protein restriction. Science 172:850–851

Zhou G, Williams RW (1999a) Mouse models for the analysis of myopia: an analysis of variation in eye size of adult mice. Optom Vis Sci 76:408–418 <nervenet.org>

Zhou G, Williams RW (1999b) Eye1 and Eye2: Gene loci that modulate eye size, lens weight, and retinal area in mouse. Invest Ophthalmol Vis Sci 40:817–825 <nervenet.org>

Zhou G, Strom RC, Giguere V, Williams RW (1998) Modulation of retinal cell populations and eye size in retinoic acid receptor knockout mice. Soc Neurosci Abst 24:1033

Genetic Interactions During Hindbrain Segmentation in the Mouse Embryo

Paul A. Trainor, Miguel Manzanares, and Robb Krumlauf[1]

1
Introduction

The process by which a simple flat sheet of cells in the early embryo proliferates and differentiates into the central nervous system (CNS) is one of the
most fascinating problems of modern biology. This is not only because of the
importance of the CNS in adult life, but also because of the enormous
quantity and complexity of cell connections that have to be established with
extraordinary precision. Large-scale genetic screens in invertebrate model
systems such as *Caenorhabditis elegans* and *Drosophila* have revealed in
significant detail the genetic components required for early CNS development
and some aspects of the basic programme appear to have been conserved
throughout evolution in the animal kingdom. In this chapter we will detail
our current understanding of the development of the early vertebrate nervous
system, with particular reference to the mouse hindbrain.

1.1
Generation of Diversity in the Developing Nervous System

The central nervous system in the mouse is derived from the embryonic
ectoderm (one of the three primary germ layers). Most of our knowledge
concerning the earliest developmental events has been derived from the
amphibian embryo. Here it was first noted that neural fate is acquired by
competent ectoderm via signals emanating from the dorsal blastopore lip
(Spemann's organiser, the equivalent of which is the node in mouse;
Nieuwkoop 1952 and 1973). This led to the proposal of the two-step activa-
tion-transformation model to describe the derivation of regionalised neural
tissue from naive embryonic ectoderm (Nieuwkoop 1952 and 1985; Toivonen
and Saxen 1968; Saxen 1989; Slack and Tannahill 1992). Firstly, activating

---
[1] Division of Developmental Neurobiology, MRC National Institute for Medical Research, The
Ridgeway, Mill Hill, London NW7 1AA, UK

signals supplied by early involuting mesoderm during gastrulation are thought to induce a default state of anterior neural differentiation, which is then modified to a more posterior character by transforming signals from later involuting mesoderm. However, it is important to note that recent studies in chick embryos have demonstrated that changes in character from posterior to anterior can also occur following neural induction, indicating that changes can occur in both anterior and posterior properties (Martinez et al. 1995; Dale et al. 1997; Foley et al. 1997). The level of a particular transforming signal at any single point along the neuraxis could be the catalyst to confer its local identity. Recent studies have also shown that preceding neural induction and the involution of mesoderm, future neural tissue has the capacity for regionalised expression of a variety of A-P markers (Doniach 1992, 1993; Doniach et al. 1992; Kintner 1992; Ruiz i Altaba 1992, 1994; Lamb and Harland 1995). Hence, in very early vertebrate embryogenesis planar and vertical signals supply critical information necessary to program future A-P regionalised markers well before the formation of lineage-restricted cellular compartments. Although at this early stage of development it is clear that anterior-posterior regionalisation of the neural plate is occurring, transplantation experiments in several species have demonstrated that the fate of individual cells is not yet irreversibly committed to an anterior or posterior character (Martinez and Alvarado-Mallart 1990; Martinez et al. 1991, 1995; Doniach 1992, 1993; Doniach et al. 1992; Kintner 1992; Ruiz i Altaba 1992, 1994; Grapin-Botton et al. 1995; Lamb and Harland 1995; Itasaki et al. 1996; Streit et al. 1997; Woo and Fraser 1997).

In the mouse, between 8.0 and 9.5 days of embryonic development, the two halves of the neural plate fold up to form a hollow tube, covered by a layer of surface ectoderm, and this process is accompanied by further regionalisation of the neural tube (reviewed in Lumsden and Krumlauf 1996). At the cephalic end, the neuraxis is partitioned by a series of swellings and constrictions into three primary vesicles that define the major brain compartments: forebrain (prosencephalon), midbrain (mesencephalon), and hindbrain (rhombencephalon). The forebrain becomes further subdivided into the anterior telencephalon and the more caudal diencephalon. The telencephalon gives rise to the cerebral hemispheres and the diencephalon will develop into the thalamic and hypothalamic brain regions. Similar to the forebrain, the hindbrain is further regionalised into the anterior metencephalon and the posterior myelencephalon. The metencephalon ultimately gives rise to the cerebellum, the part of the brain responsible for co-ordinating movements, posture and balance and the myelencephalon eventually forms the medulla oblongata, the nerves of which regulate respiratory, gastrointestinal and cardiovascular movements. In contrast to the forebrain and hindbrain, the midbrain does not subdivide further; however, its lumen will become the cerebral aqueduct. Posterior to the head region at this stage, a simple neural tube forms that tapers off towards the tail but which eventually develops into the spinal cord in association with a processes of secondary neuralation.

Regionalisation or subdivision of the cephalic part of the neural tube requires the activity of some local organising centres within the neural epithelium itself, such as the junction between the mesencephalon and metencephalon (mes-met junction or isthmus-cerebellum; Marín and Puelles 1994; Crossley et al. 1996) and also the junction between the diencephalon and mesencephalon (d/m limit; Martinez and Alvarado-Mallart 1990; Itasaki et al. 1991; Martinez et al. 1991). The influence of these organiser centres on brain patterning has been demonstrated by a number of grafting experiments performed in avian embryos which show that these organisers are responsible for producing signals that act over a relatively long range, patterning tissue both rostral and caudal to it. The consequence of these patterning events is the establishment of an organised array of differentiated brain components and segmental compartments upon which the characteristics and unique sets of neurons and accessory cells that constitute the adult CNS are subsequently modelled (Bulfone et al. 1993; Figdor and Stern 1993; Marín and Puelles 1995).

1.2
Segmental Organisation of the Hindbrain

Of all the regions of the developing nervous system, the hindbrain has attracted particular interest due to its characteristic organisation into a series of seven cell lineage restricted compartments called rhombomeres (r) (Vaage 1969; Lumsden and Keynes 1989; Fraser et al. 1990; Birgbauer and Fraser 1994; Birgbauer et al. 1995; Lumsden and Krumlauf 1996). These transverse periodic neuroepithelial bulges are clearly distinguishable by 9–9.5 days of mouse development and provide the essential ground plan for establishing subsequent features of craniofacial and CNS development (Wilkinson et al. 1989a,b; Wilkinson and Krumlauf 1990; Hunt and Krumlauf 1991; Hunt et al. 1991a,b; Couly et al. 1993; Köntges and Lumsden 1996). Not only do the rhombomeres give rise to well-defined regions of the mature adult brain (Marín and Puelles 1995; Wingate and Lumsden 1996) but its segmental organisation presages the periodic organisation of the neurons and cranial motor nerves (Lumsden and Keynes 1989; Clarke and Lumsden 1993; Keynes and Krumlauf 1994; Lumsden and Krumlauf 1996; Clarke et al. 1998) and correlates with the pathways of neural crest migration into the branchial arches (Lumsden et al. 1991; Graham et al. 1993; Sechrist et al. 1993, 1994).

The reticular neurons are the first subset of neurons to form in the hindbrain and extend axons, and they do so in an alternate manner within the confines of the newly formed rhombomeres (Lumsden and Keynes 1989; Clarke and Lumsden 1993). The first set of reticular neurons arise in rhombomere 4, followed shortly after by rhombomeres 2 and 6 (Sechrist and Bronner-Fraser 1991). Reticular neurons also develop in the odd numbered rhombomeres but at a slightly later stage. Similar to the reticular neurons, the formation and disposition of motor neurons also conforms to a two-segment periodicity pattern. Retrograde tracing of motor axons has revealed that the cell bodies of individual cranial nerves exhibit a precise relationship to specific rhombomeres (Lumsden and Keynes 1989; Marshall et al. 1992; Carpenter et al. 1993). The motor nerves of the first three branchial arches (V trigeminal, VII facio-acoustic and IX glosso-pharyngeal) are respectively derived from nuclei that are confined within rhombomeres 2, 4 and 6. Each nerve is then subsequently augmented by neurons developing in the caudally adjacent rhombomere. Therefore the segmental organisation of the hindbrain clearly underpins the metameric pattern of cranial nerves and reticular neurons (Keynes and Lumsden 1990; Keynes and Krumlauf 1994).

In addition to patterning the cranial nerves, the segmental organisation of the hindbrain also controls the patterns of cranial neural crest cell migration (Lumsden et al. 1991; Serbedzija et al. 1992; Sechrist et al. 1993; Köntges and Lumsden 1996). In vertebrate embryos, neural crest cells arise at the lateral edges of the neural plate along almost the entire neuraxis, at the junction between the neuroectoderm and ectoderm and their migration is essential for proper craniofacial morphogenesis (Le Douarin 1983; Noden 1983; Hunt et al. 1995; Couly et al. 1996, 1998; Saldivar et al. 1997). In the mouse the first population of neural crest cells to emigrate from the ridge of the neural tube do so from the caudal midbrain and rostral hindbrain at the 5–6 somite stage (8.25–8.5 dpc), long before closure of the neural tube. In contrast, the commencement of neural crest migration in the chick coincides with closure of the neural tube (Horstadius 1950; Tosney 1982; Le Douarin 1983). The patterns of neural crest cell migration in mouse and chick embryos are very similar and the duration of emigration from all axial levels typically lasts between 9 and 12 hours.

Hindbrain-derived neural crest cells in both the mouse and chick embryos migrate ventrolaterally as subectodermal streams, passing between the surface ectoderm and underlying mesoderm from the dorsal portion of the neural tube into the distal regions of the branchial arches. The pattern of hindbrain neural crest migration consists of three separate streams of cells lateral to rhombomeres r2, r4 and r6, each of which migrates into the adjacent first, second and third branchial arches respectively in keeping with their craniocaudal order (Lumsden et al. 1991; Sechrist et al. 1993). Significantly less neural crest cells delaminate from rhombomeres 3 and 5 compared to the even-numbered rhombomeres (Lumsden et al. 1991; Sechrist et al. 1993), and it has been suggested that this associated with increased cell death in these populations (Graham et al. 1993, 1994). However, recent time-lapse analyse of neural crest migration indicate that neural crest cells derived from r3 and r5 do actually migrate into the branchial arches (Sechrist et al. 1993; Kulesa and Fraser 1998). Unusually, they migrate both rostrally and caudally, joining the even-numbered rhombomere neural crest streams as they fill the branchial arches. This suggests that the paraxial environment influences the emigration of crest from odd-numbered rhombomeres. As a consequence of this, the second branchial arch contains neural crest derivatives from three consecutive rhombomeres, r3, r4 and r5 (Sechrist et al. 1993; Kulesa and Fraser 1998). Therefore it is clear that the segmental organisation of the hindbrain has a profound impact on patterning and migration of the cranial neural crest.

2
Patterns of Gene Expression
During Hindbrain Development

The actual cellular mechanisms by which the hindbrain is subdivided into rhombomeres are not well understood, but as we will see later in this chapter, we are beginning to understand the genetic cascade of events that ultimately lead to individually specified compartments. During early neuraxis development, the adoption of a neural fate by the embryonic ectoderm is reflected in the appearance of restricted domains of gene expression along the entire length of the primitive neuroectoderm (Rubenstein et al. 1994; Lumsden and Krumlauf 1996; Edlund and Jessell 1999). As development proceeds, additional genes become expressed in the neuroepithelium and their expression domains become coupled or linked to the underlying regional variation and patterns of differentiation, generating a precise map of positional information along the main A-P and D-V axes of the neural tube. Nowhere in the developing nervous system is this more evident than in the vertebrate hindbrain. Numerous genes including transcription factors, signalling molecules, membrane and nuclear receptors are expressed in a segmental manner during hindbrain development (see Fig. 1) and most of these genes exhibit extremely dynamic patterns of activity (Keynes and Krumlauf 1994; Lumsden and Krumlauf 1996). Some genes are expressed in single rhombomeres, a few are confined to pairs of rhombomeres, expressed at rhombomeres boundaries, or become activated later during neuronal differentiation. In this chapter we will discuss the roles of genes that have been shown to be necessary for the early stages of hindbrain development, when rhombomere segments and their identity are firmly established.

2.1
Hox Genes

The *Hox* gene family comprises a set of developmentally regulated transcription factors that are characterised by the presence of a sequence motif, called the homeobox, which encodes a DNA binding domain. This gene family is present in all metazoans, having been studied in detail in a wide

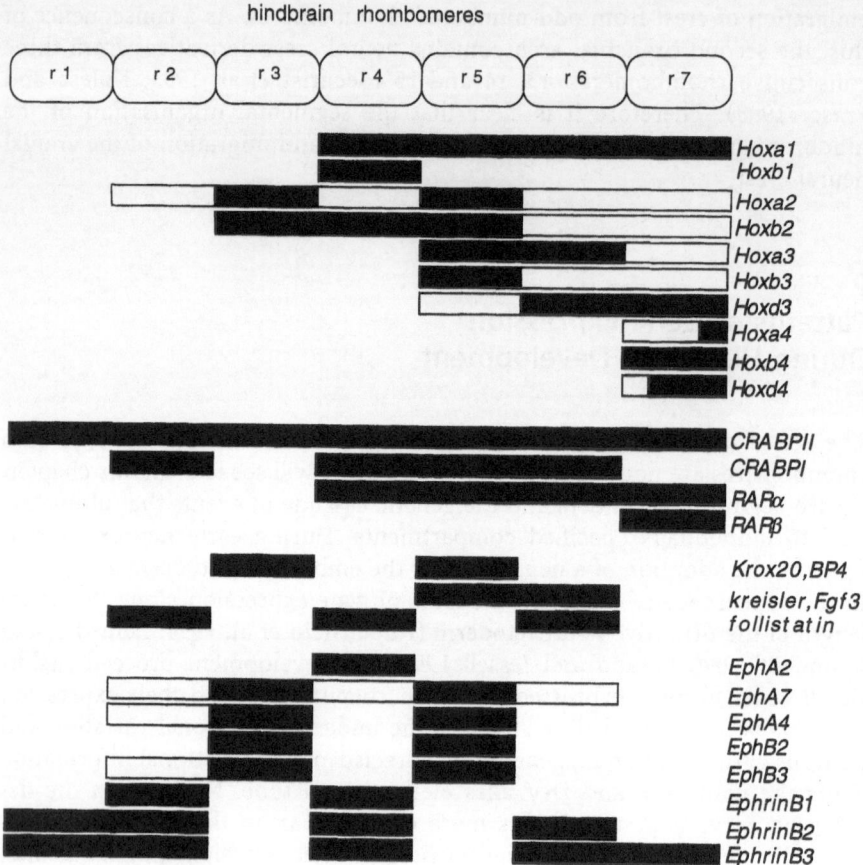

Fig. 1. Summary of segmental gene expression patterns during 7.5–10.5 dpc of hindbrain development (*solid black* and *open boxed regions* represents higher and lower levels of gene expression respectively)

variety of species including *Caenorhabditis elegans*, *Amphioxus*, *Drosophila*, fish, frog, chick, mouse and human, and it plays a crucial role in conferring axial identity to developing embryos (Kenyon and Wang 1991; Krumlauf 1992; McGinnis and Krumlauf 1992; Holland and Garcia-Fernandez 1996). The *Hox* homeotic genes are organised into a single chromosomal cluster in invertebrates. In contrast, higher vertebrates such as the mouse have 39 *Hox* (homeobox-containing) genes which are organised into four distinct chromosomal clusters (*Hoxa-Hoxd*) located on different chromosomes. However, in fish there is evidence of divergent and continuing evolution as both the number of clusters and the complement of *Hox* genes in each cluster are variable (Aparicio et al. 1997; Amores et al. 1998). This arrangement arose during evolution as a consequence of duplication and divergence of an ancestral homeobox (Kappen et al. 1989). The vertebrate *Hox* gene complexes can be aligned by the similarity of position and encoded amino acid sequence of individual gene members to a particular *Drosophila* homologue forming 13 subfamilies or "paralogous groups" (Krumlauf 1992; McGinnis and Krumlauf 1992). No single cluster contains members from all 13 groups and this is probably due to evolutionary gene loss that accompanied the duplication and divergence events.

The most striking feature of the organisation of *Hox* gene family members is that information related to gene order within a complex is transposed directly in time and space via gene expression onto the development embryo, thereby conferring positional information along the body axis (Duboule and Dolle 1989; Graham et al. 1989; Dollé et al. 1991; Izpisua-Belmonte et al. 1991; Kessel and Gruss 1991; McGinnis and Krumlauf 1992). Genes located nearer to the 3' end of the cluster are expressed earlier and more anteriorly during development than those located nearer the 5' end, such that *Hox* genes exhibit nested domains of expression along the A-P axis of the neural tube. Only the most 3' members of each complex (paralogue groups 1 to 4) are expressed in the vertebrate hindbrain (Hunt et al. 1991a) and these *Hox* genes display extraordinarily dynamic patterns of expression during the 7.5 to 10.5 dpc period of mouse embryo development (Fig. 1).

Hoxa1 and *Hoxb1* are initially expressed in the neural tube up to the presumptive r3/r4 boundaries, but as *Hoxa1* expression regresses, *Hoxb1* expression becomes confined to r4 by 8.5 dpc (Wilkinson et al. 1989b; Frohman et al. 1991a; Murphy and Hill 1991). *Hoxa2* expression is initiated in the neural epithelium at 7.5 dpc and by 8.5 dpc it is uniformly expressed up to the r1/r2 boundary (Hunt et al. 1991a; Krumlauf 1993; Prince and Lumsden 1994). Shortly thereafter *Hoxa2* becomes expressed at significantly higher levels in r3 and r5 (Nonchev et al. 1996b). In contrast, *Hoxb2* is expressed up to the r2/r3 boundary around the same time (Wilkinson et al. 1989b), but its expression levels are subsequently up regulated in r3, r4 and r5 (Sham et al. 1993; Maconochie et al. 1997). Group 3 (*Hoxa3, Hoxb3, and Hoxd3*) genes are expressed anteriorly up to the r4/r5 boundary (Wilkinson et al. 1989b; Hunt et al. 1991a), with Hoxa3 exhibiting up regulation in r5 and r6 and *Hoxb3* being similarly upregulated in r5 (Manzanares et al. 1997, 1999a). In contrast, *Hoxd3* is expressed at lower levels in r5 than more posterior regions. The group 4 genes *Hoxa4, Hoxb4* and *Hoxd4* are expressed up to a sharp r6/r7 anterior boundary by 9.5 dpc; however, the boundary of *Hoxc4* maps more posteriorly (Gaunt et al. 1989; Wilkinson et al. 1989b; Hunt et al. 1991a; Geada et al. 1992; Morrison et al. 1997).

In general, *Hox* gene expression patterns are believed to be generated in two distinct phases; establishment followed by maintenance. Furthermore, the mechanism by which *Hox* genes become expressed at higher relative levels in specific rhombomeres is partially independent from the processes that establish their more generalised expression patterns along the length of the neural tube. Current knowledge on the regulatory control of both these processes is discussed in detail below.

2.2
Upstream Regulators of *Hox* Genes

Genetic analyses of *Hox* genes largely through targeted mutations in the mouse have revealed that the segmental patterns of differential *Hox* gene expression are fundamentally important in setting positional values along the A-P axis which regulate rhombomere identity (see Sect. 4). Therefore it is important to understand the upstream signals and transcription factors that mediate segmentally restricted *Hox* expression. A broad spectrum of genes are expressed in specific rhombomeres (Lumsden and Krumlauf 1996) and comparisons with overlaps in timing and spatial distribution along with transgenic and genetic analysis have proved useful in helping to identify potential upstream regulators of *Hox* expression, such as *kreisler* (*Krml1*), *Krox20* and retinoids.

Krox20 was initially identified as an immediate early response gene in serum-stimulated fibroblasts and it encodes a protein with three C_2H_2-type zinc fingers (Chavrier et al. 1988). Cell culture co-transfection experiments have shown that *Krox20* binds to a specific DNA sequence and acts as a transcription factor (Lemaire et al. 1988; Chavrier et al. 1990; Nardelli et al. 1991; Vesque and Charnay 1992). During early embryonic development, *Krox20* is expressed in r3 by the 8.0 dpc stage followed by r5 at 9.0 dpc (Wilkinson et al. 1989a). This is important as it was the first example of a gene expressed in a two segment periodicity in the hindbrain and the timing precedes the morphological appearance of lineage restricted rhombomeric compartments in the hindbrain. By 9.5 dpc, *Krox20* expression becomes downregulated in r3 and this is quickly followed by downregulation in r5 as well. This pattern is highly conserved in other vertebrates (Wilkinson et al. 1989a; Nieto et al. 1991; Bradley et al. 1992; Oxtoby and Jowett 1993). The early and transient nature of this conserved expression pattern strongly suggested *Krox20* might be implicated in early segmental regulation, which has been confirmed by genetic and regulatory analysis in mice (see Sects. 3.2 and 4.1).

The other major candidate gene involved in controlling segmental patterning upstream of *Hox* genes is *kreisler*. *Kreisler* is a classical mouse mutant that was identified due to its circling behaviour, a phenotype that is associated with inner ear defects (Deol 1964). The gene responsible for the *kreisler* mutation has been positionally cloned, and is a member of the *Maf* oncogenic family of b-zip (basic domain-leucine zipper) transcription factors (Cordes and Barsh 1994). *Kreisler* expression is initiated at 7.5 dpc in the prospective r5 territory and is later expressed in r5 and r6 until 9.0 dpc when it is quickly downregulated in both rhombomeres (Cordes and Barsh 1994; Manzanares et al. 1999a). Interestingly, a zebrafish mutant with an altered hindbrain phenotype (*valentino*) was cloned and found to correspond to a homologue of mouse *kreisler* (Moens et al. 1996, 1998). Therefore it too appears to have a conserved role in the regulation of segmental patterning of the vertebrate hindbrain. However, there are differences between the genes in the two species, as *valentino* is expressed in r5 and r6 from the time it is activated (Moens et al. 1998).

2.3
Other Gene Families

Numerous other genes and gene families have now been described that also display segmental patterns of expression during hindbrain development (see Fig. 1). The tight correlation of expression patterns with rhombomeric organisation suggests that these genes may also participate in the pathways regulating and regulated by *Hox* genes. They include the retinoic acid nuclear receptor (*RAR*) family, cellular retinoid binding proteins, the *Eph* class of receptor tyrosine kinases and their ligands, the *ephrins*, *Fgf3*, *Wnt8*, cyclins, *follistatin*, *ErbB receptors*, neuregulin, *Ets1*, *semaphorin D*, *Fringe* genes and some of the ligands of the vertebrate Notch receptors. Furthermore, an increasing number of genes display sub-rhombomeric domains of expression that appear to correlate with both neurogenesis, differentiation and boundary formation (Cook et al. 1995; Mahmood et al. 1995; Pattyn et al. 1997; Schoorlemmer et al. 1997; Davenne et al. 1999). It is beyond the scope of this review to summarise the details and segmental patterns of all of these genes and signals, therefore we will concentrate on those that have been functionally linked with rhombomeric and neural crest patterning.

Retinoids constitute a group of vitamin-A derived signalling molecules which have been implicated in specifying and/or posteriorising the axial character of the developing CNS (Conlon 1995; Marshall et al. 1996). During embryonic development, tissues such as the node (organiser) synthesise RA from the metabolism of vitamin-A (all-trans retinol; Chen et al. 1992; Hogan et al. 1992). Exogenous retinoic acid (RA) applied to embryos during the early stages of development generally interferes with the formation of the most anterior CNS structures (forebrain and midbrain) and causes severe cranial malformations within the hindbrain and branchial arch region (Morriss and Thorogood 1978; Lammer et al. 1985; Durston et al. 1989; Conlon and Rossant 1992; Marshall et al. 1992; Hill et al. 1995; MacArthur et al. 1995; Simeone et al. 1995). These phenotypes are consistent with an increased posteriorisation of the CNS, and on this basis RA is considered one of the best candidates for a posteriorising signal in the transformation models of CNS development (Nieuwkoop 1952; Toivonen and Saxen 1968; Saxen 1989).

In addition to the nuclear retinoic acid receptors (*RARs* and *RXRs*), the mechanism of RA action is thought to involve cellular retinol and retinol binding proteins (*CRBP1* and *II*, *CRABP I* and *II*). These proteins are believed to help control the regional variation in RA concentration along the AP axis by controlling availability of free versus bound retinoids (Mangelsdorf et al.

1995). *CRBP* expression is confined primarily to the floor plate, where it binds retinol taken up from the blood. *CRABPI* is expressed in the mouse hindbrain in r4–r6 and at lower levels in r2 at 9.5 dpc (Dencker et al. 1990; Maden et al. 1992; Ruberte et al. 1992; Lyn and Giguere 1994). *CRABPII* is also expressed at high levels in the hindbrain and extends posteriorly along the length of the neural tube (Ruberte et al. 1992; Lyn and Giguere 1994). *CRABPs* therefore may be responsible for sequestering RA and thus limiting the amount of RA available to bind to the nuclear retinoic acid (*RARs* and *RXRs*). This could then control the balance or a switch between the ability of the receptors to function as repressors in the absence of ligand versus their roles as activators in the presence of ligands (Mangelsdorf and Evans 1995; Mangelsdorf et al. 1995). However, it should be noted that combined loss of function alleles for both *CRABP* genes results in mice that are essentially normal (Lampron et al. 1995), leaving the role of these proteins unclear. Another newly emerging area of study involves the expression of genes associated with the synthesis and degradation of retinoids. Many of these appear to have interesting patterns of expression in the CNS and would be an important control point for modulating the relative levels of retinoids available for signalling.

The retinoid nuclear receptor family comprises sequence specific DNA binding proteins that form heteromeric complexes by recognising direct repeat motifs called retinoic acid response elements (RAREs) located within target genes (Mangelsdorf and Evans 1995; Mangelsdorf et al. 1995). Although there are three genes in each of the *RAR* and *RXR* families (α, β, and γ), not all the members are expressed in the hindbrain. *RARα* is expressed up to the r3/r4 border while *RARβ* is expressed only to an anterior limit in the neural tube coinciding with the r6/r7 border (Mendelsohn et al. 1991, 1994). *RXRα* and β are expressed fairly uniformly throughout the entire hindbrain neuroepithelium (Dollé et al. 1994). The presence of retinoids together with their binding proteins and nuclear receptors within the CNS in patterns overlapping with the expression of *Hox* genes is suggestive of RA playing roles in both regulating the *Hox* genes and in the positional specification of the hindbrain and neural crest. This has been confirmed by functional and regulatory analysis (see Sects. 3.1 and 4.2), highlighting the retinoid signalling pathways as key components of patterning in the vertebrate CNS.

Gene expression studies during embryogenesis identified a large family of receptor tyrosine kinases (*Eph*) and associated ligands (*ephrins*) that were segmentally expressed in the developing hindbrain (Gilardi-Hebenstreit et al. 1992; Nieto et al. 1992a,b; Becker et al. 1994; Ruiz and Robertson 1994; Flenniken et al. 1996; Gale et al. 1996a,b; Flanagan and Vanderhaeghen 1998). The *ephrins* comprise a family of proteins that are anchored in the plasma membrane, either through a glucosyl-phosphatidyl inositol (GPI) linkage (*ephrin-A* type ligands) or by having a transmembrane domain and a cytoplasmic region (*ephrin-B* type ligands) (Flanagan and Vanderhaeghen 1998). Similarly, the *Eph* receptors can also be subdivided into two groups based on their binding specificities (Gale et al. 1996b; Flanagan and Vanderhaeghen

1998). In general, receptors of the *EphA* type only interact with GPI linked ligands while receptors of the *EphB* type only interact with transmembrane bound ligands. Interestingly, while *EphB* receptors are expressed in odd-numbered rhombomeres during the 8.0 to 10.5 dpc period of embryonic development, their corresponding *ephrinB* ligands are expressed in the adjacent even-numbered rhombomeres (Lumsden and Krumlauf 1996). This raises the possibility that *Eph* receptors and *ephrins* are involved in processes of cell repulsion, which by preventing cell intermingling between even and odd rhombomere cells essentially maintains proper inter-rhombomeric boundaries (Xu et al. 1995, 1999; Mellitzer et al. 1999). In addition to the *EphB* receptors, the receptors *EphA2* and *EphA4* are also segmentally expressed in the hindbrain (Nieto et al. 1992a; Becker et al. 1994; Irving et al. 1996). *EphA4* expression is initiated in presumptive r3 and r4 at around 7.25 dpc. Up regulation of expression occurs in pre-r3 at 7.75 dpc and later in r5 at 8.5 dpc (five somites). Concurrently there is a downregulation in r4 and by 8.75 dpc (12 somite stage) *EphA4* expression has become sharply restricted to definitive r3 and r5 and also at lower levels in r2. A dominant negative approach to interfere with the *EphA4* receptor function led to the disruption of spatially restricted gene expression in r3 and r5 (Xu et al. 1995, 1999). This and other related investigations (Robinson et al. 1997; Smith et al. 1997; Mellitzer et al. 1999) implicate members of the *Eph* family and the *ephrin* ligands in contributing to the establishment of the basic organisation of the hindbrain and craniofacial structures.

3
Genetic Control of Hindbrain Patterning

Genetic and expression analysis in several vertebrates has shown that *Hox* genes are an integral part of the process specifying regional variation in the developing hindbrain. Therefore it has been important to investigate the cascade of events that work upstream to establish and regulate their expression during the process of segmentation itself. Considerable effort therefore has been placed upon identifying upstream regulatory factors that impose segment-restricted expression of *Hox* genes as a means of unravelling the molecular mechanisms underlying the generation and specification of axial segments. This has largely depended upon a bottom up-approach, with the utilisation of transgenic analysis to map and characterise *cis*-acting regulatory elements from the *Hox* genes, and identify the factors interacting with these elements. Due to the extensive conservation in both the *Hox* expression patterns and their gene organisation in most vertebrates, evolutionary comparisons and cross-species functional analysis have played a pivotal role in helping to unravel some of the aspects of the *Hox* regulatory cascade in hindbrain segmentation.

3.1
Retinoic Acid Pathways

As noted in Sect. 2.3, RA has the potential to change the A-P character of the CNS and participates as an overall mediator of nested *Hox* gene expression. Excess RA causes both an anterior shift in *Hox* gene expression and an anterior to posterior transformation of regional fate (Durston et al. 1989; Simeone et al. 1990, 1991; Kessel and Gruss 1991; Papalopulu et al. 1991a,b; Conlon and Rossant 1992; Marshall et al. 1992; Moroni et al. 1993; Hill et al. 1995; Papalopulu and Kintner 1996; Morrison et al. 1996, 1997). Conversely, the suppression of RA signalling by the expression of dominant-negative retinoid receptors results in anteriorisation (Kolm and Sive 1995; Blumberg et al. 1997; Kolm et al. 1997; Sharpe and Goldstone 1997) and quail embryos generated on a retinoid free diet have severe deletions of hindbrain regions (Maden et al. 1996). All of this has helped to illustrate that RA signalling pathways seem to play a key direct and/or indirect role in modulating *Hox* expression.

In relation to this, an increasing number of studies have also found retinoic acid response elements (RAREs) in *Hox* control regions, enabling them to respond directly to retinoid signalling (Langston and Gudas 1992; Moroni et al. 1993; Pöpperl and Featherstone 1993; Marshall et al. 1994; Studer et al. 1994, 1998; Ogura and Evans 1995a,b; Langston et al. 1997; Gould et al. 1998; Huang et al. 1998; Packer et al. 1998). Perhaps the best characterised are those associated with the group 1 paralogues *Hoxa1* and *Hoxb1*. In the 3' flanking sequences two RAREs have been found in the *Hoxb1* locus (Marshall et al. 1994; Ogura and Evans 1995b; Langston et al. 1997) and a single RARE in the *Hoxa1* locus (Langston and Gudas 1992). Transgenic analysis has strongly suggested that these RAREs play a role in regulating the expression of the respective genes (Marshall et al. 1994; Frasch et al. 1995; Huang et al. 1998) and alteration of the elements by targeted mutagenesis in ES cells has confirmed the key role of these 3' RAREs in regulating endogenous *Hoxa1* and *Hoxb1* (Dupé et al. 1997; Gavalas et al. 1998; Studer et al. 1998). In addition, another RARE has been identified in the 5' flanking region of the *Hoxb1* gene and shown to play a role in restricting its expression to r4, by repressing expression in r3 and r5 (Stachel et al. 1993; Studer et al. 1994; Ogura and Evans 1995a). These studies reveal that individual *Hox* genes can have multiple RAREs that co-ordinate and integrate the distinct patterning influences of retinoids.

Together, these transgenic and genetic mutational analyses have helped to build an emerging picture of how retinoids are involved in establishing group 1 *Hox* expression (Morrison 1998). *Hoxa1* and *Hoxb1* are among the first *Hox* genes to be activated during development and their expression in neural ectoderm is initiated during gastrulation in response to RA through the 3' RAREs. This domain extends up to a sharp anterior limit at the presumptive rhombomere 3/4 hindbrain border. As this early *Hoxa1* and *Hoxb1* expression subsides, retinoids make a second input into *Hoxb1* regulation by helping to refine its domain of expression to a single rhombomere segment (r4). This second input involves a negative pathway whereby the 5' RARE serves to abolish *Hoxb1* expression in cells migrating into the neighbouring odd rhombomeres, r3 and r5. Therefore RA functions to both activate and repress *Hoxb1* expression in different regions through different RAREs.

In line with this model for regulation in the mouse, studies in the frog embryos have also shown that RA has a role in regulating another group 1 member, *Hoxd1* (Kolm and Sive 1995; Kolm et al. 1997). In mouse and chick embryos *Hoxd1* is not expressed in the CNS (Hunt et al. 1991a; Frohman and Martin 1992) however, in *Xenopus Hoxd1* is the primary group 1 member expressed in the early CNS during gastrulation (Cho and De Robertis 1990; Kolm and Sive 1995). Furthermore, using dominant-negative *RARs* it has been shown that activation of *Hoxd1* in frog embryos requires RA signalling (Kolm and Sive 1995; Kolm et al. 1997). This suggests that all three vertebrate *Hox* group 1 members have RAREs that activate their early expression and that these might have been lost or mutated in the higher vertebrates in the case of *Hoxd1*.

Another case where RA signalling has been clearly shown to be involved in establishing early domains of *Hox* expression in the CNS comes from analysis of the group 4 paralogues *Hoxa4*, *Hoxb4* and *Hoxd4*. These genes are also capable of responding to exogenous RA, but they do so in a manner that is different to the more 3' genes such as *Hoxa1* and *Hoxb1* (Conlon and Rossant 1992; Marshall et al. 1992, 1996; Morrison et al. 1996, 1997; Packer et al. 1998). Whereas RA treatment at 7.25–7.75 dpc rapidly induces the most 3' genes such as *Hoxa1*, *Hoxa2* and *Hoxb1* expression expanding their expression, into more anterior regions, there is no response of the group 4 genes at this stage. However, if RA is administered between 8.5 and 9.5 dpc the group 4 genes rapidly respond while the more 3' genes are no longer activated (Morrison et al. 1996, 1997). Uncharacterised mechanisms serve to restrict the timing and degree of response of *Hox* genes to RA. Recent evidence has illustrated that the response of at least one of the group 4 genes, *Hoxb4*, is direct and that a 3' RARE neural enhancer is necessary for regulating early expression up to the r6/r7 boundary (Gould et al. 1998). This RARE is stimulated by a mesodermal signal and helps to trigger the proper early domain of *Hoxb4* expression, which is later maintained at the correct boundary by a second autoregulatory element in the locus (Gould et al. 1998). Since there is evidence of auto- and cross-regulatory interactions between the group 4 genes (Pöpperl and Featherstone 1992; Gould et al. 1997, 1998; Packer et al. 1998), it is not clear whether RA plays a similar direct role in establishing *Hoxd4* and *Hoxa4* expression or if this is mediated indirectly by cross-regulatory influences of *Hoxb4*.

To date, it remains unclear whether the colinear RA response of other *Hox* genes is also directly mediated by RAREs and retinoid receptors as illustrated for the group 1 and group 4 genes. Some of these same elements might be shared between adjacent genes (Gould et al. 1997), allowing them to exert an influence over multiple genes in the clusters. It will be important to characterise more genes in the complexes to determine how general the direct activation mechanisms are on a larger scale. In addition to the roles of auto- and cross-regulatory interactions in maintaining the proper limits of *Hox* expression, vertebrate homologues of the *Polycomb* and *Trithorax* group genes are also important in maintaining the correct domains through chromatin mediated mechanisms (Paro 1993; Gould 1997; Pirrotta 1997a,b). Mutations in some of these genes result in anterior shifts of segmental expression in the hindbrain (Takihara et al. 1997), and they also appear to be able to alter the response of Hox genes to retinoic acid (Bel et al. 1998). Hence the PcG, and TrxG gene families may have an important role in mediating or modulating the indirect influences of RA on *Hox* genes and their ability to respond to RA. Regardless of the specific mechanisms involved, an important area of future work will be to define the multiple means by which RA signalling establishes and maintains axial patterning processes through the *Hox* genes and other positional cues.

The node is a rich source of RA (Chen et al. 1992, 1996; Hogan et al. 1992), and consistent with its role as an organiser tissue, it produces increasing amounts of RA during regression. It is possible that the nested domains of *Hox* gene expression could be controlled either by a posterior-anterior gradient of RA diffusing from the node or by an increasing exposure to RA as cells pass the node in A to P progression. Alternatively, a constant level of RA might be translated into a graded response through interactions with modulating factors. It has yet to be shown that RA normally forms a graded signal or that a gradient is indeed necessary; however the local activity of retinoids could be modified by co-activators and co-repressors of retinoid signalling (Mangelsdorf and Evans 1995; Mangelsdorf et al. 1995; Morrison et al. 1997). The work showing that RAREs are essential for proper *Hoxb1* or *Hoxb4* expression might reflect a need for an RA signalling component and does not necessarily imply that RA provides an instructive influence on axial expression. However, in specificity swap experiments where RARE sequences in a *Hoxb4* enhancer were changed to that of *Hoxb1*, this resulted in an earlier activation of expression and a more anterior boundary, demonstrating that RA can indeed provide instructive signals that set precise *Hox* expression pattern (Gould et al. 1998).

3.2
Krox20 Targets

Krox20 has a pivotal role in controlling hindbrain segmentation (Schneider-Maunoury et al. 1993; Swiatek and Gridley 1993) and the specific upregulation of *Hoxb2* and *Hoxa2* and of *EphA4* in r3 and r5 suggested a possible link or overlap with that of *Krox20*. The regulatory roles of *Krox20* have been clearly demonstrated via transgenic and mutational analyses. Analysis of the

5' 2 kb flanking region immediately upstream of the *Hoxb2* gene led to the identification of a 569 bp fragment that was important for restricting domains of expression in r3 and r5 in response to ectopically expressed *Krox20* (Sham et al. 1993). This 569 bp fragment contains three Krox20 binding sites which very closely resemble a 9 nt consensus Krox20 binding site 5'-GCGGGGGCG-3' (Nardelli et al. 1991). Single G–C mutations in any of the three putative binding sites dramatically reduce its complex formation with Krox20 in in vitro binding assays. Alone, the 569 bp fragment is insufficient to drive expression in r3 and r5, however the addition of a further 122 bp of 5' flanking region renders the enhancer capable of restricting transgene expression to r3 and r5. The additional 122 bp does not contain any extra Krox20 binding sites but must provide some other as yet unclassified *cis*-acting elements that serve to potentiate or enhance *Krox20*-dependent activity of this enhancer (Sham et al. 1993). Further analysis revealed that a second motif positioned adjacent to the Krox20 binding sites in *Hoxb2* is also required for activity and shows that other factors in the minimal enhancer are also needed to potentiate *Krox20* activity (Vesque et al. 1996).

Similar transgenic assays were performed on *Hoxa2*. Analyses of over 22 kb spanning the three genes *Hoxa1* to *Hoxa3* defined an 809 bp enhancer in the 5' flanking region of the *Hoxa2* gene that contains two putative Krox20 binding sites (Nonchev et al. 1996a,b). These two co-operative sites are contained within a 257 bp subfragment and are essential for upregulation of *Hoxa2* expression in r3 and r5. They are insufficient on their own to establish expression in r3 and r5 as they require interaction with as yet unidentified cofactors (Nonchev et al. 1996a,b). Hence, both *Hoxa2* and *Hoxb2* are direct targets for *Krox20* mediated up regulation in r3 and r5. Expression of *Hoxb3* is lost in a *Krox20* mutant and it may also be directly regulated by *Krox20* (Seitanidou et al. 1997).

The overlapping expression patterns of *Krox20* and *EphA4* also led to the suggestion that transcriptional control and cell–cell signalling could be coupled together. Therefore transgenic regulatory analysis was performed on a 7.5 kb of 5' flanking sequence which is able to reconstitute the *EphA4* endogenous expression pattern in r3 and r5. *Cis*-acting regulatory sequences important for controlling *Eph44* gene expression in r3 and r5 and were mapped by deletion analysis and localised to a 470 bp element that contains eight putative Krox20 binding sites (Theil et al. 1998). Mutations in these sites abolish r3/r5 enhancer activity and ectopic *Krox20* expression leads to the direct activation of this enhancer. Together, this shows that *EphA4* is a direct target of *Krox20* and provides a common link for co-ordination of genes involved in segmental identity (*Hox*) and cell lineage restrictions (*Eph*).

These transgenic and mutational analyses indicate that *Krox20* is an essential component of the upstream regulatory cascade governing the transcriptional control of *Hoxa2*, *Hoxb2*, *Hoxb3* and *EphA4* during hindbrain segmentation. The results also suggest that the identity and movement of cells during hindbrain development are coupled in order to generate sharply re-

stricted and specified segmental domains and this is further substantiated in
the phenotype of *Krox20* null mutants, which is discussed below.

3.3
Kreisler Targets

Previous analyses in *kreisler* mutants showed that *Hox* gene expression is
altered or lost in the hindbrain (Frohman et al. 1993; McKay et al. 1994, 1997;
Manzanares et al. 1997b). The cloning of the *kreisler* gene and the analysis of
its expression showed that it was strongly expressed in r5 and r6 (Cordes and
Barsh 1994), suggesting it might play a direct role in the transcriptional
regulation of *Hox* genes important for hindbrain development. Transgenic
analyses of the regulatory regions of *Hoxb3* in mouse and chick and *Hoxa3* in
mouse have shown that both these genes are under the direct control of the
product of the *kreisler* gene (Manzanares et al. 1997, 1999a).

A 650 bp element upstream of the mouse *Hoxb3* gene is capable of directing reporter gene expression specifically to r5 (Manzanares et al. 1997).
Sequence comparison of this region with a functionally equivalent 800bp
from the chick gene revealed that only two short stretches of sequences were
conserved, one of 20 and one of 45 bp. Each of these blocks contains binding
sites for the *kreisler* protein that are necessary for expression and sufficient
for a *kreisler* response (Manzanares et al. 1997). Despite the dependence of
this enhancer on *kreisler* activity, it only mediates expression in r5 and not r6,
showing that other factors may serve to restrict its potential. Further deletion
and mutation analysis identified a second *cis*-element positioned adjacent to
one of the *kreisler* sites. This element corresponds to an activation site for
Ets-related transcription factors (ERAS) and is necessary to potentiate and
restrict *kreisler*-dependent activity of the enhancer to r5 (Manzanares et al.
1997). Studies in other systems have shown that Ets proteins and Maf proteins like *kreisler* physically interact and play synergistic roles in regulating
cell lineage decisions (Sieweke et al. 1996). This suggests that Ets and Maf
proteins may have multiple roles in controlling cell fate and identity. Other
cis-elements in the *Hoxb3* enhancer may also contribute to its restricted
expression in r5 and further analysis is required.

As described above with *Krox20*, it is possible that *kreisler* can regulate
multiple *Hox* genes and a similar regulatory analysis has been performed with
another group 3 paralogue, *Hoxa3*. *Hoxa3* is upregulated in both r5 and r6,
and regulatory analysis of its genomic locus identified a 600 bp element
responsible for expression in these rhombomeres (Manzanares et al. 1999a).
Sequence analysis of this fragment revealed the presence of a unique *kreisler*-
binding site, mutations in which abolish activity. When oligomerised and
tested with a reporter gene the transcriptional readout of these sites mirrored
that of endogenous *kreisler*, showing strong expression in r5 and r6. This
demonstrates that both *Hoxa3* and *Hoxb3* are direct targets of *kreisler* in the

hindbrain, but that there are distinct differences in the nature of this response as other factors serve to restrict its ability to stimulate *Hoxb3* to r5 (Manzanares et al. 1999a). These types of studies have defined clear roles for *Krox20* and *kreisler* in directly regulating multiple *Hox* genes. However, they themselves overlap in controlling aspect of r5 patterning, and synergy or combinatorial activities of these two proteins on targets in r5 have not yet been examined.

3.4
Hox Gene Auto- and Cross-Regulation

Regulatory interactions between the Hox genes themselves play an important role in controlling and maintaining their segmental expression. This is clearly displayed by the *Hoxb1* gene which, after its early RA-mediated establishment phase, maintains high levels of expression in r4 (Murphy et al. 1989; Wilkinson et al. 1989b). Rhombomere 4 grafts have demonstrated that this is achieved through an autonomous mechanism (Guthrie et al. 1992; Kuratani and Eichele 1993). The continued expression or maintenance phase of *Hoxb1* in r4 does not directly involve RA, but is achieved through a conserved auto-regulatory loop (Pöpperl et al. 1995). Three related sequence motifs located 5' to the *Hoxb1* gene are required for r4 activity of the enhancer and are highly conserved among mouse, chicken and pufferfish *Hoxb1* genes (Pöpperl et al. 1995). These repeats represent a bipartite recognition site comprised of overlapping sites for *Hoxb1* and for a mouse homologue of the *Exd* homeobox gene, *Pbx*. *Hoxb1* together with a mouse *Pbx* family protein is essential for enhancer activity. Mutational analyses have clearly demonstrated that early expression of *Hoxb1* and of *Hoxa1*, stimulated by RA, is able to activate this auto-regulatory loop (Studer et al. 1996, 1998; Dupé et al. 1997; Gavalas et al. 1998) enabling *Hoxb1* to regulate r4 identity. The activation is restricted to r4 through the influence of the RARE-dependent repressor elements that block the auto-regulatory loop in r3 and r5 (Studer et al. 1994).

Hoxb1 also exerts a cross-regulatory role on another *Hox* gene. Despite the similarities in expression and regulation between various *Hox* gene paralogues, there is frequently variation in their relative levels within specific segments. For example, of the group 2 homologues, *Hoxb2* is upregulated in r4, but in contrast *Hoxa2* is not. This type of differential expression suggests that in even-numbered rhombomeres these two genes have distinct modes of regulation. Transgenic deletion analyses of 5' flanking regions have again been crucial for identifying *cis*-acting elements responsible for upregulation of *Hoxb2* gene expression in r4 (Maconochie et al. 1997). Within the 5' flanking region of the *Hoxb2* gene a 181 bp element was identified that was capable of mediating the upregulation of *Hoxb2* expression in r4. Sequence analysis demonstrated that this element contained no consensus RAREs; however, interestingly it revealed a single motif that was highly related to the autoregulatory motifs identified in the *Hoxb1* locus. An alignment of the site in the *Hoxb2* gene with the three sites from the *Hoxb1* gene indicated that the motif is related to a bipartite *Pbx/Hox* consensus (5'TGATCG-3') sequence, although it is not identical to any of the *Hoxb1* motifs (Maconochie et al. 1997). Consequently, it was shown that in vitro *Hoxb1* is able to bind to the *Hoxb2* motif in an *exd*-dependent manner. The *Hoxb2* motif was also shown to be able to distinguish between *Hoxb2* and group 1 proteins, implying that r4-specific expression of *Hoxb2* is not a consequence of its own auto-regulation but results from cross-regulatory interaction with group 1 genes (Maconochie et al. 1997). Deletions of the *Pbx/Hox* site in the *Hoxb2* en-hancer confirmed that it is required for normal in vivo r4 activity. These results were confirmed by the absence of *Hoxb2* upregulation in r4 in *Hoxb1* null mutants (Studer et al. 1996; Maconochie et al. 1997) and demonstrates that *Hoxb2* is a direct target of *Hoxb1* (discussed below).

Another demonstration of the importance of *Hox* auto- and cross-regulation has come from analyses of the *cis*-acting regions, in group 4 pa-ralogues (Gould et al. 1997, 1998). *Hoxb4* and *Hoxd4* have domains of expression that partially overlap and map to the r6/r7 junction. Enhancers capable of setting these r6/r7 anterior limits of neural expression have been found in the 3' flanking sequences of these genes (Whiting et al. 1991; Aparicio et al. 1995; Morrison et al. 1995, 1997; Gould et al. 1997, 1998). The neural enhancer of the *Hoxb4* gene (region A) is highly conserved in the *Hoxb4* gene of other vertebrates, and although it lies 3' of the *Hoxb4* gene, it sits directly adjacent to a distal promoter of the *Hoxb3* gene and this *cis*-element is shared by the *Hoxb4* and *Hoxb3* genes (Sham et al. 1992; Aparicio et al. 1995; Morrison et al. 1995; Gould et al. 1997). Sequence comparisons of region A from pufferfish, chicken and mouse identified a highly conserved region (CR3) which is alone capable of directing expression with a sharp r6/r7 anterior boundary. Similar to *Hoxb1*, *Hoxb4* and *Hoxd4* are also capable of auto-regulation (Gould et al. 1997). The auto-regulatory mechanism is largely mediated through two TAAT/ATTA motifs (designated HS1 and HS2) con-tained within the CR3 fragment and mutations in these sites abolish normal expression patterns. Analysis of CR3 in *Drosophila* embryos revealed that it exhibits an evolutionary conserved response to the *Dfd*-related group 4 *Hox* genes (Chan et al. 1997; Gould et al. 1997). The CR3 element therefore ap-pears to be involved in the maintenance but not the establishment of *Hoxb4* and *Hoxd4* gene expression via auto- and cross-regulatory mechanisms.

In a manner analogous to that of *Hoxb1*, it has been shown that the early phase of *Hoxb4* expression is mediated by a 3' RARE (Gould et al. 1998). In response to a secreted signal from somites, RA signalling directly establishes an early and transient r6/r7 domain that triggers the *Hoxb4* autoregulatory loop. This then maintains the proper domains of expression in later stages. Therefore, both group 1 and group 4 genes use similar RA-mediated means of activating Hox genes in specific domains and then later maintain those through cross-regulatory interactions. Further analysis is needed to determine whether this biphasic model extends to many of the other paralogy groups.

Although *Hoxa4* regulation differs from that of other group 4 paralogues, there are similarities (Behringer et al. 1993; Morrison et al. 1997). A 3' enhancer has been identified that directs neural expression with an r6/r7 anterior boundary but only at later stages. There is a 5' enhancer that also mediates r6/r7 expression, but it does so at earlier stages, suggesting dual control of proper *Hoxa4* expression in the hindbrain and temporal differences in expression. Hence, the elements that regulate *Hoxa4* in the hindbrain appear to be slightly different from those of its paralogues.

The above studies highlight the general importance of auto-regulatory and cross-regulatory mechanisms for the functional maintenance of *Hox* gene expression during vertebrate hindbrain development. This clearly complements any of the roles that the *PcG* and *TrxG* genes have in maintaining expression and show how changing the expression of one *Hox* gene can be translated to global changes in other genes. The fact that some elements are shared has important implications for maintaining the *Hox* complexes themselves. It is possible that removing these genes or regions might also alter control regions in a particular complex and hence they would be required for proper regulation of multiple genes. This would provide a basis for keeping genes clustered in order to maintain appropriate expression patterns and this necessity is highlighted in the analyses of null mutations in *Hox* genes detailed below.

Finally, the identification of high-affinity bipartite binding sites for *Hox* proteins and their cofactors of in vivo relevance has greatly helped the identification of downstream target genes. While data are rapidly emerging on the types of control elements used by *Hox* genes, much more work needs to be directed at the nature of the mesodermal and axial signals that activate and maintain these CNS patterns. This will be essential to determine how the *Hox* genes integrate such a wide array of signalling information to regulate morphogenesis.

4
Mutational Analyses of Gene Function

4.1
Segmentation Genes

Null mutations provide an important mechanism for observing a particular gene's overall role during embryonic development. Mice homozygous for a targeted mutation in *Krox20* die shortly after birth and exhibit fusions of the trigeminal ganglion with facial and vestibular ganglia as a consequence of the profound perturbation of hindbrain morphogenesis (Schneider-Maunoury et al. 1993, 1997; Swiatek and Gridley 1993). Although the presumptive territories of r3 and r5 do form during the very early stages of hindbrain development, these two rhombomeres are not maintained and their structural derivatives are subsequently eliminated. Consistent with the interactions unveiled by transgenic analyses between *Krox20* and *Hoxa2*, *Hoxb2*, *Hoxb3* and *EphA4*, no upregulation of these three genes was observed in *Krox20* mutants (Seitanidou et al. 1997).

Interestingly, the mutational analyses also revealed interactions between *Krox20* and other genes in the regulatory cascade controlling hindbrain segmentation. During normal mouse hindbrain development *follistatin* is expressed in r2, r4 and r6. In *Krox20−/−* embryos however, *follistatin* is expressed in a continuous pattern in r2, r3 and r4 (Schneider-Maunoury et al. 1997). This suggests that *Krox20* inactivation results in a specific activation of *follistatin* expression in r3 but does not affect its regulation in r5 or even-numbered rhombomeres. In normal development therefore the absence of *follistatin* expression in r3 may be due to *Krox20*-dependent repression. If this repression occurs in a direct manner it means that *Krox20* can act as both a positive and a negative transcriptional regulator within the same cells. The differential activity may require the interaction of *Krox20* with distinct co-factors which have been found to alter its activity (Svaren et al. 1998).

In addition to these roles *Krox20*, has recently been shown to synergise with *Hoxa1* and regulate patterning in r3 (Helmbacher et al. 1998). Even though r3 is outside the normal domain of *Hoxa1* expression this suggests that it may indirectly influence r3 patterning through altering inter-rhombomeric interactions that interface with *Krox20*. Overall therefore there are at least five known regulatory genes (*EphA4*, *Hoxa2*, *Hoxb2*, *Hoxb3* and *follistatin*) belonging to different families that are under the direct or indirect control of *Krox20* in r3 and/or r5. This transcription factor therefore is a key regulator of gene expression in the developing hindbrain that is essential to maintain the proper segmental domains of r3 and r5.

The classical mutant *kreisler* was recovered from an X-ray induced mutagenesis and identified by its circling behaviour (Deol 1964) which is typical of inner ear and vestibular defects. The physical nature of the mutation is a micro-inversion that maps approximately 50 kb away from the coding regions of the gene (Cordes and Barsh 1994). Apart from the hindbrain, *kreisler* is expressed in other tissues, most of which are not affected in the classical mutant. This indicates that the *kreisler* mutation is not a null allele, but a regulatory mutation affecting those elements responsible for the hindbrain domain of expression of the gene (Cordes and Barsh 1994). In addition to inner ear abnormalities, *kreisler* mutants also exhibit defects in neural crest derived skeletal elements such as the hyoid (Frohman et al. 1993; Deol 1964).

When examined at early embryonic stages, the primary defect in *kreisler* mutants was identified as being a disruption of segmentation in the otic region of the hindbrain (Deol 1964; Frohman et al. 1993; McKay et al. 1994,

1996; Manzanares et al. 1999b). The rhombomeric borders that normally demarcate r4, r5 and r6 are absent. The normal expression domains of *Fgf3* and *CRABPI* and the upregulation of *Hoxa3* in r6, are absent and although *Krox20* expression in presumptive r3 is present, the posterior band associated with r5 is missing. Similarly, the normal expression domains of *Hoxb2*, *Hoxb3* and *Hoxb4* in r5 are also abolished. Analysis of patterns of expression of *Eph7A* and *ephrinB2* indicate that only a single segment correlating with r5 is missing (Manzanares et al. 1999b). Therefore the segmentation defect associated with the *kreisler* mutant is a specific loss of r5 and although an r6 territory does form (contrary to earlier reports) it never properly matures (Manzanares et al. 1999b). Cell miscibility studies in the mouse indicate that an even character has been imparted to r6 and this is further supported by the fact that r5-derived FBM neurons which migrate ventrally into r5 before turning laterally in r6 in normal embryos, undertake a truncated lateral migration pathway immediately after emerging from r4 in *kreisler* embryos. *Kreisler* therefore regulates multiple steps in hindbrain segmentation. It functions early in segmentation to mediate the proper formation of r5, and thus is a true segmentation gene in the mouse (Manzanares et al. 1999b). In addition, through direct regulation of *Hox* genes (*Hoxa3* and *Hoxb3*), *kreisler* also controls segmental identity, and this is further evidenced by the recent finding that ectopic expression of *kreisler* in r3 transforms it to an r5 identity (Theil et al. 1999). Hence like *Krox20*, *kreisler*, rather than being restricted to a single patterning aspect, is involved at several different levels including establishment and maintenance of A-P identity and segment specification.

A zebrafish homologue (*valentino*) of *kreisler* has recently been cloned from the analysis of a mutant identified in a screen (Moens et al. 1996, 1998). While the expression pattern of *valentino* is also highly upregulated in r5 and r6, its role in hindbrain patterning has been interpreted in a very different way from that of *kreisler* (Moens et al. 1998; Manzanares et al. 1999b). Instead of regulating the formation of r5 independent of r6, it is suggested that *valentino* functions to subdivide a pro-rhombomeric like territory into mature r5 and r6. While this could reflect genuine differences in how these genes function in the two species, it is also possible that the genes actually function in a similar manner. Differences in the types and stages of analysis and markers used could account for the two divergent models and further experiments will be required to distinguish between these possibilities.

In addition to *Krox20* and *kreisler*, there is also some evidence that suggests *Hox* genes may play a role in early segmentation. Two different null alleles of *Hoxa1* have been generated (Lufkin et al. 1991; Chisaka et al. 1992) and they display phenotypes in r5. In one case they show only a partial deletion of r5 markers and territories (Dollé et al. 1993; Mark et al. 1993) while in the other there is a complete absence of r5 (Carpenter et al. 1993). Hence, *Hoxa1* could be involved in maintaining or possibly establishing r5 in addition to its other roles, described below.

4.2
Segment Identity Genes

By analogy to their *Drosophila* counterparts it was predicted that *Hox* genes played a role in regulating the identity of hindbrain segments (Wilkinson et al. 1989b; Hunt et al. 1991a,b). This role has been documented for some of the 3' members of the *Hox* genes. Ectopic expression of *Hoxa1* in mouse and fish embryos results in the transformation of r2 to an r4 identity (Zhang et al. 1994; Alexandre et al. 1996). These gain of function experiments suggest that *Hoxa1* has a normal role in regulating segmental identity. *Hoxa1* mouse mutants die at birth from anoxia, and they exhibit marked defects in the inner ear and in specific cranial nerve components (Lufkin et al. 1991; Chisaka et al. 1992). The embryonic phenotype is characterised by a reduction in the size of r4 and the complete or partial deletion of r5, as revealed by the diminished expression of *Hoxb1* in r4, and of *Krox20* and *Fgf3* in r5. This may account for the absence of facial nerve and abducens nerve motor neurons. In addition, patches of cells with an r2 like character appear in r3, which probably cause the r3 motor nerves to migrate in a pattern typical of an even-numbered rhombomere (Helmbacher et al. 1998). Hence, in the single *Hoxa1* mutants there appear to be only minor defects in r4.

In contrast to the *Hoxa1* mutants, *Hoxb1* mutants exhibit no segmentation defects, but they do show changes in segmental identity (Goddard et al. 1996; Studer et al. 1996, 1998). In *Hoxb1* mutant embryos, molecular analyses indicate that the patterning of r4 is initiated properly but not maintained (Studer et al. 1996, 1998). Although r4 markers such as *Wnt8c* and *CRABP1* fail to be upregulated, the associated ectopic expression of r2 markers (*EphA4*) suggests that r4 adopts an altered identity. Cellular analyses by dye tracing show that the r4-specific facial branchiomotor neurons (FBM) and contralateral vestibular acoustic (CVA) efferent neurons are incorrectly specified (Studer et al. 1996). Both sets of neurons fail to migrate into the correct position and subsequently there is a loss of the facial motor nerve. These results demonstrate that as part of its role in maintaining rhombomere identity, *Hoxb1* is involved in controlling migratory properties of motor neurons in the hindbrain.

The phenotypes of individual *Hoxa1* and *Hoxb1* loss of function (null) mutations suggest that these genes play distinct roles in hindbrain development. Genetic analyses of double mutants (*Hoxa1/Hoxb1*) exhibit a surprising range of phenotypes that are absent from either individual mutant (Gavalas et al. 1998; Studer et al. 1998). This suggests that there is extensive synergy between *Hoxa1* and *Hoxb1* in the patterning of cranial nerves VII–XI and in the establishment of r4 identity. In the absence of both genes, a territory appears in the region of r4; however, the earliest r4 marker, that is the *Eph* tyrosine kinase receptor *EphA2*, fails to be activated. This suggests a failure to initiate rather than maintain the specification of r4 identity and indicates that *EphA2* lies downstream of *Hoxa1* and *Hoxb1* in the genetic cascade that regulates hindbrain segmentation and specification. The double mutants also revealed that *Hoxa1* and *Hoxb1* work synergistically in initiating the r4-restricted expression of *Hoxb1*. *Hoxa1* has a normal role in activating the r4 enhancer of *Hoxb1* through 3' RARE para-regulatory interactions during the establishment phase of *Hox* gene expression, but it is unable to participate in the long-term maintenance of *Hoxb1* expression because it is not expressed at these later stages (Studer et al. 1998). Therefore the para-regulatory interactions between *Hoxa1* and *Hoxb1* together with RA ensure that *Hoxb1* expression is generated in r4, while the *Hoxb1* auto-regulatory function ensures that high levels are maintained (Pöpperl et al. 1995; Studer et al. 1996, 1998). Furthermore, *Hoxb1* works in a direct cross-regulatory manner to upregulate *Hoxb2* expression in r4. Therefore *Hoxb1* utilises auto-, para- and cross-regulatory mechanisms as part of its role in maintaining r4 identity and regulating facial motor neuron patterning.

Recent analysis of *Krox20/Hoxa1* double mutants has revealed that further synergistic interactions are involved in hindbrain patterning (Helmbacher et al. 1998). The *Hoxa1* mutant is characterised by the presence of r2-like cells in r3 combined with abnormal motor axon navigation. These phenotypic manifestations become more severe with the additional activation of one *Krox20* allele and demonstrate that *Hoxa1* and *Krox20* synergise in a dosage-dependent manner to specify r3 identify and odd versus even rhombomere characteristics. Control of r3 development therefore may not be autonomous but dependent on interactions with *Hoxa1* expressing cells.

Hoxa2 is the most anteriorly expressed *Hox* gene and the only *Hox* gene to be expressed in r2 (Krumlauf 1993; Prince and Lumsden 1994). The targeted inactivation of *Hoxa2* results in lethality at birth and homeotic transformation of the second arch neural crest derived elements into first arch derivatives (Rijli et al. 1993). In *Hoxa2* mutant embryos, the hindbrain is also affected, with the segmental identities of r2 and r3 being altered (Gavalas et al. 1997). The alar territories of r2 and r3 are reduced and there is a concomitant expansion of r1. *EphA4* expression is selectively abolished in r2, which suggests that *Hoxa2* is required to maintain *EphA4* expression in r2. It also suggests that there is a change in the identity of r2 since *Hoxa2* is the only *Hox* gene expressed in this rhombomere. Cellular analysis by retrograde dye tracing supports this hypothesis by showing that r2 and r3 trigeminal motor axons turn caudally and exit the hindbrain from the r4 facial nerve exit point and not from their normal exit point in r2. Dorsal r2-r3 patterning is particularly affected with the loss of cochlea nuclei and the enlargement of the lateral part of the cerebellum. *Hoxa2* therefore not only acts as a selector gene for second arch mesenchymal neural crest cells but also plays a fundamental role in rostral hindbrain patterning by establishing r2 and to a lesser extent r3 segmental identities (Rijli et al. 1993; Gavalas et al. 1997). In addition, *Hoxa2* plays a crucial role in the control of r2-r3 motor axon guidance and its absence causes homeotic transformations in the alar plates of these rhombomeres.

In contrast to *Hoxa2* null mutants, the targeted mutation in the *Hoxb2* locus apparently does not result in segmental abnormalities in the developing hindbrain (Barrow and Capecchi 1996). However, this is a compound mutant that also affects the expression of other members of the *HoxB* complex in *cis*, so it is difficult to determine its precise role in segmental patterning. However, recently a new allele of *Hoxb2* has been generated and this has been used in double mutant analysis with *Hoxa2* (Davenne et al. 1999). In the single *Hoxb2* mutants there were changes in the identity of r4 at later stages, showing that it has a role in maintaining segmental identity in line with the finding that it is a direct target of *Hoxb1* in r4. Despite the absence of *Hoxb2* upregulation in r3 and r5 these rhombomeres develop normally.

In addition to the A-P changes, phenotypic changes were also evident in the D-V patterning of the hindbrain neurons (Davenne et al. 1999). Specific alterations to *Mash1* (a marker for a subset or neuronal precursors), *Math3*, *Nkx2.2* and *Phox2b* expression were observed in r4, which now more closely resembled the wild type patterns typical of r2 and suggests that there may be incorrect specification of a subset of ventral motorneuron precursors. *Hoxb2* therefore plays an important role in regulating a genetic cascade that may control the generation and/or fate of subsets of r4 motorneuron progenitors. Retrograde dye tracing of r4 branchiomotor neurons (which form the somatic motor component of the facial nerve and migrate caudally into r5) revealed that this population was greatly reduced and in addition, a few laterally migrating r3 neurons incorrectly projected to the r4 exit point. Therefore the development of the somatic motor component of the facial nerve is impaired in *Hoxb2−/−* mutants, which is reminiscent of the phenotype described in the *Hoxb1* mutants.

The generation of double mutants (*Hoxa2/Hoxb2*) revealed that *Hoxa2* and *Hoxb2* are coupled to both A-P and D-V patterning but play distinct roles in the cascades that mediate neurogenesis in the developing hindbrain (Davenne et al. 1999). The morphological and cellular analyses of single mutants support the idea that *Hoxa2* controls development in the alar and dorsal basal plates of r2 and r3 whereas *Hoxb2* is essential for motorneuron development in the ventral basal plate of r4. Several *Pax* genes including *Pax3* and *Pax6* are expressed in D-V domains of the developing hindbrain where they function as determinants of dorsal and ventral differentiation pathways of neural progenitors. In the double mutants, selective alterations in the D-V restriction of *Pax3* and *Pax6* were observed in r2 and r3 where *Hoxa2* and *Hoxb2* are the only *Hox* genes expressed. The absence of a particular ventral interneuron subtype in r3 in the double mutants but not in either of the single mutants suggests that there is functional synergy between *Hoxa2* and *Hoxb2* in r3. In addition, despite the fact that the normal number of hindbrain segments form in the double mutants, there appears to be a requirement for *Hoxa2* and *Hoxb2* in patterning inter-rhombomeric boundaries, because boundary formation between r1–r4 is non-existent. These results indicate that *Hoxa2* and *Hoxb2* differentially control alar and basal plate development within distinct rhombomeres and provide a link between *Hox*-mediated patterning along the A-P axis and neurogenesis along the D-V axis. It also provides a mechanism for the generation of neurons at reproducible positions within hindbrain segments.

With respect to AP patterning in the double *Hoxa2/Hoxb2* mutants, r3 and r5 were still present (Davenne et al. 1999). However, abnormalities were detected at rhombomere boundaries. The fact that r3 and r5 are maintained in the double mutants shows that the role of *Krox20* in maintaining these segments must be mediated through alternative targets.

A target mutation in the *Hoxa3* gene leads to two classes of defects in the formation of the IXth cranial nerve. In the first class, there was an apparent deletion of the proximal portion of the IXth (glossopharyngeal) cranial ganglion (Chisaka and Capecchi 1991). These embryos appear to have at least some part of the IXth cranial ganglion, but it was not connected to the hindbrain. This phenotype may represent the deletion of neural crest derived cells that contribute to the formation of the superior ganglion. The second class of mutant phenotype seen in the *Hoxa3* mutant embryos was a fusion of the IXth and Xth cranial ganglia. Examination of the cranial nerves in the *Hoxb3−/−* embryos showed that both classes of cranial ganglion phenotype were also present in these mutants, but at a lower penetrance than in the *Hoxa3* mutants (Condie and Capecchi 1993, 1994; Manley and Capecchi 1997, 1998). The cranial nerves in the *Hoxd3−/−* embryos however, have been shown to be completely normal. Therefore at least two of the group 3 paralogues play essential roles in formation of the IXth cranial ganglion. Double mutants (*Hoxa3/Hoxb3*, *Hoxa3/Hoxd3* and *Hoxb3/Hoxd3*) were subsequently generated in order to determine whether *Hoxd3* plays a synergistic role in development of the cranial ganglia. In *Hoxb3/Hoxd3* double mutants there was a marked increase in the penetrance of the IXth cranial ganglion defects relative to that observed in the *Hoxb3* single mutant. This increase in penetrance shows that *Hoxd3* does indeed play a synergistic role with *Hoxb3* in cranial nerve development even though the *Hoxd3* single mutant does not show a defect in these structures.

Together these studies illustrate that *Hox* genes function in many steps of segmental identity and patterning of craniofacial structures of neural crest origin. It confirms their important role as mediators of positional information and in the future it will be necessary to generate conditional or partial loss of function alleles to explore their specific roles in individual tissues. It will also be critical to understand the nature of the cellular changes and associated molecular mechanisms and target genes through which they function.

5
Mechanisms of Hindbrain Segmentation

The restriction of intermingling between rhombomeres is crucial for the establishment of segment identity and the maintenance of organised patterns during hindbrain and craniofacial development. Classical studies in which distinct cell populations were mixed together, such as in the case of odd and even rhombomeric populations, demonstrated that there was a preferential association of cells with similar adhesive properties (Guthrie and Lumsden 1991; Guthrie et al. 1993; Manzanares et al. 1999b). This led to the hypothesis that a hierarchy of cell adhesion molecules could facilitate the segregation of distinct cell populations (Wizenmann and Lumsden 1997); however to date, an adhesion molecule with alternating segmental expression in the hindbrain has not been identified. In contrast, the *Eph* receptors and their membrane-bound ligands (the *ephrins*) are expressed alternately in complementary rhombomeres which implicated them in mediating cell repulsion at rhombomere boundaries (Gilardi-Hebenstreit et al. 1992; Nieto et al. 1992a; Becker et al. 1994; Gale et al. 1996b; Lumsden and Krumlauf 1996; Flanagan and Vanderhaeghen 1998). Recent experiments now suggest that *Eph* receptor tyrosine kinases are required for the segmental restriction of cell intermingling during hindbrain development (Xu et al. 1995, 1999). The mosaic activation of *Eph* receptors such as *EphA4* in the hindbrain leads to the sorting of cells to the boundaries of odd rhombomeres. In contrast, the mosaic activation of *ephrins* such as *ephrinB2* leads to sorting at the boundaries in even rhombomeres. The finding that activation of *Eph* receptors or *ephrins* each leads to a cell sorting response suggests that bidirectional signalling at rhombomere boundaries restricts cell intermingling (Mellitzer et al. 1999). The above analyses were carried out at the 12–15 somite stage in zebrafish embryos, yet presumptive r3/r5 can be detected at the 2 somite stage by the expression of *Krox20*. It seems therefore that the restrictions in cell mixing are established during the later stages of hindbrain segmentation, perhaps as a consequence of *EphA4* being a downstream target of *Krox20*. These studies reveal the key roles *Eph* receptors and their ligands play in controlling cellular restrictions and movements in the hindbrain. They have also been shown to have important functions in regulating the migration of cranial neural crest cells and axon guidance (Smith et al. 1997; Flanagan and Vanderhaeghen 1998).

Krox20 regulates *EphA4* in r3 and r5, revealing that it exerts its functions through both the *Hox* genes and *Eph* receptors. However, *Hox* genes themselves have been implicated in regulating *Eph* receptors (Taneja et al. 1996; Chen and Ruley 1998). The *EphA2* gene is expressed at a high level in r4 and transgenic analysis identified a 12 kb region able to direct reporter expression to r4 (Chen and Ruley 1998). Sequence analysis revealed that there were multiple bipartite Hox/Pbx binding sites, nearly identical to those found in the *Hoxb1* and *Hoxb2* r4 enhancers. Further analysis revealed that *Hoxb1* can activate expression from these elements, and that *EphA2* expression is lost in double *Hoxb1/Hoxa1* mutants (Studer et al 1996, 1998; Chen and Ruley 1998; Gavalas et al. 1998). These studies show that *EphA2* is a direct target of *Hoxa1* and *Hoxb1* in r4. Expression studies in *Hoxa2* mutants have shown that the conserved rhombomere-specific expression of *EphA7* (*Mdk1*) is dependent upon *Hoxa2*. Therefore, in addition to their roles in controlling segmental identity, *Hox* genes along with *Krox20* participate in regulating lineage restrictions and cell movements in the hindbrain through the *Eph* receptors.

6
Conclusions

Neuraxial patterning is a continuous process that extends over many days during embryonic development and requires an extraordinarily complex set of genetic interactions for proper specification (Fig. 2). During gastrulation a crude A-P pattern is conferred on naive ectoderm by signals emanating from the underlying mesoderm. This coarse-grained pattern which is detectable by broad domains of gene expression is then refined and reinforced by locally acting mechanisms to generate the characteristic morphological features of the vertebrate hindbrain. It is clear that RA plays a crucial role (primarily a posteriorising one) in establishing the broad domains of *Hox* gene expression in the neural epithelium which are vital for the acquisition of A-P polarity in the neuraxis. Subsequent position-specific expression of developmental control genes dictates regional specialisation and the overall plan of the CNS. With the exception of *Hoxb1*, RA is only involved in the establishment phase of *Hox* gene expression. The maintenance and refinement of *Hox* gene expression into discrete nested domains relies specifically upon interactions with segmentation genes such as *Krox20* and *kreisler* and also on highly conserved auto-, cross- and para-regulatory mechanisms within the *Hox* gene complexes. Clearly, hindbrain segmentation in the form of cell lineage restrictions and compartmentalisation is coupled with acquisition of specific cellular identities. The intricate spatial order of differentiated neurons, essential to the subsequent formation of functional circuits, is crucially dependent on correct regional specification. Loss of function mutations, ectopic gene expression and transgenic analyses as described in this chapter critically disrupt the precise regional specification of the neuraxis which leads to the absence, transformation or mismigration of particular neuronal populations. The key element still missing is proof that *Hox* gene are involved in regulating downstream genes that play key roles in the structural organisation of the hindbrain and the generation of specific cellular subtypes. These issues will be the focus of future analyses of vertebrate hindbrain development.

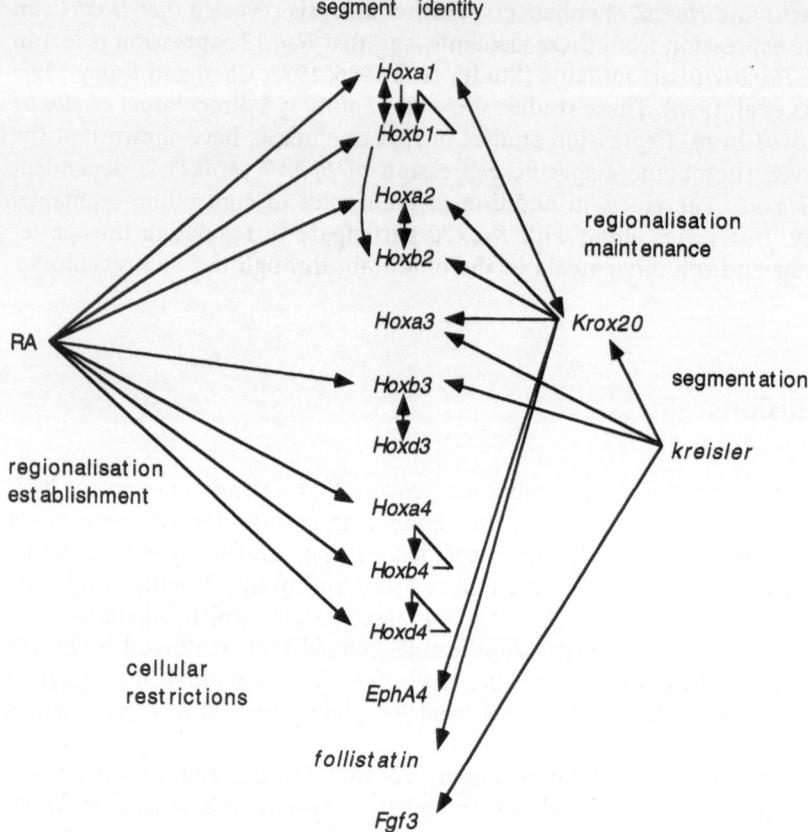

Fig. 2. Complex regulation of segmental patterning during hindbrain development (*single head*, *double head* and *bent arrows* represent direct, synergistic and auto-regulatory interactions respectively)

References

Alexandre D, Clarke J, Oxtoby E, Yan Y-L, Jowett T, Holder N (1996) Ectopic expression of *Hoxa-1* in the zebrafish alters the fate of the mandibular arch neural crest and phenocopies a retinoic acid-induced phenotype. Development 122:735–746

Amores A, Force A, Yan Y-L, Joly L, Amemiya C, Fritz A, Ho R, Langeland J, Prince V, Wang Y-L, Westerfield M, Ekker M, Postlehwait J (1998) Zebrafish *hox* clusters and vertebrate genome evolution. Science 282:1711–1714

Aparicio S, Morrison A, Gould A, Gilthorpe J, Chaudhuri C, Rigby PWJ, Krumlauf R, Brenner S (1995) Detecting conserved regulatory elements with the model genome of the Japanese puffer fish *Fugu rubripes*. Proc Natl Acad Sci USA 92:1684–1688

Aparicio S, Hawker K, Cottage A, Mikawa Y, Zuo L, Chen E, Krumlauf R, Brenner S (1997) Organisation of the *Fugu rubripes Hox* clusters, evidence for continuing evolution of vertebrate *Hox* complexes. Nat Genet 16:79–84

Barrow J, Capecchi M (1996) Targeted disruption of the *Hoxb2* locus in mice interferes with expression of *Hoxb1* and *Hoxb4*. Development 122:3817–3828

Becker N, Seitanidou T, Murphy P, Mattei M-G, Topilko P, Nieto MA, Wilkinson DG, Charnay P, Gilardi-Hebenstreit P (1994) Several receptor tyrosine kinase genes of the *Eph* family are segmentally expressed in the developing hindbrain. Mech Dev 47:3–18

Behringer R, Crotty DA, Tennyson VM, Brinster R, Palmiter R, Wolgemuth D (1993) Sequences 5′ of the homeobox of the *Hox-1.4* gene direct tissue-specific expression of *lacZ* during mouse development. Development 117:823–833

Bel S, Coré N, Djabali M, Kieboom K, Van der Lugt N, Alkema M, Van Lohuizen M (1998) Genetic interactions and dosage effects of *Polycomb* group genes in mice. Development 125:3543–3551

Birgbauer E, Fraser SE (1994) Violation of cell lineage restriction compartments in the chick hindbrain. Development 120:1347–1356

Birgbauer E, Sechrist J, Bronner-Fraser M, Fraser S (1995) Rhombomeric origin and rostro-caudal reassortment of neural crest cells revealed by intravital microscopy. Development 121:935–945

Blumberg B, Bolado J, Moreno T, Kintner C, Evans R, Papalopulu N (1997) An essential role for retinoid signaling in anteroposterior neural patterning. Development 124:373–379

Bradley LC, Snape A, Bhatt S, Wilkinson DG (1992) The structure and expression of the *Xenopus Krox-20* gene: conserved and divergent patterns of expression in the rhombomeres and neural crest. Mech Dev 40:73–84

Bulfone A, Puelles L, Porteus M, Frohman M, Martin G, Rubenstein J (1993) Spatially-restricted expression of *Dlx-1*, *Dlx-2* (*Tes-1*), *GBx-2* and *Wnt-3* in the embryonic day 12.5 mouse forebrain defines potential transverse and longitudinal segmental boundaries. J Neurosci 13:3155–3172

Carpenter EM, Goddard JM, Chisaka O, Manley NR, Capecchi MR (1993) Loss of *Hoxa-1* (*Hox-1.6*) function results in the reorganization of the murine hindbrain. Development 118:1063–1075

Chan S-K, Ryoo H-D, Gould A, Krumlauf R, Mann R (1997) Switching the in vivo specificity of a minimal *Hox*-responsive element. Development 124:2007–2014

Chavrier P, Zerial M, Lemaire P, Almendral J, Bravo R, Charnay P (1988) A gene encoding a protein with zinc fingers is activated during G0/G1 transition in cultured cells. EMBO J 7:29–35

Chavrier P, Vesque C, Galliot B, Vigneron M, Dollé P, Duboule D, Charnay P (1990) The segment-specific gene *Krox-20* encodes a transcription factor with binding sites in the promoter of the *Hox 1.4* gene. EMBO J 9:1209–1218

Chen J, Ruley H (1998) AN enhancer element in the *EphA2* (*Eck*) gene sufficient for rhombomere-specific expression is activated by *Hoxa1* and *Hoxb1*. J Biol Chem 273:24670–24675

Chen Y, Dong D, Kostetskii I, Zile MH (1996) Hensen's node from vitamin A-deficient quail embryo induces chick limb bud duplication and retains its normal asymmetric expression of *Sonic hedgehog* (*Shh*). Dev Biol 173:256–264

Chen YP, Huang L, Russo AF, Solursh M (1992) Retinoic acid is enriched in Hensen's node and is developmentally regulated in the early chick embryo. Proc Natl Acad Sci USA 89:10056–10059

Chisaka O, Capecchi M (1991) Regionally restricted developmental defects resulting from targeted disruption of the mouse homeobox gene *Hox1.5*. Nature 350:473–479

Chisaka O, Musci T, Capecchi M (1992) Developmental defects of the ear, cranial nerves and hindbrain resulting from targeted disruption of the mouse homeobox gene *Hox-1.6*. Nature 355:516–520

Cho K, De Robertis E (1990) Differential activation of *Xenopus* homeobox genes by mesoderm-inducing growth factors and retinoic acid. Genes Dev 4:1910–1917

Clarke JD, Lumsden A (1993) Segmental repetition of neuronal phenotype sets in the chick embryo hindbrain. Development 118:151–162

Clarke JDW, Erskine L, Lumsden A (1998) Differential progenitor dispersal and the spatial origin of early neurons can explain the predominance of single-phenotype clones in the chick hindbrain. Dev Dyn 212:14–26

Condie B, Capecchi M (1993) Mice homozygous for a targeted disruption of Hoxd-3 (Hox-4.1) exhibit anterior transformations of the first and second cervical vertebrate, the atlas and axis. Development 119:579–595

Condie BG, Capecchi MR (1994) Mice with targeted disruptions in the paralogous genes Hoxa-3 and Hoxd-3 reveal synergistic interactions. Nature 370:304–307

Conlon RA (1995) Retinoic acid and pattern formation in vertebrates. TIG 11:314–319

Conlon RA, Rossant J (1992) Exogenous retinoic acid rapidly induces anterior ectopic expression of murine Hox-2 genes in vivo. Development 116:357–368

Cook M, Gould A, Brand N, Davies J, Strutt P, Shaknovich R, Licht J, Waxman S, Chien Z, Gluecksohn-Waelsch S, Krumlauf R, Zelent A (1995) Expression of the zinc-finger gene PLZF at rhombomere boundaries in the vertebrate hindbrain. Proc Natl Acad Sci USA 92:2249–2253

Cordes SP, Barsh GS (1994) The mouse segmentation gene kr encodes a novel basic domain-leucine zipper transcription factor. Cell 79:1025–1034

Couly G, Grapin-Botton A, Coltey P, Ruhin B, Le Douarin NM (1998) Determination of the identity of the derivatives of the cephalic neural crest: incompatibility between Hox gene expression and lower jaw development. Development 128:3445–3459

Couly GF, Coltey PM, Le Douarin NM (1993) The triple origin of skull in higher vertebrates – a study in quail-chick chimeras. Development 117:409–429

Couly GF, Grapin-Bottom A, Coltey P, Le Douarin NM (1996) The regeneration of the cephalic neural crest, a problem revisited: the regenerating cells originate from the contralateral or from the anterior and posterior neural folds. Development 122:3393–3407

Crossley PH, Martinez S, Martin GR (1996) Midbrain development induced by FGF8 in the chick embryo. Nature 380:66–68

Dale J, Vesque C, Lints T, Sampath K, Furley A, Dodd J, Placzek M (1997) Cooperation of BMP7 and SHH in the induction of forebrain ventral midline cells by prechordal mesoderm. Cell 90:257–269

Davenne M, Maconochie M, Neun R, Brunet J-F, Chambon P, Krumlauf R, Rijli F (1999) Hoxa2 and Hoxb2 control dorsoventral patterns of neuronal developments in the rostral hindbrain. Neuron 22:677–691

Dencker L, Annerwall E, Busch C, Eriksson U (1990) Localization of specific retinoid-binding sites and expression of cellular retinoic-acid-binding protein (CRABP) in the early mouse embryo. Development 110:343–352

Deol MS (1964) The abnormalities of the inner ear in kreisler mice. J Embryol Exp Morph 12:475–490

Dollé P, Izpisùa-Belmonte J, Brown J, Tickle C, Duboule D (1991) Hox-4 genes and the morphogenesis of mammalian genitalia. Genes Dev 5:1767–1776

Dollé P, Lufkin T, Krumlauf R, Mark M, Duboule D, Chambon P (1993) Local alterations of Krox-20 and Hox gene expression in the hindbrain suggest lack of rhombomeres 4 and 5 in homozygote null Hoxa-1 (Hox-1.6) mutant embryos. Proc Natl Acad Sci USA 90:7666–7670

Dollé P, Fraulob V, Kastner P, Chambon P (1994) Developmental expression of murine retinoid X receptor (RXR) genes. Mech Dev 45:91–104

Doniach T (1992) Induction of anteroposterior neural pattern in Xenopus by planar signals. Dev Suppl:183–193

Doniach T (1993) Planar and vertical induction of anteroposterior pattern during the development of the amphibian central nervous system. J Neurobiol 24:1256–1275

Doniach T, Phillip C, Gerhart J (1992) Planar induction of anteroposterior pattern in the developing central nervous system of Xenopus laevis. Science 257:542–545

Duboule D, Dolle P (1989) The structural and functional organization of the murine Hox gene family resembles that of Drosophila homeotic genes. EMBO J 8:1497–1505

Dupé V, Davenne M, Brocard J, Dollé P, Mark M, Dierich A, Chambon P, Rijli F (1997) In vivo functional analysis of the *Hoxa1* 3' retinoid response element (3' RARE). Development 124:399–410

Durston A, Timmermans J, Hage W, Hendriks H, de Vries N, Heideveld M, Nieuwkoop P (1989) Retinoic acid causes an anteroposterior transformation in the developing central nervous system. Nature 340:140–144

Edlund T, Jessell T (1999) Progression from extrinsic to intrinsic signaling in cell fate specification: a view from the nervous system. Cell 96:211–224

Figdor MC, Stern CD (1993) Segmental organisation of embryonic diencephalon. Nature 363:630–634

Flanagan JG, Vanderhaeghen P (1998) The ephrins and Eph receptors in neural development. Ann Rev Neurobiol 21:309–345

Flenniken AM, Gale NW, Yancopoulos GD, Wilkinson DG (1996) Distinct and overlapping expression of ligands for Eph-related receptor tyrosine kinases during mouse embryogenesis. Dev Biol 179:382–401

Foley A, Storey K, Stern C (1997) The prechordal region lacks neural inducing ability, but can confer anterior character to more posterior neuroepithelium. Development 124:2983–2996

Frasch M, Chen X, Lufkin T (1995) Evolutionary-conserved enhancers direct region-specific expression of the murine *Hoxa-1* and *Hoxa-2* loci in both mice and *Drosophila*. Development 121:957–974

Fraser S, Keynes R, Lumsden A (1990) Segmentation in the chick embryo hindbrain is defined by cell lineage restrictions. Nature 344:431–435

Frohman M, Martin G (1992) Isolation and analysis of embryonic expression of *Hox4.9*, a member of the murine labial-like gene family. Mech Dev 38:55–67

Frohman MA, Boyle M, Martin GR (1990) Isolation of the mouse *Hox2.9* gene; analysis of embryonic expression suggests that positional information along the anterior-posterior axis is specified by mesoderm. Development 110:589–607

Frohman MA, Martin GR, Cordes S, Halamek LP, Barsh GS (1993) Altered rhombomere-specific gene expression and hyoid bone differentiation in the mouse segmentation mutant *kreisler* (*kr*). Development 117:925–936

Gale N, Flenniken A, Compton D, Jenkins N, Copeland N, Gilbert D, Davis S, Wilkinson D, Yancopoulos G (1996a) Elk-L3, a novel transmembrane ligand for the Eph family of receptor tyrosine kinases, expressed in embryonic floor plate, roof plate and hindbrain segments. Oncogene 13:1343–1352

Gale NW, Holland SJ, Valenzuela DM, Flenniken A, Pan L, Ryan TE, Henkemeyer M, Strebhardt K, Hirai H, Wilkinson DG, Pawson T, Davis S, Yancopoulos GD (1996b) Eph receptors and ligands comprise two major specificity subclasses and are reciprocally compartmentalized during embryogenesis. Neuron 17:9–19

Gaunt SJ, Krumlauf R, Duboule D (1989) Mouse homeo-genes within a subfamily, *Hox-1.4, -2.6 and -5.1*, display similar anteroposterior domains of expression in the embryo, but show stage- and tissue-dependent differences in their regulation. Development 107:131–141

Gavalas A, Davenne M, Lumsden A, Chambon P, Rijli F (1997) Role of *Hoxa-2* in axon pathfinding and rostral hindbrain patterning. Development 124:3693–3702

Gavalas A, Studer M, Lumsden A, Rijli F, Krumlauf R, Chambon P (1998) *Hoxa1* and *Hoxb1* synergize in patterning the hindbrain, cranial nerves and second pharyngeal arch. Development 125:1123–1136

Geada AMC, Gaunt SJ, Azzawi M, Shimeld SM, Pearce J, Sharpe PT (1992) Sequence and embryonic expression of the murine *Hox-3.5* gene. Development 116:497–506

Gilardi-Hebenstreit P, Nieto A, Frain M, Mattei M, Chestier A, Wilkinson D, Charnay P (1992) An EPH-related receptor protein-tyrosine kinase gene segmentally-expressed in the developing mouse hindbrain. Oncogene 7:2499–2506

Goddard J, Rossel M, Manley N, Capecchi M (1996) Mice with targeted disruption of *Hoxb1* fail to form the motor nucleus of the VIIth nerve. Development 122:3217–3228

Gould A (1997) Functions of mammalian *Polycomb*-group and *trithorax*-group related genes. Curr Opin Genet Dev 7:488–494

Gould A, Morrison A, Sproat G, White R, Krumlauf R (1997) Positive cross-regulation and enhancer sharing: two mechanisms for specifying overlapping *Hox* expression patterns. Genes Dev 11:900–913

Gould A, Itasaki N, Krumlauf R (1998) Initiation of rhombomeric *Hoxb4* expression requires induction by somites and a retinoid pathway. Neuron 21:39–51

Graham A, Papalopulu N, Krumlauf R (1989) The murine and *Drosophila* homeobox clusters have common features of organisation and expression. Cell 57:367–378

Graham A, Heyman I, Lumsden A (1993) Even-numbered rhombomeres control the apoptotic elimination of neural crest cells from odd-numbered rhombomeres in the chick hindbrain. Development 119:233–245

Graham A, Francis-West P, Brickell P, Lumsden A (1994) The signalling molecule BMP4 mediates apoptosis in the rhombencephalic neural crest. Nature 372:684–686

Grapin-Botton A, Bonnin M-A, Ariza-McNaughton L, Krumlauf R, LeDouarin NM (1995) Plasticity of transposed rhombomeres: *Hox* gene induction is correlated with phenotypic modifications. Development 121:2707–2721

Guthrie S, Lumsden A (1991) Formation and regeneration of rhombomere boundaries in the developing chick hindbrain. Development 112:221–229

Guthrie S, Muchamore I, Kuroiwa A, Marshall H, Krumlauf R, Lumsden A (1992) Neuroectodermal autonomy of *Hox-2.9* expression revealed by rhombomere transpositions. Nature 356:157–159

Guthrie S, Prince V, Lumsden A (1993) Selective dispersal of avian rhombomere cells in orthotopic and heterotopic grafts. Development 118:527–538

Helmbacher F, Pujades C, Desmarquet C, Frain M, Rijli F, Chambon P, Charnay P (1998) *Hoxa1* and *Krox20* synergize to control the development of rhombomere 3. Development 125:4739–4748

Hill J, Clarke JDW, Vargesson N, Jowett T, Holder N (1995) Exogenous retinoic acid causes specific alterations in the development of the midbrain and hindbrain of the zebrafish embryo including positional respecification of the Mauthner neuron. Mech Dev 50:3–16

Hogan BLM, Thaller C, Eichle G (1992) Evidence that Hensen's node is a site of retinoic acid synthesis. Nature 359:237–241

Holland PWH, Garcia-Fernandez J (1996) *Hox* genes and chordate evolution. Dev Biol 173:382–395

Horstadius S (1950) The neural crest. Oxford University Press, London

Huang D, Chen S, Langston A, Gudas L (1998) A conserved retinoic acid responsive element in the murine *Hoxb-1* gene is required for expression in the developing gut. Development 125:3235–3246

Hunt P, Krumlauf R (1991) Deciphering the *Hox* code: clues to patterning the branchial region of the head. Cell 66:1075–1078

Hunt P, Gulisano M, Cook M, Sham M, Faiella A, Wilkinson D, Boncinelli E, Krumlauf R (1991a) A distinct *Hox* code for the branchial region of the head. Nature 353:861–864

Hunt P, Whiting J, Muchamore I, Marshall H, Krumlauf R (1991b) Homeobox genes and models for patterning the hindbrain and branchial arches. Development 112 (Suppl: Molecular and cellular basis of pattern formation):187–196

Hunt P, Ferretti P, Krumlauf R, Thorogood P (1995) Restoration of normal Hox code and branchial arch morphogenesis after extensive deletion of hindbrain neural crest. Dev Biol 168:584–597

Irving C, Nieto M, Das Gupta R, Charnay P, Wilkinson D (1996) Progressive spatial restriction of Sek1 and Krox20 gene expression during hindbrain segmentation. Dev Biol 173:26–38

Itasaki N, Ichijo H, Hama C, Matsuno T, Nakamura H (1991) Establishment of rostrocaudal polarity in tectal primordium: engrailed expression and subsequent tectal polarity. Development 113:1133–1144

Itasaki N, Sharpe J, Morrison A, Krumlauf R (1996) Reprogramming *Hox* expression in the vertebrate hindbrain: influence of paraxial mesoderm and rhombomere transposition. Neuron 16:487–500

Izpisua-Belmonte J-C, Tickle C, Dolle P, Wolpert L, Duboule D (1991) Expression of homeobox *Hox-4* genes and the specification of position in chick wing development. Nature 350: 585–589

Kappen C, Schugart K, Ruddle F (1989) Two steps in the evolution of Antennapedia-class vertebrate homeobox genes. Proc Natl Acad Sci USA 86:5459–5463

Kenyon C, Wang B (1991) A cluster of Antennapedia-class homeobox genes in a nonsegmented animal. Science 253:516–517

Kessel M, Gruss P (1991) Homeotic transformations of murine prevertebrae and concomitant alteration of Hox codes induced by retinoic acid. Cell 67:89–104

Keynes R, Krumlauf R (1994) Hox genes and regionalization of the nervous system. Annu Rev Neurosci 17:109–132

Keynes R, Lumsden A (1990) Segmentation and the origins of regional diversity in the vertebrate central nervous system. Neuron 4:1–9

Kintner C (1992) Molecular bases of early neural development in *Xenopus* embryos. Annu Rev Neurosci 15:251–284

Kolm P, Sive H (1995) Regulation of the *Xenopus* labial homeodomain genes, *HoxA1* and *HoxD1*: activation by retinoids and peptide growth factors. Dev Biol 167:34–49

Kolm P, Apekin V, Sive H (1997) *Xenopus* hindbrain patterning requires retinoid signaling. Dev Biol 192:1–16

Köntges G, Lumsden A (1996) Rhombencephalic neural crest segmentation is preserved throughout craniofacial ontogeny. Development 122:3229–3242

Krumlauf R (1992) Evolution of the vertebrate Hox homeobox genes. Bioessays 14:245–252

Krumlauf R (1993) *Hox* genes and pattern formation in the branchial region of the vertebrate head. TIG 9:106–112

Kulesa P, Fraser S (1998) Neural crest cell dynamics revealed by time-lapse video microscopy of whole chick explant cultures. Dev Biol 204:327–344

Kuratani SC, Eichele G (1993) Rhombomere transposition repatterns the segmental organization of cranial nerves and reveals cell-autonomous expression of a homeodomain protein. Development 117:105–117

Lamb TM, Harland RM (1995) Fibroblast growth factor is a direct neural inducer, which combined with noggin generates anterior-posterior neural pattern. Development 121: 3627–3636

Lammer E, Chen D, Hoar R, Agnish A, Benke P, Braun J, Curry C, Fernhoff P, Grix A, Lott I, Richard J, Sun S (1985) Retinoic acid embryopathy. New Engl J Med 313:837–841

Lampron C, Rochette-Egly C, Gorry P, Dolle P, Mark M, Lufkin T, LeMeur M, Chambon P (1995) Mice deficient in cellular retinoic acid binding protein II (CRABPII) or in both CRABPI and CRABPII are essentially normal. Development 121:539–548

Langston A, Thompson J, Gudas L (1997) Retinoic acid-responsive enhancers located 3′ of the HoxA and the HoxB gene clusters. J Biol Chem 272:2167–2175

Langston AW, Gudas LJ (1992) Identification of a retinoic acid responsive enhancer 3′ of the murine homeobox gene *Hox-1.6*. Mech Dev 38:217–228

Le Douarin N (1983) The Neural Crest. Cambridge University Press, Cambridge

Lemaire P, Revelant O, Bravo R, Charnay P (1988) Two mouse genes encoding potential transcription factors with identical DNA-binding domains are activated by growth factors in cultured cells. Proc Natl Acad Sci USA 85:4691–4695

Lufkin T, Dierich A, LeMeur M, Mark M, Chambon P (1991) Disruption of the *Hox-1.6* homeobox gene results in defects in a region corresponding to its rostral domain of expression. Cell 66:1105–1119

Lumsden A, Keynes R (1989) Segmental patterns of neuronal development in the chick hindbrain. Nature 337:424–428

Lumsden A, Krumlauf R (1996) Patterning the vertebrate neuraxis. Science 274:1109–1115

Lumsden A, Sprawson N, Graham A (1991) Segmental origin and migration of neural crest cells in the hindbrain region of the chick embryo. Development 113:1281–1291

Lyn S, Giguere V (1994) Localisation of CRABP-I and CRABP-II mRNA in the early mouse embryo by whole-mount in situ hybridisation: implications for teratogenesis and neural development. Dev Dyn 199:280–291

MacArthur CA, Lawshe A, Xu J, Santos-Ocampo S, Heikinheimo M, Chellaiah AT, Ornitz DM (1995) FGF-8 isoforms activate receptor splice forms that are expressed in mesenchymal regions of mouse development. Development 121:3603–3613

Maconochie M, Nonchev S, Studer M, Chan S-K, Pöpperl H, Sham M-H, Mann R, Krumlauf R (1997) Cross-regulation in the mouse HoxB complex: the expression of Hoxb2 in rhombomere 4 is regulated by Hoxb1. Genes Dev 11:1885–1896

Maden M, Horton C, Graham A, Leonard L, Pizzey J, Siegenthaler G, Lumsden A, Eriksson U (1992) Domains of cellular retinoic acid-binding protein I (CRABP I) expression in the hindbrain and neural crest of the mouse embryo. Mech Dev 37:13–23

Maden M, Gale E, Kostetskii I, Zile M (1996) Vitamin A deficient quail embryos have half a hindbrain and other neural defects. Curr Biol 6:417–426

Mahmood R, Kiefer P, Guthrie S, Dickson C, Mason I (1995) Multiple roles for FGF-3 during cranial neural development in the chicken. Development 121:1399–1410

Mahmood R, Kiefer P, Guthrie S, Dickson C, Mason I (1995) Multiple roles for FGF-3 during cranial neural development in the chicken. Development 121:1399–1410

Mangelsdorf DJ, Evans RM (1995) The RXR Heterodimers and Ophan Receptors. Cell 83:841–850

Mangelsdorf DJ, Thummel C, Beato M, Herrlich P, Schütz G, Umesono K, Blumberg B, Kastner P, Mark M, Chambon P, Evans RM (1995) The nuclear receptor superfamily: the second decade. Cell 83:835–839

Manley N, Capecchi M (1997) Hox group 3 paralogous genes act synergistically in the formation of somitic and neural crest-derived structures. Developmental Biology 192:274–288

Manley N, Capecchi M (1998) Hox group 3 paralogs regulate the development and migration of the thymus, thyroid and parathyroid glands. Dev Biol 195:1–15

Manzanares M, Cordes S, Kwan C-T, Sham M-H, Barsh G, Krumlauf R (1997) Segmental regulation of Hoxb3 by kreisler. Nature 387:191–195

Manzanares M, Cordes S, Ariza-McNaughton L, Sadl V, Maruthainar K, Barsh G, Krumlauf R (1999a) Conserved and distinct roles of kreisler in regulation of the paralogous Hoxa3 and Hoxb3 genes. Development 126:759–769

Manzanares M, Trainor P, Nonchev S, Ariza-mcNaughton L, Brodie J, Gould A, Marshall H, Morrison A, Kwan C-T, Sham M-H, Wilkinson D, Krumlauf R (1999b) The role of kreisler in segmentation during hindbrain development. Dev Biol 211:220–237

Marín F, Puelles L (1994) Patterning of the embryonic avian midbrain after experimental inversions: a polarizing activity from the isthums. Dev Biol 161:19–37

Marín F, Puelles L (1995) Morphological fate of rhombomeres in quail/chick chimeras: a segmental analysis of hindbrain nuclei. Eur J Neurosci 7:1714–1738

Mark M, Lufkin T, Vonesch J-L, Ruberte E, Olivo J-C, Dollé P, Gorry P, Lumsden A, Chambon P (1993) Two rhombomeres are altered in Hoxa-1 mutant mice. Development 119:319–338

Marshall H, Nonchev S, Sham MH, Muchamore I, Lumsden A, Krumlauf R (1992) Retinoic acid alters hindbrain Hox code and induces transformation of rhombomeres 2/3 into a 4/5 identity. Nature 360:737–741

Marshall H, Studer M, Pöpperl H, Aparicio S, Kuroiwa A, Brenner S, Krumlauf R (1994) A conserved retinoic acid response element required for early expression of the homeobox gene Hoxb-1. Nature 370:567–571

Marshall H, Morrison A, Studer M, Pöpperl H, Krumlauf R (1996) Retinoids and Hox genes. FASEB 10:969–978

Martinez S, Alvarado-Mallart RM (1990) Expression of the homeobox Chicken gene in chick/ quail chimeras with inverted mes-metencephalic grafts. Dev Biol 139:432–436

Martinez S, Wassef M, Alvarado-Mallart RM (1991) Induction of a mesencephalic phenotype in the 2-day-old chick prosencephalon is preceded by the early expression of the homeobox gene engrailed. Neuron 6:971-981

Martinez S, Marin F, Nieto MA, Puelles L (1995) Induction of ectopic engrailed expression and fate change in avian rhombomeres: intersegmental boundaries as barriers. Mech Dev 51:289–303

McGinnis W, Krumlauf R (1992) Homeobox genes and axial patterning. Cell 68:283-302

McKay I, Lewis J, Lumsden A (1996) The role of FGF-3 in early inner ear development: an analysis in normal and kreisler mutant mice. Dev Biol:370–378

McKay I, Lewis J, Lumsden A (1997) Organization and development of facial motor neurons in the kreisler mutant mouse. Eur J Neurosci 9:1499-1506

McKay IJ, Muchamore I, Krumlauf R, Maden M, Lumsden A, Lewis J (1994) The kreisler mouse: a hindbrain segmentation mutant that lacks two rhombomeres. Development 120:2199–2211

Mellitzer G, Xu Q, Wilkinson D (1999) Restriction of cell intermingling and communication by Eph receptors and ephrins. Nature 400:77–81

Mendelsohn C, Ruberte E, Le Meur M, Morriss-Kay G, Chambon P (1991) Developmental analysis of the retinoic acid-inducible RAR-beta 2 promoter in transgenic animals. Development 113:723–734

Mendelsohn C, Larkin S, Mark M, LeMeur M, Clifford J, Zelent A, Chambon P (1994) RAR isoforms: distinct transcriptional control by retinoic acid and specific spatial patterns of promoter activity during mouse embryonic development. Mech Dev 45:227–241

Moens CB, Yan Y-L, Appel B, Force AG, Kimmel CB (1996) valentino: a zebrafish gene required for normal hindbrain segmentation. Development 122:3981-3990

Moens CB, Cordes SP, Giorgianni MW, Barsh GS, Kimmel CB (1998) Equivalence in the genetic control of hindbrain segmentation in fish and mouse. Development 125:381–391

Moroni M, Vigano M, Mavilio F (1993) Regulation of the human HoxD4 gene by retinoids. Mech Dev 44:139–154

Morrison A (1998) 1 + 1 = 4 and much much more. Bioessays 20:794-797

Morrison A, Chaudhuri C, Ariza-McNaughton L, Muchamore I, Kuroiwa A, Krumlauf R (1995) Comparative analysis of chicken Hoxb-4 regulation in transgenic mice. Mech Dev 53:47–59

Morrison A, Moroni M, Ariza-McNaughton L, Krumlauf R, Mavilio F (1996) In vitro and transgenic analysis of a human HoxD4 retinoid-responsive enhancer. Development 122:1895–1907

Morrison A, Ariza-McNaughton L, Gould A, Featherstone M, Krumlauf R (1997) HoxD4 and regulation of the group 4 paralog genes. Development 124:3135–3146

Morriss GM, Thorogood PV (1978) An approach to cranial neural crest migration and differentiation in mammalian embryos. In: Johnson MH (ed) Development in mammals. vol 3. Elsevier North-Holland, Amsterdam, pp 363–411

Murphy P, Hill RE (1991) Expression of the mouse labial-like homeobox-containing genes, Hox 2.9 and Hox 1.6, during segmentation of the hindbrain. Development 111:61–74

Murphy P, Davidson DR, Hill RE (1989) Segment-specific expression of a homeobox-containing gene in the mouse hindbrain. Nature 341:156–159

Nardelli J, Gibson T, Vesque C, Charnay P (1991) Base sequence discrimination by zinc-finger DNA-binding domains. Nature 349:175–178

Nieto MA, Bradley LC, Wilkinson DG (1991) Conserved segmental expression of Krox-20 in the vertebrate hindbrain and its relationship to lineage restriction. Development Suppl 2:59–62

Nieto MA, Gilardi HP, Charnay P, Wilkinson DG (1992a) A receptor protein tyrosine kinase implicated in the segmental patterning of the hindbrain and mesoderm. Development 116:1137–1150

Nieto MA, Gilardi-Hebenstreit P, Charnay P, Wilkinson D (1992b) A receptor protein tyrosine kinase implicated in the segmental patterning of the hindbrain and mesoderm. Development 116:1137–1150

Nieuwkoop P (1952) Activation and organisation of the central nervous system in amphibians. J Exp Zool 120:1–108

Nieuwkoop P (1973) The organisation centre of the amphibian embryo: its origin, spatial organisation and morphogenetic action. Adv in Morphog 10:1–39

Nieuwkoop P (1985) Inductive interactions in early amphibian development and their general nature. J Embryol Exp Morph 89 (suppl.):333–347

Noden D (1983) The role of the neural crest in patterning of avian cranial skeletal, connective, and muscle tissues. Dev Biol 96:144–165

Nonchev S, Maconochie M, Vesque C, Aparicio S, Ariza-McNaughton L, Manzanares M, Maruthainar K, Kuroiwa A, Brenner S, Charnay P, Krumlauf R (1996a) The conserved role of Krox-20 in directing Hox gene expression during vertebrate hindbrain segmentation. Proc Natl Acad Sci USA 93:9339–9345

Nonchev S, Vesque C, Maconochie M, Seitanidou T, Ariza-McNaughton L, Frain M, Marshall H, Sham MH, Krumlauf R, Charnay P (1996b) Segmental expression of Hoxa-2 in the hindbrain is directly regulated by Krox-20. Development 122:543–554

Ogura T, Evans R (1995a) Evidence for two distinct retinoic acid response pathways for Hoxb-1 gene regulation. Proc Natl Acad Sci USA 92:392–396

Ogura T, Evans R (1995b) A retinoic acid-triggered cascade of Hoxb-1 gene activation. Proc Natl Acad Sci USA 92:387–391

Oxtoby E, Jowett T (1993) Cloning of the zebra fish Krox-20 gene (Krx-20) and its expression during hindbrain development. Nucleic Acids Res 21:1087–1095

Packer A, Crotty D, Elwell V, Wolgemuth D (1998) Expression of the murine Hoxa4 gene requires both autoregulation and a conserved retinoic acid response element. Development 125:1991–1998

Papalopulu N, Kintner C (1996) A posteriorising factor, retinoic acid, reveals that anteroposterior patterning controls the timing of neuronal differentiation in Xenopus neuroectoderm. Development 122:3409–3418

Papalopulu N, Clarke J, Bradley L, Wilkinson D, Krumlauf R, Holder N (1991a) Retinoic acid causes abnormal development and segmental patterning of the anterior hindbrain in Xenopus embryos. Development 113:1145–1159

Papalopulu N, Lovell-Badge R, Krumlauf R (1991b) The expression of murine Hox-2 genes is dependent on the differentiation pathway and displays collinear sensitivity of retinoic acid in F9 cells and Xenopus embryos. Nucleic Acids Res 19:5497–5506

Paro R (1993) Mechanisms of heritable gene repression during development of Drosophila. Curr Opin Cell Biol 5:999–1005

Pattyn A, Morin X, Cremer H, Goridis C, Brunet J-F (1997) Expression and interactions of the two closely related homeobox genes Phox2a and Phox2b during neurogenesis. Development 124:4065–4075

Pirrotta V (1997a) Chromatin-silencing mechanisms in Drosophila maintain patterns of gene expression. Trends Genet 13:314–318

Pirrotta V (1997b) PcG complexes and chromatin silencing. Curr Opin Genet Dev 7:249–258

Pöpperl H, Featherstone M (1992) An autoregulatory element of the murine Hox-4.2 gene. EMBO J 11:3673–3680

Pöpperl H, Featherstone M (1993) Identification of a retinoic acid response element upstream of the murine Hox-4.2 gene. Mol Cell Biol 13:257–265

Pöpperl H, Bienz M, Studer M, Chan S, Aparicio S, Brenner S, Mann R, Krumlauf R (1995) Segmental expression of Hoxb1 is controlled by a highly conserved autoregulatory loop dependent upon exd/Pbx. Cell 81:1031–1042

Prince V, Lumsden A (1994) Hoxa-2 expression in normal and transposed rhombomeres: independent regulation in the neural tube and neural crest. Development 120:911–923

Rijli FM, Mark M, Lakkaraju S, Dierich A, Dolle P, Chambon P (1993) A homeotic transformation is generated in the rostral branchial region of the head by disruption of Hoxa-2, which acts as a selector gene. Cell 75:1333–1349

Robinson V, Smith A, Flenniken AM, Wilkinson DG (1997) Role of Eph receptors and ephrins in neural crest pathfinding. Cell Tissue Res 290:265–274

Rubenstein JLR, Martinez S, Shimamura K, Puelles L (1994) The embryonic vertebrate forebrain: the prosomeric model. Science 266:578–580

Ruberte E, Friederich V, Morriss-Kay G, Chambon P (1992) Differential distribution patterns of CRABP-I and CRABP-II transcripts during mouse embryogenesis. Development 115:973–989

Ruiz i Altaba A (1992) Planar and vertical signals in the induction and patterning of the *Xenopus* nervous system. Development 116:67–80

Ruiz i Altaba A (1994) Pattern formation in the vertebrate neural plate. TINS 17:233–243

Ruiz J, Robertson E (1994) The expression of the receptor-protein tyrosine kinase gene, eck, is highly restricted during early mouse development. Mech Dev 46:87–100

Saldivar JR, Sechrist JW, Krull CE, Ruffin S, Bronner-Fraser M (1997) Dorsal hindbrain ablation results in the rerouting of neural crest migration and the changes in gene expression, but normal hyoid development. Development 124:2729–2739

Saxen L (1989) Neural induction. Int J Dev Biol 33:21–48

Schneider-Maunoury S, Topilko P, Seitanidou T, Levi G, Cohen-Tannoudji M, Pournin S, Babinet C, Charnay P (1993) Disruption of *Krox-20* results in alteration of rhombomeres 3 and 5 in the developing hindbrain. Cell 75:1199–1214

Schneider-Maunoury S, Seitanidou T, Charnay P, Lumsden A (1997) Segmental and neuronal architecture of the hindbrain of *Krox-20* mouse mutants. Development 124:1215–1226

Schoorlemmer J, Marcos-Gutiérrez, Were F, Martínez R, García E, Satijn D, Otte A, Vidal M (1997) Ring1A is a transcriptional repressor that interacts with polycomb-M33 and is expressed at rhombomere boundaries in the mouse hindbrain. EMBO J 16:5930–5942

Sechrist J, Bronner-Fraser M (1991) Birth and differentiation of reticular neurons in the chick hindbrain: ontogeny of the first neuronal population. Neuron 7:947–963

Sechrist J, Serbedzija GN, Scherson T, Fraser SE, Bronner-Fraser M (1993) Segmental migration of the hindbrain neural crest does not arise from its segmental generation. Development 118(3):691–703

Sechrist J, Scherson T, Bronner-Fraser M (1994) Rhombomere rotation reveals that multiple mechanisms contribute to segmental pattern of hindbrain neural crest migration. Development 120:1777–1790

Seitanidou T, Schneider-Manunoury S, Desmarquet C, Wilkinson D, Charnay P (1997) *Krox20* is a key regulator of rhombomere-specific gene expression in the developing hindbrain. Mech Dev 65:31–42

Serbedzija G, Fraser S, Bronner-Fraser M (1992) Vital dye analysis of cranial neural crest cell migration in the mouse embryo. Development 116:297–307

Sham M-H, Hunt P, Nonchev S, Papalopulu N, Graham A, Boncinelli E, Krumlauf R (1992) Analysis of the murine *Hox-2.7* gene: conserved alternative transcripts with differential distributions in the nervous system and the potential for shared regulatory regions. EMBO J 11:1825–1836

Sham MH, Vesque C, Nonchev S, Marshall H, Frain M, Das Gupta R, Whiting J, Wilkinson D, Charnay P, Krumlauf R (1993) The zinc finger gene *Krox-20* regulates *Hoxb-2* (*Hox2.8*) during hindbrain segmentation. Cell 72:183–196

Sharpe C, Goldstone K (1997) Retinoid receptors promote primary neurogenesis in *Xenopus*. Development 124:515–523

Sieweke M, Tekotte H, Frampton J, Graf T (1996) MafB is an interaction partner and repressor of Ets-1 that inhibits erythroid differentiation. Cell 85:49–60

Simeone A, Acampora D, Arcioni L, Andrews PW, Boncinelli E, Mavilio F (1990) Sequential activation of *Hox2* homeobox genes by retinoic acid in human embryonal carcinoma cells. Nature 346:763–766

Simeone A, Acampora D, Nigro V, Faiella A, D'Esposito M, Stornaiuolo A, Mavilio F, Boncinelli E (1991) Differential regulation by retinoic acid of the homebox genes of the four *Hox* loci in human embryonal carcinoma cells. Mech Dev 33:215–227

Simeone A, Avantaggiato V, Moroni MC, Mavilio F, Arra C, Cotelli F, Nigro V, Acampora D (1995) Retinoic acid induces stage-specific antero-posterior transformation of rostral central nervous system. Mech of Dev 51:83–98

Slack JMW, Tannahill D (1992) Mechanism of anteroposterior axis specification in vertebrates. Lessons from the amphibians. Development 114:285–302

Smith A, Robinson V, Patel K, Wilkinson DG (1997) The *EphA4* and *EphB1* receptor tyrosine kinases and *ephrin-B2* ligand regulate targeted migration of branchial neural crest cells. Curr Biol 7:561–570

Stachel SE, Grunwald DJ, Myers PZ (1993) Lithium perturbation and *goosecoid* expression identify a dorsal specification pathway in the pregastrula zebrafish. Development 117:1261–1274

Streit A, Sockanathan S, Pérez L, Rex N, Scotting P, Sharpe P, Lovell-Badge R, Stern C (1997) Preventing the loss of competence for neural induction: role of HGF/SF, L% and *Sox-2*. Development 124:1191–1202

Studer M, Pöpperl H, Marshall H, Kuroiwa A, Krumlauf R (1994) Role of a conserved retinoic acid response element in rhombomere restriction of *Hoxb-1*. Science 265:1728–1732

Studer M, Lumsden A, Ariza-McNaughton L, Bradley A, Krumlauf R (1996) Altered segmental identity and abnormal migration of motor neurons in mice lacking *Hoxb-1*. Nature 384:630–635

Studer M, Gavalas A, Marshall H, Ariza-McNaughton L, Rijli F, Chambon P, Krumlauf R (1998) Genetic interaction between *Hoxa1* and *Hoxb1* reveal new roles in regulation of early hindbrain patterning. Development 125:1025–1036

Svaren J, Sevetson B, Golda T, Stanton J, Swirnoff A, Milbrandt J (1998) Novel mutants of NAB corepressors enhance activation by Egr transactivators. EMBO J 17:6010–6019

Swiatek PJ, Gridley T (1993) Perinatal lethality and defects in hindbrain development in mice homozygous for a targeted mutation of the zinc finger gene *Krox-20*. Genes Dev 7:2071–2084

Takihara Y, Tomotsune D, Shirai M, Katoh-Fukui Y, Nishii K, Motaleb M, Nomura M, Tsuchiya R, Fujita Y, Shibata Y, Higashinakagawa T, Shimada K (1997) Targeted disruption of the mouse homologue of the *Drosophila polyhomeotic* gene leads to altered anteroposterior patterning and neural crest defects. Development 124:3673–3682

Taneja R, Thisse B, Rijli FM, Thisse C, Bouillet P, Dolle P (1996) The expression pattern of the mouse receptor tyrosine kinase gene MDK1 is conserved through evolution and requires Hoxa-2 for rhombomere-specific expression in mouse embryos. Dev Biol 177:397–412

Theil T, Frain M, Gilardi-Hebenstreit P, Flenniken A, Charnay P, Wilkinson D (1998) Segmental expression of the *EphA4* (*Sek-1*) receptor tyrosine kinase in the hindbrain is under the direct transcriptional control of *Krox20*. Development 125:443–452

Theil T, Ariza-McNaughton L, Manzanares M, Krumlauf R, Wilkinson D (1999) *kreisler* regulates rostrocaudal identity in the hindbrain. Development (in press)

Toivonen S, Saxen L (1968) Morphogenetic interaction of presumptive neural and mesodermal cells mixed in different ratios. Science 158:539–540

Tosney K (1982) The segregation and early migration of cranial neural crest cells in the avian embryo. Dev Biol 89:13–24

Vaage S (1969) The segmentation of the primitive neural tube in chick embryos (*Gallus domesticus*). Adv Anat Embryol Cell Biol 41:1–88

Vesque C, Charnay P (1992) Mapping the functional regions of the segment specific transcription factor Krox-20. Nuclei Acids Res 10:2485–2492

Vesque C, Maconochie M, Nonchev S, Ariza-McNaughton L, Kuroiwa A, Charnay P, Krumlauf R (1996) *Hoxb-2* transcriptional activation by *Krox-20* in vertebrate hindbrain requires an evolutionary conserved *cis*-acting element in addition to the Krox-20 site. EMBO J 15: 5383–5896

Whiting J, Marshall H, Cook M, Krumlauf R, Rigby PWJ, Stott D, Allemann RK (1991) Multiple spatially specific enhancers are required to reconstruct the pattern of *Hox-2.6* gene expression. Genes Dev 5:2048–2059

Wilkinson D, Krumlauf R (1990) Molecular approaches to the segmentation of the hindbrain. Trends Neurosci 13:335–339

Wilkinson DG, Bhatt S, Chavrier P, Bravo R, Charnay P (1989a) Segment-specific expression of a zinc-finger gene in the developing nervous system of the mouse. Nature 337:461–465

Wilkinson DG, Bhatt S, Cook M, Boncinelli E, Krumlauf R (1989b) Segmental expression of *Hox-2* homeobox-containing genes in the developing mouse hindbrain. Nature 341:405–409

Wingate R, Lumsden A (1996) Persistence of rhombomeric organisation in the postsegmental avian hindbrain. Development 122:2143–2152

Wizenmann A, Lumsden A (1997) Segregation of rhombomeres by differential chemoaffinity. Mol Cell Neurosci 9:448–459

Woo K, Fraser S (1997) Specification of the zebrafish nervous system by non-axial signals. Science 277:254–257

Xu Q, Alldus G, Holder N, Wilkinson DG (1995) Expression of truncated *Sek-1* receptor tyrosine kinase disrupts the segmental restriction of gene expression in the *Xenopus* and zebrafish hindbrain. Development 121:4005–4016

Xu Q, Mellitzer G, Robinson V, Wilkinson D (1999) In vivo cell sorting in complementary segmental domains mediated by *Eph* receptors and *ephrins*. Nature 399:267–271

Zhang M, Kim H-J, Marshall H, Gendron-Maguire M, Lucas AD, Baron A, Gudas LJ, Gridley T, Krumlauf R, Grippo JF (1994) Ectopic *Hoxa-1* induces rhombomere transformation in mouse hindbrain. Development 120:2431–2442

Neurogenetic Compartments of the Mouse Diencephalon and some Characteristic Gene Expression Patterns

Salvador Martínez and Luis Puelles[1]

1
Introduction

In the last 10 years our concept of the developing diencephalon has changed dramatically. This is a consequence of an increasing number of morphological, chemoarchitectural, gene expression and experimental data that resist a satisfactory interpretation within the usual morphological schema suggested by textbooks, represented by the so-called *columnar* view of the vertebrate forebrain. The columnar paradigm was instaurated by Herrick (1910), who divided the vertebrate diencephalon into four superposed columns separated by ventricular sulci. This schema was later supported by numerous adherents, among which Kuhlenbeck played a singular role (Kuhlenbeck 1973, and earlier work reviewed therein). The four columns were called epithalamus, thalamus dorsalis, thalamus ventralis and hypothalamus (from dorsal to ventral). They were held to be longitudinal parts of the neural tube, though this view is only possible by arbitrary disregard of the notorius axial bending of the rostral neural tube at the cephalic flexure (Keyser 1972; Puelles and Rubenstein 1993; Puelles 1995). Among other problems (see Puelles 1995), this schema typically dealt poorly with the pretectum, causing many authors to fail to distinguish it adequately from the dorsal thalamus, the epithalamus or the midbrain roof. Various descriptive embryologists noticed over the years the difficulties of the columnar approach, favoring a segmental paradigm, but did not achieve a substantial impact with their alternative interpretations (Rendahl 1924; Tello 1934; Bergquist 1954; Coggeshall 1964; Keyser 1972; Gribnau and Geijsberts 1985). These, nevertheless, finally constituted the base of the present conceptions, together with parallel work on non-mammalian vertebrates.

It turned out that molecular and experimental data on neural tube regionalization, which started to accumulate in the late 1980s, showed a

---

[1] Department of Morphological Sciences, University of Murcia, 30100 Murcia, Spain (salvador@fcu.um.es; puelles@fcu.um.es)

Results and Problems in Cell Differentiation, Vol. 30
Goffinet and Rakic (Eds.): Mouse Brain Development
© Springer-Verlag Berlin Heidelberg 2000

general lack of relevance of the ventricular sulci to strict regional delimitation of primary molecularly defined areas in the neural wall (sulci being themselves secondary morphogenetic phenomena). Simultaneously, interest was rekindled in the axial curvatures of the neuraxis, since it became clear that most developmental genes reflect the axial incurvations in their expression patterns (Puelles and Rubenstein 1993; Puelles 1995; Shimamura et al. 1995). Molecular boundaries do correlate with cytoarchitectonic and glial limits observed in the forebrain wall. The observed boundary patterns are more in agreement with the forementioned, alternative *segmental* views of the forebrain, which divided the developing diencephalon into several transverse segments or neuromeres (reviewed in Puelles et al. 1987; Puelles and Rubenstein 1993). The resulting prosomeric model embodies the most advanced proposal for understanding prosencephalic regionalization and morphogenesis consistently with experimental data on early neural induction and subsequent spatial patterning (Figdor and Stern 1993; Rubenstein et al. 1994; Puelles 1995; Nieuwenhuys 1998).

A forebrain segment or prosomere is formed by four superposed longitudinal domains, called, from ventral to dorsal: floor plate, basal plate, alar plate and roof plate; the mutual limits of these domains are defined by molecular patterns and cellular fates commonly established by dorsoventral patterning of the forebrain wall, irrespective of sulci. These fundamental neurogenetic elements are metameric, that is, are repeatedly represented across the diverse diencephalic segments, and are serially continuous with similar longitudinal subdivisions in the midbrain, hindbrain (rhombomeres) and spinal cord (myelomeres) (Vaage 1969). Neuromeres are thus serially iterated regions transverse to the brain axis that encompass a common system of longitudinal domains. The interneuromeric limits are also primarily molecular and related in their origin to anteroposterior patterning; sometimes they may be overtly marked by constrictions separating bulging parts of the neural wall (this occurs mainly in the alar plate and is in any case a transient phenomenon related to differential growth). Neuromeric units appear sequentially during early development in a complex spatiotemporal pattern by subdivision or differential growth of previous simpler 'proneuromeres'. The final number of neuromeres seems to be constant in all vertebrates (Vaage 1969; Nieuwenhuys 1998; Pombal and Puelles 1999).

There is definitely a correspondence of localized gene expression domains with the site-characteristic patterns of neurogenesis, neuronal migration and glial differentiation of neuroepithelial domains (Redies 1995). Each metameric subdomain of the basal and alar neuroepithelial regions forms a distinct histogenetic area where radial migration brings diverse neuronal populations into the mantle zone. These histogenetic fields can be considered as more or less self-contained morphogenetic units, though this scenario may be complicated by the eventual superposition of tangential neuronal migrations, which can occur both dorsoventrally and rostrocaudally. However, in the diencephalon such cell movements seem to be restricted to the interior of single prosomeres. The subsequent progressive subdivision and territorial specification of the neural complexes formed within each segmental field – the neuromere-derived domains – leads to the final histological diversity of the neural wall.

At the early neural plate stages, the prosencephalon, which represents its anterior, ventrally bent and expanded region (Fig. 1a), is subdivided during the process of neurulation in two transversal proneuromeric regions: the caudal proneuromere corresponds to the diencephalon proper (excluding most of the hypothalamus; this is essentially in agreement with an old schema proposed by His (1893), and the rostral proneuromere represents the secondary prosencephalon (Fig. 1b,c). The latter contains the telencephalon, the eye vesicles, the hypothalamus and some prethalamic areas (Rubenstein et al. 1994, 1998; Shimamura et al. 1995). These prosencephalic proneuromeres are subsequently subdivided into smaller transversal domains which are described as prosomeres (Puelles et al. 1987; Bulfone et al. 1993; Puelles and Rubenstein 1993; Rubenstein et al. 1994, 1998; Puelles 1995). The diencephalon proper is subdivided into four transversal units which, from caudal to rostral, contain in their alar domains the pretectum (prosomere 1), the dorsal thalamus and epithalamus (prosomere 2), the ventral thalamus (prosomere 3) and the eminenthia thalami (prosomere 4) (Fig. 1c).

There is an increasing body of data concerning genes that are expressed during development and show a prosomere-related pattern that suggests a

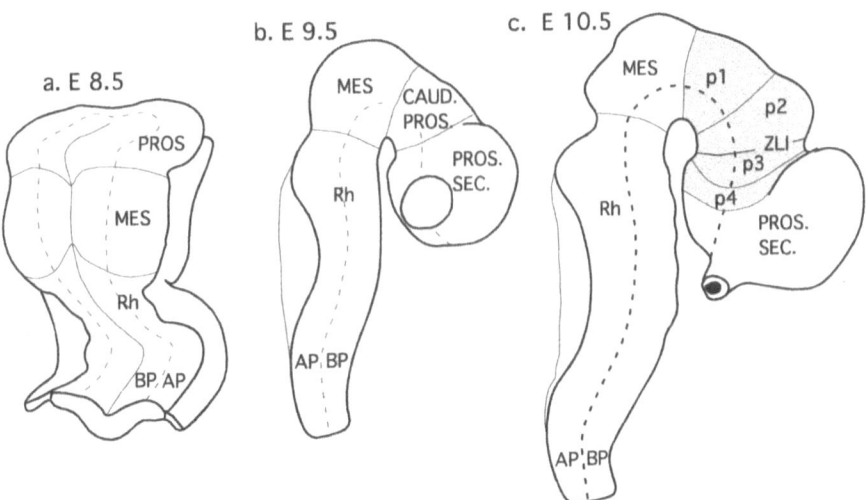

Fig. 1. Prosencephalic segmentation. Diagrams of E8.5 (**a**), E9.5 (**b**) and E10.5 (**c**) mouse embryo neural plate and tube. The prosencephalic region of the neural plate at E8.5 is progressively regionalized in two prosencephalic proneuromers at E9.5 (**b**) and in four diencephalic prosomers at E10.5 (**c**). *AP* alar plate; *BP* basal plate; *Rh* rhombencephalon; *ZL* zona limitans intrathalamica

role in diencephalic morphogenesis, i.e., *Gbx2* (Bulfone et al. 1993; Miya-shita-Lin et al. 1999), *Otx1/2* and *Otp* (Simeone et al. 1993, 1994), *Six3* (Oliver et al. 1995), *PLZF* (Avantaggiato et al. 1995), *Pax6* (Stoykova et al. 1996; Grindley et al. 1997), *Ebf-1/2/3* (Garel et al. 1997), and *Otlx2/Brx1* (Mucchielli et al. 1996; Kitamura et al. 1997). A detailed map of diencephalic develop-ment in mouse may be important to aid gene expression pattern descriptions and establish a common conceptual scaffold for comparisons between dif-ferent studies. In addition, a definition of the representative nuclei or other landmarks inside the prosomeres will be useful to recognize structural phe-notypes resulting from experimental mutagenesis in mice. We present here a description of the development of the mouse diencephalon at early stages of development, with special emphasis on anatomical landmarks and specific structures that can be characterized as markers for a specific domain inside a given prosomere. We present data on classical Nissl staining and im-munohistochemistry for calcium-binding proteins (calbindin, CB, and calretinin, CR), in order to have markers for specific cell populations and better detect differences between some diencephalic regions. This material was selected in order to obtain maximal information with simple histological techniques, easily reproducible in any laboratory (see also the Paxinos et al. 1994, 1999; Foster 1998; Jacobowitz et al. 1998, atlases).

2
Origin and Definition of Diencephalon

At the initial stages of mouse forebrain development (E8) the neural plate does not show morphological clues as to the future diencephalic re-gionalization, though molecular markers already identify diverse nested zones positive i.e. for *Otx-2*, *Pax-6*, *BF-1* or *Six-3* (Simeone et al. 1992; Oliver et al. 1995). A lateral transverse furrow in the neural folds at the cephalic neural plate is usually taken to approximately identify the limit between the prospective prosencephalic vesicle, rostrally, and the prospective mesence-phalic vesicle, caudally (Fig. 1a), though the caudal limit of the *Pax-6* domain seems a more exact reference.

After neurulation, occurring between E8 and E9.5, the prosencephalon starts to be separated into the prospective diencephalon, which initially contains mainly what we may call 'caudal diencephalon', and the rostral secondary prosencephalon, from which the 'rostral diencephalon' has yet to segregate by differential growth (Fig. 1b,c). At this stage the limit between diencephalon and mesencephalon appears as a slight constriction that co-incides with the caudal boundary of expression of the gene *Pax-6* (Stoykova and Gruss 1994). We shall refer to this boundary as the *mes-diencephalic limit*. The alar plate expression domain of *Pax-6* incipiently shows downregulation of this signal in a transverse inverted-V-shaped area corresponding to the prospective zona limitans intrathalamica (or simply 'zona limitans'). This appears at the rostral end of the caudal diencephalon, the extent of which is clearly marked as well by the expression of *Irx-1/2/3* in the alar plate (Bosse et al. 1997; note these authors misinterpret p2 as pretectum), *Wnt-13* at the roof plate (already present at E8.5; Zakin et al. 1998), *DM-20* at the basal plate (Timsit et al. 1991) and *HNF3β* at the floor plate (Sasaki and Hogan 1993; Grindley et al. 1997). No other limit is evident morphologically more rostrally (Fig. 1b), though 1 day later, at E10.5, the expression of the gene *Otp* around the optic stalks ends caudally with a sharp transverse boundary corresponding to the prospective definitive diencephalo-prosencephalic limit (p4/p5, Fig. 1c; Simeone et al. 1994). The gene *Sim-1* seems to coincide here with *Otp* (Fan et al. 1996).

During these stages the secondary prosencephalon becomes subdivided into two territories: rostrally, the dorsalmost part of the alar plate generates the evagination of the telencephalic vesicle, whereas caudally the two rostral diencephalic prosomeres are successively incorporated into the diencephalon (rostral to the zona limitans), conforming the definitive four diencephalic segments; these extend to include in p4 the mammillary pouch, ventrally, and the eminentia thalami, dorsally (Fig. 1c). There are advantages and disadvantages in placing p4 in the diencephalon proper, against the previous option of placing it in the secondary prosencephalon (Puelles and Rubenstein 1993; Puelles 1995). For instance, genes expressed in the basal plate – i.e., *Shh*, *Sim-1* – seem to suggest some sort of unity of the mammillary region with the retromammillary area and the prerubral tegmentum in the caudal diencephalon. It is suggested that epichordal properties of the floor- and basal plates may end at the mammillary pouch (compare with Kuhlenbeck 1973). In the alar plate, expression of *Dlx* genes also unifies p3 and p4 (Bulfone et al. 1993). On the other hand, the genes *Tbr-1* (Bulfone et al. 1995) and *AP-2.2* (Chazaud et al. 1996) characterize at E10.5 the prospective telencephalic pallium in p5, together with the eminentia thalami in p4. Gene patterns also exist that unify the complex formed by p3-p6, like the gene *Arx* (Miura et al. 1997). Thus, it must be concluded that various molecular determinants of fate seem to appear in a nested fashion across these prosomeres, making the distinction by this means of a di-prosencephalic boundary a matter of convenience.

Simultaneously, the signals of *AP-2* (Chazaud et al. 1996) and *Ebf-1/2/3* (Garel et al. 1997) in the pretectum and the expression of *Gbx-2* (Bulfone et al. 1993) in the dorsal thalamus clearly establish the molecular boundary p1/p2, whereas *TCF-4* signal seems to unify the whole p1 and p2 alar domain of the caudal diencephalon (Cho and Dressler 1998). In the rostral diencephalon, the gene *Gsh-2* expressed in alar p3 (Hsieh-Li et al. 1995) apparently distinguishes p3 from p4, together with *Tbr-1* and *AP-2.2*, which are both negative in p3 (cited above). Finally, the p5 and p6 prosomeres postulated within the secondary prosencephalon (Puelles and Rubenstein 1993) may be delimited by the respective molecular boundaries of *Six-3* (Oliver et al. 1995) and *Punc* (Salbaum 1998). Curiously, this suggests that the *Six-3*-positive neurohypophysis may actually lie in p6, rather than in p5 as was thought before (Puelles 1995).

We thus redefine the diencephalon proper as the neuroepithelial region that is limited caudally by the mes-diencephalic limit, and rostrally by the definitive diprosencephalic limit. The general morphology of diencephalic segments is irregular, since the caudal segments (p1 and p2) show larger alar plate than basal plate domains, while rostral segments (p3 and p4) are more homogeneous or have an expanded basal plate (mammillary pouch in p4). Apart from the latter, the diencephalic basal plate contains the retro-mammillary area (p3), posterior tuberculum (p2) and prerubral tegmental zone (p1) and arches around the cephalic flexure, just dorsal to the end of the notochord (Fig. 1b).

3
Diencephalic Segmentation

The process of morphologic segmentation in the diencephalon starts at E9.5 and follows during the next 2–3 days. We have previously described that the zona limitans intrathalamica (ZL) appears as a transverse ventricular ridge opposed to a surface constriction or furrow, between prosomeres p2 and p3. The ZL shows during the early stages of neural development important molecular properties (Rubenstein et al. 1994; Shimamura et al. 1996) and functional properties (Martinez et al. 1991, 1999; Crossley et al. 1996; Shimamura and Rubenstein 1997). The diencephalic region appears clearly divided into four neuromeres at E10, after the appearence of two new transverse constrictions at both sides of the ZL (Fig. 1c). E10 and E11 are the stages of mouse embryo development in which the diencephalic prosomeres are morphologically apparent. Afterwards, the progress in differential growth of the diencephalic wall, caused by heterogeneous neurogenetic and migra-tion processes in each prosomere, as well as by the establishment of multiple axonal pathways, leads to a progressive hiding of the limits between segments (which persist in the form of deformed radial glia patterns) and to the loss of the initial multivesicular shape.

Starting out from the mesencephalon, we find a ventricular ridge that marks the limit between mesencephalic and diencephalic; this is the mes-diencephalic limit (Fig. 2a–d,i,j). The first diencephalic prosomere (p1) contains the presumptive pretectal region, and is more extensive dorsally than ventrally. The basal plate will develop as tegmental area and the alar plate will generate the pretectal region (Fig. 2a–e,i,j). Rostrally this prosomere

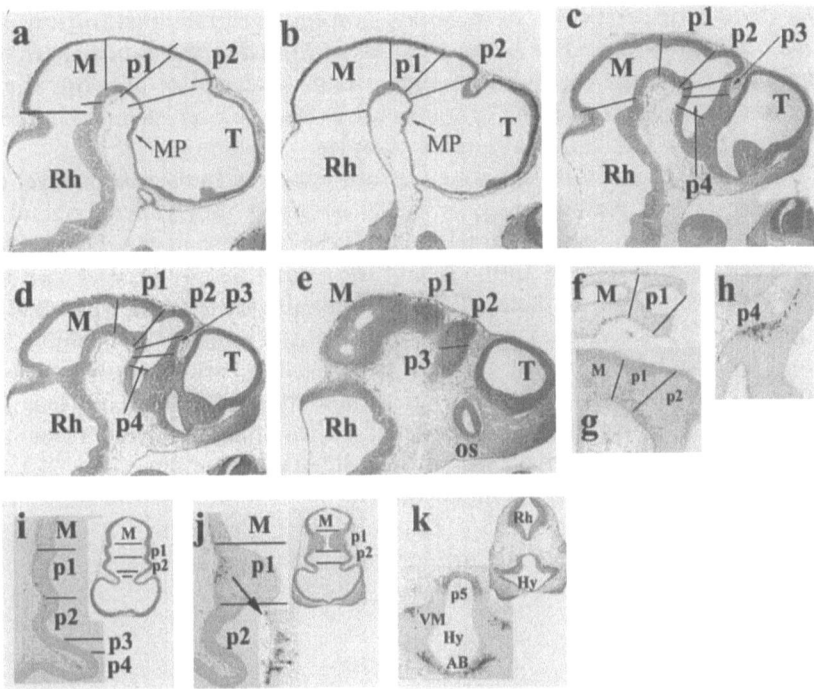

Fig. 2a-k. Diencephalic segmentation at E10.5. **a-e** Sagittal sections of mouse embryo cephalic region. From medial (**a**) to lateral (**e**), the prosencephalic segments are clearly visible in Nissl-stained series. **f-h** Details of anti-CR immunostained series showing the presence of post-mitotic CR-positive cells in the alar plate of p1 and p2 (**f,g**) and in the lateral levels of eminentia thalami, p4 (**h**). **i-k** Horizontal sections processed by anti-CR immunostaining; *inserts* show parallel Nissl-stained sections, showing CR-positive neurons in the mantle layer of p1 and p2 alar plates (**i,j**), the positive cells in the ventro medial (*VM*) and antero basal (*AB*) areas of p5 and p6 respectively (**k**) and immunopositive epithelial cells in the infundibular region, p5 (**k**). *AB* antero basal nucleus; *Hy* hypothalamus; *M* mesencephalon; *MP* mammillary pouch; *p1-p4* prosomers; *Rh* rhombencephalon; *T* telencephalon; *VM* ventro-medial nucleus

is separated from p2 by the p1/p2 limit (Fig. 2a–e,i,j). Prosomere 2 lies between this limit and the ZL. Shorter than p1 in the antero-posterior direction, p2 shows a narrow basal plate, which will form the tegmental tuberculum posterior (prominent inside the ventricle; Fig. 2c,d), and a more extense alar plate, which will develop as the dorsal thalamus and epithalamus complex (Fig. 2). Rostral to the ZL, the two rostral diencephalic prosomeres, p3 and p4, are visible. P3 is also narrow at the basal plate, where it forms the retromammillary area just caudal to the mammillary pouch (MP; Fig. 2a,b); it expands dorsally into the anterior region of the diencephalic alar plate, which corresponds to the ventral thalamus proper (note that a number of atlases and other publications confusingly add elements corresponding to p4 to the 'ventral thalamus'). P4 is the region that bulges ventricularly at the dorsal transition between diencephalon and prosencephalon; it is limited rostrally by the di/prosencephalic limit, which descends ventrally just in front of the mammillary pouch. The basal plate of p4 contains the caudal part of the hypothalamus, the mammillary region, and its alar plate develops the eminentia thalami formation (ET) transiently, which is later included in the stria terminalis/stria medullaris nuclear complex.

The first CR and CaBP positive neurons appeared in the mantle layer of the p1 and p2 alar plates already in E10.5 embryos; they generally send their growing axonal processes caudalwards. These cells also have been detected by means of anti-b-tubulin antibody and were described by Mastick and Easter (1996) as the origin of the medial longitudinal fascicle (flm). The basal plate of p4 and p3 shows the pioneering axons of the tract of the postoptic commissure, coming from the positive cells localized in the prospective ventromedial hypothalamic nucleus in p5 and the anterobasal nucleus in p6 (VM; AB; Fig. 2k). The floor plate of p5, or tuberal region, presents CR immunopositive epithelial cells, at both sides of the midline (Fig. 2k). More dorsally, the alar plate of p4 incipiently becomes populated by a characteristic CR-immunoreactive cell group (Fig. 2h), whereas a nearby CB-positive area in p5 may correspond to the primordium of the paraventricular nucleus.

4
Diencephalic Histogenetic Differentiation

At E12.5 the diencephalon presents a more complex structure. Often we can distinguish several anteroposterior, ventrodorsal or sometimes oblique cytoarchitectonic domains within each segment, as well as different concentric neuronal strata in most of these areas (normally, periventricular, intermediate and superficial strata, some of which may yet become divided into sublayers, as occurs in the lateral geniculate nucleus). This secondary regionalization, which is most marked in the alar plate, obviously must be a consequence of further molecular specification of the ventricular zone, the mantle zone, or both, affecting the development, migration and layering of separate sets of neuron types in increasing interaction with their afferents. One expects the observation of gene expression patterns that precede and accompany the morphological appearance of these additional intraneuromeric areas and boundaries. In some cases, this phase of histogenesis is preceded by gradiental expression of some genes across the primary domain, which may lead to differential upregulation of the secondary patterns, i.e., the gradient of *Wnt3* described by Bulfone et al. (1993), or the dorsoventrally nested expression domains of the *Gli* or *BMP* genes (Furuta et al. 1997; Platt et al. 1997; Ruiz i Altaba 1998). Gradiental expression of members of the family of Eph receptors and ligands possibly correlates with some subdivisions and their involvement in topographically ordered connections (Mellitzer et al. 1999), whereas differential combinatorial expression of cadherins (Redies and Takeichi 1996; Korematsu and Redies 1997) has been proposed to correlate with subsequent formation of sets of nuclei across different brain segments, which participate in a given functional pathway or system.

One result of the significant development of the mantle layer is the appearance of some transversely or longitudinally coursing axonal tracts. These tracts can be easily detected even in Nissl-stained material and represent the most important structural landmarks for morphological orientation at more advanced embryonic and postnatal stages, due to their stereotyped topological positions within the prosomeric schema. The rationale for this is based on accumulated observations that growing axons usually either are attracted/repelled by the floor plate (or roof plate) of the brain, which causes them to course in a transversal plane, or tend to follow longitudinal courses within one of the main longitudinal zones (basal or alar plates). The fact that neuronal or glial localized sources that spread out or contain chemical determinants of ulterior axonal navigational decisions are positioned within the neuromeric framework and act at close range leads to a large impact of neuromeric topology on the whole axonal network (changes in course, emission of collateral branches, terminal arborization and synaptogenesis). Axonal tracts thus tend to form orthogonal (perpendicular) to the primary transverse or longitudinal units postulated in the prosomeric schema (Fig. 3b), and we can roughly adscribe them to one of these two categories: longitudinal tracts and transverse tracts (this refers to the places where they are easily detected due to their compact fasciculation; most tracts actually have transverse/longitudinal portions of their complete course, which are only observable experimentally).

5
Alar Plate Domains at E12.5

The alar plate of p1 (pretectum) is not very well understood in mammals, a situation that we feel is not helped by assigning it to the midbrain, either in part (Kuhlenbeck 1973) or in whole (Swanson 1992). According to Rendahl (1924) and Keyser (1972) it may be subdivided at least into two cytoarchitectonic domains. The caudal one is related to the posterior commissure, a compact transverse tract visible at the transition between the periventricular and intermediate strata, and is called the 'commissural pretectum'. The existence of specific grisea forming the intermediate and superficial strata excentrically to the commissural fibers is not yet unequivocally demonstrated in mammals, though cells deep to the nucleus of the optic tract may be part of it (Caballero-Bleda et al. 1992). Such grisea, nevertheless, are well known in

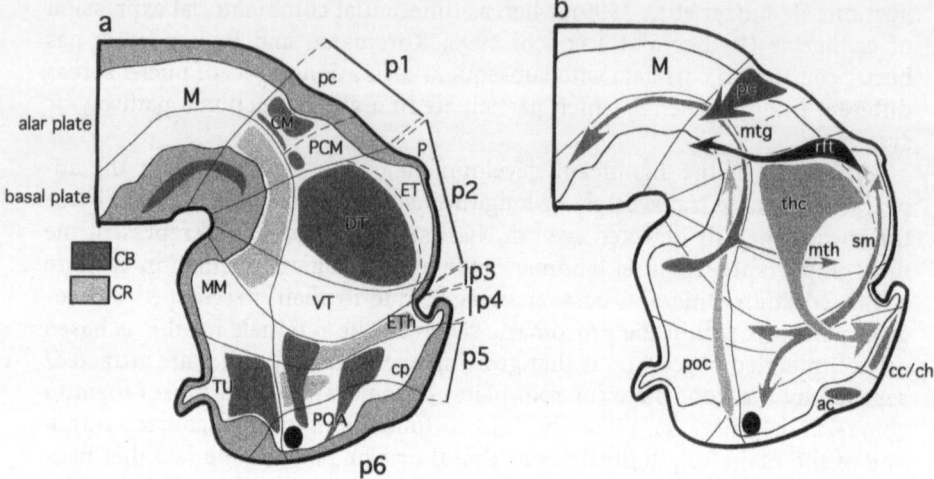

Fig. 3a,b. Segmental organization of the diencephalon. **a** The prosomeric model in relation to anti-CB and anti-CR immunopositive domains. **b** Topology of main diencephalic tracts in relation to the transversal and longitudinal domains of the prosomeric model: landmarks for diencephalic prosomeres. *ac* anterior commissure; *cc/ch* corpus callosum and hippocampal commissure; *CM* commissural pretectum; *cp* commissural plate; *DT* dorsal thalamus; *ET* epithalamus; *Eth* eminentia thalami; *M* mesencephalon; *MM* mammillary region; *mtg* mammillo-tegmental tract; *mth* mammillo-thalamic tract; *PCM* precommissural pretectum; *POA* preoptic area; *poc* postoptic commissure tract; *sm* stria medullaris; *thc* thalamo-cortical tract; *TU* tuberal region; *VT* ventral thalamus

reptiles and birds (the principal pretectal and subpretectal nuclei), so that their identification in mammals probably awaits an appropriate investigation. The rostral part of p1 is named the 'precommissural pretectum'; it lies in front of the posterior commissure and behind the retroflex tract, which courses already within p2 (Fig. 3b). This area later builds the anterior pretectal nucleus and a deep periaqueductal layer (Fig. 4). All sauropsids further show a third set of pretectal nuclei intercalated between the 'commissural' and 'precommissural' pretectal regions (n. of the posterior commissure in reptiles, or spiriform nuclei in birds). Reports on the rabbit pretectum by Caballero-Bleda et al. (1992) and Lagares et al. (1994) showed that such an entity may also be found in mammals, conforming to the definition by our group of an intercalated 'juxtacommissural pretectum'. Likewise, our CB and

--→

Fig. 4a–p. Diencephalon at E12.5. Sagittal sections of E12.5 mouse brain: **a–d** Nissl-stained series; **e–h** anti-calbindin (*CB*) immunostained series; **i–l** anti-calretinin (*CR*) immunostained series. Parallel consecutive sections are presented in *columns* (**a,e,i; d,f,j;...**). **m–p** Details of anti-CB (**m,o**) and anti-CR (**n,p**) processed consecutive sections. *CM* commissural pretectum; *MGE* medial ganglionic eminence; *PCM* precommissural pretectum; *p1–p6* prosomeres; *rft* retroflex tract; *sm* stria medullaris; *T* telencephalon; *zl* zona limitans

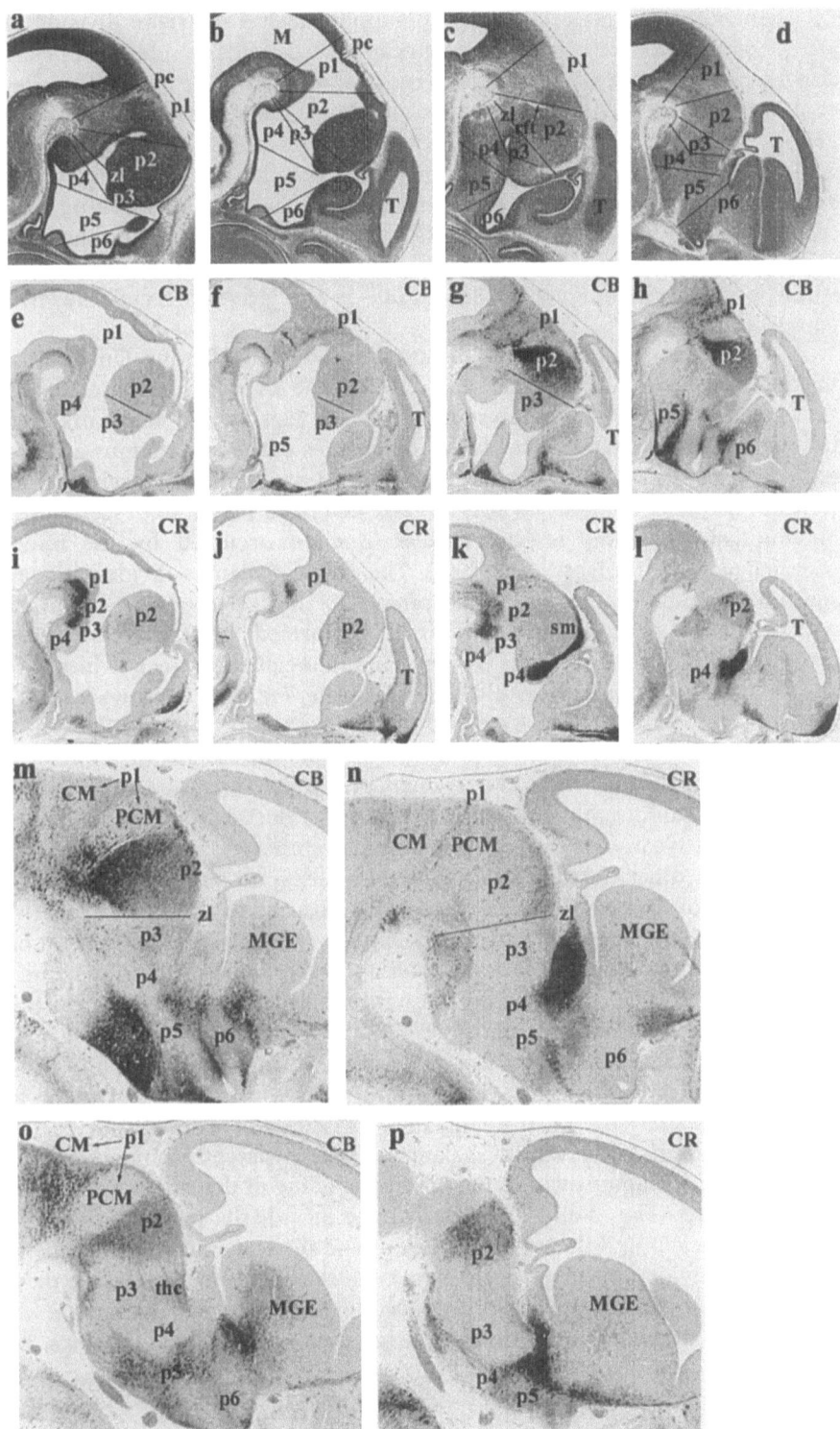

CR immunoreacted material seems to support such a tripartite division of alar p1 in the mouse (Fig. 4m,n). Pretectal anteroposterior subdivisions are also supported by the expression patterns of some genes, like *Pax-6* (Grindley et al. 1997), *Sax-1* (Schubert et al. 1995), *PLZF* (Avantaggiato et al. 1995) and *Ebf-1/2/3* (Garel et al. 1997). Finally, note that the so-called posterior pretectal nucleus of mammals, whose deeper neuronal populations are strongly CB-positive and serve to distinguish it from the superior colliculus lacking such cells, lies strictly caudal to the posterior commissure, that is, behind the mesdiencephalic limit, and is therefore a primarily mesencephalic formation, whose probable homologue in sauropsids is the 'griseum tectale' layered complex.

Alar p2 separates soon into epithalamus and dorsal thalamus. Only dorsal thalamus expresses *Gbx-2* (Bulfone et al. 1993) and a ventrodorsal gradient of CB (Puelles et al. 1992; for present results see Figs. 3 and 4a,g,h,m). The habenular nuclei contain the neuronal derivatives of the epithalamus and are the origin of the retroflex tract, which courses strictly transversely just in front of the p1/p2 boundary across the alar and basal plates, to descend then longitudinally into the isthmic median domain occupied by the interpeduncular nucleus (Figs. 3b and 4c). The dorsal thalamus divides at later stages into numerous nuclei, which project to the telencephalon in varied patterns. The mammillo-thalamic tract is formed by transverse, ventrodorsally coursing collaterals of the mammillotegmental tract (which has itself a longitudinal course within the basal plate; Fig. 3b). It grows into the anterior nuclear complex of the dorsal thalamus after E15, appearing slightly caudal to the ZL. That is, it lies inside alar p2 and is not at its rostral boundary, as is often indicated confusingly in the literature. The intergeniculate leaflet formation, whose origin in either dorsal or ventral thalamus has not been clearly defined in previous literature, apparently originates, together with other 'retroreticular' cells, in a separate anteroventral histogenetic area of alar p2 which is placed just adjacent to the ZL (Caballero-Bleda et al. 1993). These cells distinctly express *Nkx-2.2*, whereas the rest of the dorsal thalamus doesn't (Kitamura et al. 1997). Some of these cells, immunoreactive for CR, seem to invade tangentially at the brain surface the lateral and medial geniculate nuclei (Fig. 4l,n,p). Such a migration was also observed in the homologous cell population in birds, forming the perirotundic area and the interstitial nucleus of the optic tract (Puelles et al. 1991; Martínez et al. 1991; De Castro et al. 1998; Uchikawa et al. 1999).

Alar p3 comprises the ventral thalamus, whose apparent position 'ventral' to the 'dorsal' thalamus owes to the marked bending of the length axis at the cephalic flexure (Figs. 3 and 4). Its derivatives include the superficial ventral geniculate nucleus and the reticular nucleus and the zona incerta; all of them seem to subdivide into distinct parts (see the relevant gene expression data reported by Stoykova and Gruss 1994; Mucchielli et al. 1996; Kitamura et al. 1997). The ventral thalamus is distinctly negative for both CB and CR (Figs. 3a and 4). It strongly expresses *Dlx* genes (Robinson et al. 1991; Bulfone et al. 1993; Price 1993), except perhaps at its dorsalmost part, near the choroidal roof.

Alar p4 is easily distinguished in CR-immunoreacted material by the massive expression of this protein in practically all its mantle zone, which fills in the bulge of the eminentia thalami. Many of these cells seem to project into the stria medullaris (therefore also CR-positive), which courses longitudinally across p4, p3 and p2, to cross the midline in the habenular commissure at the roof plate of p2, just rostral to the pineal stalk (Figs. 3a,b and 4k,l,n,p). The expression of CB is complementary over the p4/p5, or di-prosencephalic boundary, appearing strongly expressed in the anterior part of the bed nucleus of the stria terminalis (Fig. 3m,o). The eminentia thalami persists into the adult with strong expression of *Tbr-1* (Bulfone et al. 1995), lacks expression of *Dlx* and shows ventricular expression of *Pax-6* (Bulfone et al. 1993; Stoykova and Gruss 1994).

Acknowledgement. Work supported by EC contract BIO4-CT96-042 and Spanish DGICYT PB98-397 (LP)

References

Altman J, Bayer SA (1978) Development of the diencephalon in the rat. I. Autoradiographic study of the time of origin and settling patterns of neurons of the hypothalamus. J Comp Neurol 182:945–972

Altman J, Bayer SA (1979) Development of the diencephalon in the rat. IV. Quantitative study of the time of origin of neurons and the internuclear chronological gradients in the thalamus. J Comp Neurol 188:455–472

Angevine JB (1970) Time of neuron origin in the diencephalon of the mouse. An autoradiographic study. J Comp Neurol 139:129–188

Avantaggiato V, Pandolfi PP, Ruthardt M, Hawe N, Acampora D, Pelicci PG, Simeone A (1995) Developmental analysis of murine *Promyelocyte Leukemia Zinc Finger* (*PLZF*) gene expression: implications for the neuromeric model of the forebrain organization. J Neurosci 15:4927–4942

Bergquist H (1954) Ontogenesis of diencephalic nuclei in vertebrates. A comparative study. K Fysiogr Sallsk Lund Handl 6:1–34

Bosse A, Zulch A, Becker MB, Torres M, Gomez-Skarmeta JL, Modolell J, Gruss P (1997) Identification of the vertebrate *Iroquois* homeobox gene family with overlapping expression during early development of the nervous system. Mech Dev 69:169–181

Bulfone A, Puelles L, Porteus MH, Frohman MA, Martin GR, Rubenstein JLR (1993) Spatially restricted expression of *Dlx-1*, *Dlx-2* (*Tes-1*), *Gbx-2*, and *Wnt-3* in the embryonic day 12.5 mouse forebrain defines potential transverse and longitudinal segmental boundaries. J Neurosci 13:3155–3172

Bulfone A, Smiga SM, Shimamura K, Puelles L, Peterson A, Rubenstein JLR (1995) *T-brain-1*: a homolog of *Brachyury* whose expression defines molecularly distinct domains within the cerebral cortex. Neuron 15:63–78

Caballero-Bleda M, Fernandez B, Puelles L (1992) The pretectal complex of the rabbit: distribution of AChE and NADPH-diaphorase activities. Acta Anat 144: 7–16

Caballero-Bleda M, Lagares C, Fernández B, Puelles L (1993) A chemoarchitectonically similar internal extension connects the rabbit intergeniculate leaflet to midline dorsal thalamic nuclei. J Hirnforsch 34: 33–40

Chazaud C, Oulad-Abdelghani M, Bouillet P, Decimo D, Chambon P, Dolle P (1996) AP-2.2, a novel gene related to AP-2, is expressed in the forebrain, limbs and face during mouse embryogenesis. Mech Dev 54: 83–94

Cho EA, Dressler GR (1998) TCF-4 binds beta-catenin and is expressed in distinct regions of the embryonic brain and limbs. Mech Dev 77: 9–18

Coggeshall RE (1964) A study of diencephalic development in the albino rat. J Comp Neurol 122: 241–269

Crossley PH, Martínez S, Martin GR (1996) Midbrain development induced by FGF8 in the chick embryo. Nature 380: 66–68

de Castro F, Cobos I, Puelles L, Martinez S (1998) Calretinin in pretecto- and olivocerebellar projections in the chick: immunohistochemical and experimental study. J Comp Neurol 397: 149–162

Fan C-M, Kuwana E, Bulfone A, Fletcher CF, Copeland NG, Jenkins NA, Crews S, Martínez S, Puelles L, Rubenstein JLR, Tessier-Lavigne M (1996) Expression patterns of two murine homologs of Drosophila single-minded suggest possible roles in embryonic patterning and in the pathogenesis of Down syndrome. Mol Cell Neurosci 7: 1–16

Figdor MC, Stern CD (1993) Segmental organization of embryonic diencephalon. Nature 363: 630–634

Foster GA (1998) Chemical neuroanatomy of the prenatal rat brain. A developmental atlas. Oxford University Press, Oxford New York Tokyo

Furuta Y, Piston DW, Hogan BL (1997) Bone morphogenetic proteins (BMPs) as regulators of dorsal forebrain development. Development 124: 2203–2212

Garel S, Marin F, Mattei MG, Vesque C, Vincent A, Charnay P (1997) Family of Ebf/Olf-1-related genes potentially involved in neuronal differentiation and regional specification in the central nervous system. Dev Dyn 210: 191–205

Gribnau AAM, Geijsberts LGM (1985) Morphogenesis of the brain in staged Rhesus monkey embryos. Adv Anat Embryol Cell Biol 91: 1–69

Grindley JC, Hargett LK, Hill RE, Ross A, Hogan BL (1997) Disruption of PAX6 function in mice homozygous for the Pax6/Sey-1/Neu mutation produces abnormalities in the early development and regionalization of the diencephalon. Mech Dev 64: 111–126

Herrick CJ (1910) The morphology of the forebrain in amphibia and reptilia. J Comp Neurol 20: 413–545

His W (1893) Vorschläge zur Einteilung des Gehirns. Arch Anat Entwicklungsgesch 17: 172–179

Hsieh-Li HM, Witte DP, Szucsik JC, Weinstein M, Li H, Potter SS (1995) Gsh-2, a murine homeobox gene expressed in the developing brain. Mech Dev 50: 177–186

Jacobowitz DM, Abbott LC (1997) Chemoarchitectonic atlas of the developing mouse brain. CRC Press, Boca Raton

Keyser A (1972) The development of the diencephalon of the chinese hamster. Acta Anatomica 83: 1–181

Kitamura K, Miura H, Yanazawa M, Miyashita T, Kato K (1997) Expression patterns of Brx1 (Rieg gene), Sonic hedgehog, Nkx2.2, Dlx1 and Arx during zona limitans intrathalamica and embryonic ventral lateral geniculate nuclear formation. Mech Dev 67: 83–96

Korematsu K, Redies C (1997) Restricted expression of cadherin-8 in segmental and functional subdivisions of the embryonic mouse brain. Dev Dyn 208: 178–189

Kuhlenbeck H (1973) The central nervous system of vertebrates. Vol 3 Part II. S Karger, Berlin

Lagares C, Caballero-Bleda M, Fernández B, Puelles L (1994) Reciprocal connections between the rabbit suprageniculate pretectal nucleus and the superior colliculus: tracer study with horseradish peroxidase and Fluorogold. Visual Neurosci 11:347–353

Lumsden A, Krumlauf R (1996) Patterning the vertebrate neuraxis. Science 274:1109–1115

Martínez S, Wassef M, Alvarado-Mallart RM (1991) Induction of a mesencephalic phenotype in the 2-day-old chick prosencephalon is preceded by the early expression of the homeobox gene *En2*. Neuron 6:971–981

Martínez S, Crossley PH, Cobos I, Rubenstein JLR, Martin GR (1999) FGF-8 induces an isthmic organizer and isthmocerebellar development in the caudal forebrain via a repressive effect on *Otx2* expression. Development 126:1189–1200

Mastick GS, Easter SE (1996) Initial organization of neurons and tracts in the embryonic mouse fore- and midbrain. Dev Biol 173:79–94

Mellitzer G, Xu Q, Wilkinson DG (1999) Eph receptors and ephrins restricted cell intermingling and communication. Nature 400:77–81

Miura H, Yanazawa M, Kato K, Kitamura K (1997) Expression of a novel *aristaless* related homeobox gene '*Arx*' in the vertebrate telencephalon, diencephalon and floor plate. Mech Dev 65:99–109

Miyashita-Lin EM, Hevner R, Wassarman KM, Martinez S, Martin GR, Rubenstein JLR (1999) Neocortical regionalization is preserved in the absence of thalamic innervation in newborn *Gbx-2* mutant mice. Science 285:906–909

Mucchielli ML, Martínez S, Pattyn A, Goridis C, Brunet JF (1996) *Otlx-2*, an *Otx*-related homeobox gene expressed in the pituitary gland and in a restricted pattern in the forebrain. Mol Cell Neurosci 8:258–271

Nieuwenhuys R (1998) Morphogenesis and general structure. In: Nieuwenhuys R, ten Donkelaar HJ, Nicholson C (eds) The Central Nervous System of Vertebrates, Vol I; chapter 4, pp 159–228. Springer, Berlin Heidelberg New York

Oliver G, Mailhos A, Wehr R, Copeland NG, Jenkins NA, Gruss P (1995) *Six3*, a murine homologue of the *sine oculis* gene, demarcates the most anterior border of the developing neural plate and is expressed during eye development. Development 121:4045–4055

Paxinos G, Ashwell KS, Törk Y (1994) Atlas of the developing rat nervous system. 2nd edn. Academic Press, London

Paxinos G, Kus L, Ashwell KS, Watson C (1999) Chemoarchitectonic Atlas of the Rat Brain. Academic Press, London

Platt KA, Michaud J, Joyner AL (1997) Expression of the mouse *Gli* and *Ptc* genes is adjacent to embryonic sources of hedgehog signals, suggesting a conservation of pathways between flies and mice. Mech Dev 62:121–135

Pombal MA, Puelles L (1999) A prosomeric map of the lamprey forebrain based on calretinin immunocytochemistry, Nissl stain and ancillary markers. J Comp Neurol (in press)

Price M, Lemaistre M, Pischetola M, Lauro RD, Duboule D (1991) A mouse gene related to *Distal-less* shows a restricted expression in the developing forebrain. Nature 351:748–751

Puelles L (1995) A segmental morphological paradigm for understanding vertebrate forebrains. Brain Behav Evol 46:319–337

Puelles L, Rubenstein JLR (1993) Expression patterns of homeobox and other putative regulatory genes in the embryonic mouse forebrain suggest a neuromeric organization. TINS 16:472–479

Puelles L, Amat JA, Martinez-de-la-Torre M (1987) Segment-related, mosaic neurogenetic pattern in the forebrain and mesencephalon of early chick embryos: I. Topography of AChE-positive neuroblasts up to stage HH18. J Comp Neurol 266:247–268

Puelles L, Guillén M, Martínez de la Torre M (1991) Observations on the fate of nucleus superficialis magnocellularis of Rendahl in the avian diencephalon, bearing on the organization and nomenclature of neighboring retinorecipient nuclei. Anat Embryol 183:221–233

Puelles L, Sanchez MP, Spreafico R, Fairen A (1992) Prenatal development of calbindin immunoreactivity in the dorsal thalamus of the rat. Neuroscience 46:135–147

Redies C (1995) Cadherin expression in the developing vertebrate brain: from neuromeres to brain nuclei and neural circuits. Exp Brain Res 220:243–256

Redies C, Takeichi M (1996) Cadherins in the developing central nervous system: an adhesive code for segmental and functional subdivisions. Dev Biol 180:413–423

Rendahl H (1924) Embryologische und morphologische Studien über das Zwischenhirn beim Huhn. Acta Zool Stockh 5:241–344

Robinson GW, Wray S, Mahon KA (1991) Spatially restricted expression of a member of a new family of murine distal-less homeobox genes in the developing forebrain. New Biologist 3:1183–1194

Rubenstein JLR, Martínez S, Shimamura K, Puelles L (1994) The embryonic vertebrate forebrain: the prosomeric model. Science 266:578–580

Rubenstein JLR, Shimamura K, Martínez S, Puelles L (1998) Regionalization of the prosencephalic neural plate. An Rev Neurosci 21:445–477

Ruiz I, Altaba A (1998) Neural patterning. Deconstructing the organizer. Nature 391:748–749

Salbaum JM (1998) Punc, a novel mouse gene of the immunoglobulin superfamily, is expressed predominantly in the developing nervous system. Mech Dev 71:201–204

Sasaki H, Hogan BL (1993) Differential expression of multiple fork head related genes during gastrulation and axial pattern formation in the mouse embryo. Development 118:47–59

Schubert FR, Fainsod A, Gruenbaum Y, Gruss P (1995) Expression of the novel murine homeobox gene Sax-1 in the developing nervous system. Mech Dev 51:99–114

Shimamura K, Rubenstein JLR (1997) Inductive interactions direct early regionalization of the mouse forebrain. Development 124:2709–2718

Shimamura K, Hartigan DJ, Martínez S, Puelles L, Rubenstein JLR (1995) Longitudinal organization of the anterior neural plate and neural tube. Development 121:3923–3933

Simeone A, Acampora D, Gulisano M, Stornaiuolo A, Boncinelli E (1992) Nested expression domains for homeobox genes in developing rostral brain. Nature 358:687–690

Simeone A, Acampora D, Mallaci A, Stornaiuolo A, D'Apice MR, Nigro V, Boncinelli E (1993) A vertebrate gene related to orthodenticle contains a homeodomain of the bicoid class and demarcates anterior neuroectoderm in the gastrulating mouse embryo. EMBO J 12:2735–2747

Simeone A, D'Apice MR, Nigro V, Casanova J, Graziani G, Acampora D, Avantaggiato V (1994) Orthopedia, a novel gene homeobox-containing gene expressed in the developing central nervous system of both mouse and Drosophila. Neuron 13:83

Stoykova A, Gruss P (1994) Roles of Pax-genes in developing and adult brain as suggested by expression patterns. J Neurosci 14:1395–1412

Stoykova A, Fritsch R, Walther C, Gruss P (1996) Forebrain patterning defects in small eye mutant mice. Development 122:3453–3465

Swanson LW (1992) Brain Maps: Structure of the rat brain. Elsevier, Amsterdam

Tello JF (1934) Les differenciations neurofibrillaires dans le prosencephale de la souris de 4 a 15 millimetres. Trav Lab Rech Biol 29:339–396

Timsit S, Martínez S, Allinquant B, Peyron F, Puelles L, Zalc B (1991) Oligodendrocytes originate in a restricted zone of the embryonic ventral neural tube defined by DM-20 mRNA expression. J Neurosci 15:1012–1024

Uchikawa M, Kamachi Y, Kondo H (1999) Two distinct subgroups of Group B Sox genes for transcriptional activators and repressors: their expression during embryonic organogenesis of the chicken. Mech Dev 84:103–120

Vaage S (1969) The segmentation of the primitive neural tube in chick embryos (Gallus domesticus). Ergeb Anat Entwicklungsgesch 41:1–88

Zakin LD, Mazan S, Maury M, Martin N, Guenet JL, Brulet P (1998) Structure and expression of Wnt13, a novel mouse Wnt2 related gene. Mech Dev 73:107–116

Neuronogenesis and the Early Events of Neocortical Histogenesis

V. S. Caviness Jr[1], T. Takahashi[1,2], and R. S. Nowakowski[3]

1
Introduction

The neocortex is central to the most highly evolved processing functions of the mammalian brain. These functions are intimately related to the architecture and massive scale of the neocortex and its regional organization as components of distributed neural systems. The neurons of the mammalian neocortex, represented by a great diversity of distinct classes and subclasses (Lorente de No 1938; Cajal 1952) arise from a proliferative pseudostratified ventricular epithelium (PVE) at the surface of the embryonic ventricular cavities (His 1889; Sauer 1935; Sauer 1936; Boulder Committee 1970) (Fig. 1A). Various lines of evidence now provide a glimpse of the workings of the proliferative process which reveal its pervasive integration with the succession of histogenetic events that follow. Thus, there is now evidence to suggest that neurons are specified with respect to class in the course of neuronogenesis in the PVE (Parnavelas et al. 1991; Mione et al. 1994; Mione et al. 1997) and that the regional map of the neocortex is foreshadowed by a corresponding regional map of PVE (Rakic 1988). That is, the earliest events of specification of cell class and of regional specification appear to occur coordinately with neuronogenesis within the proliferative cells of the PVE (Takahashi et al. 1999a).

The perspective, that the elementary proliferative mechanisms of the PVE are broadly integrated with its larger histogenetic agenda, will be the central theme of the presentation to follow. Formal development of this theme will be preceded by an overview of the properties of the epithelium which support the proliferative process. The theme will then be developed from three perspectives. These are, at the most elementary level, the parameters which govern for any defined founder population a quantitative model for the rate at which

[1] Department of Neurology, Massachusetts General Hospital, Harvard Medical School, Boston, MA, 02114, USA
[2] Department of Pediatrics, Keio University School of Medicine, Tokyo 160, Japan
[3] Department of Neuroscience and Cell Biology, UMDNJ-Robert Wood Johnson Medical School, Piscataway, NJ 08854, USA

Results and Problems in Cell Differentiation, Vol. 30
Goffinet and Rakic (Eds.): Mouse Brain Development
© Springer-Verlag Berlin Heidelberg 2000

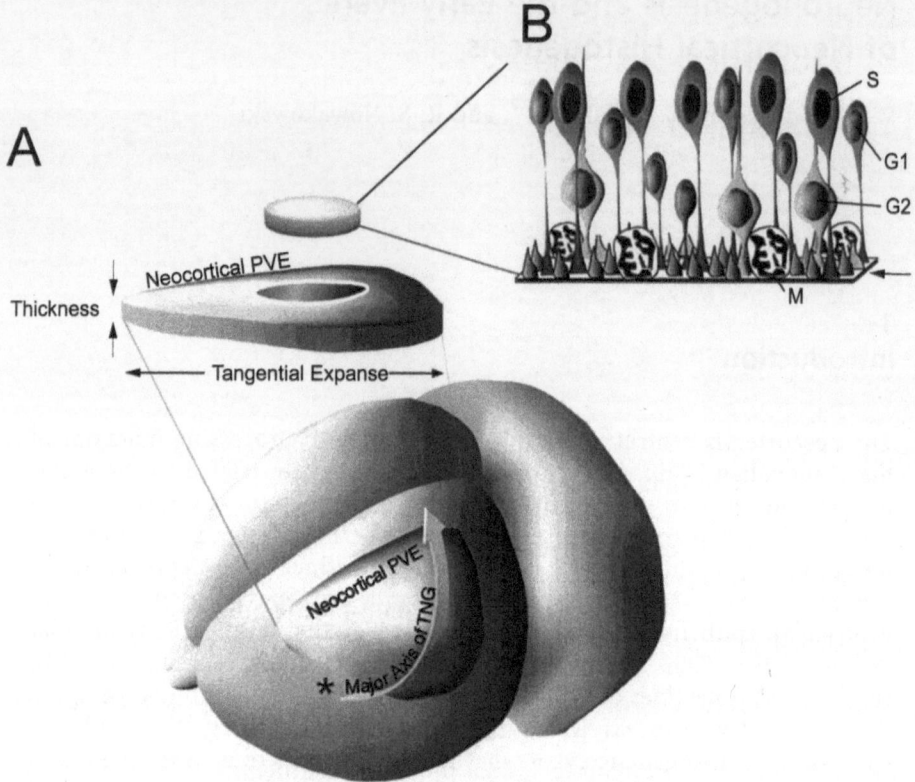

Fig. 1A,B. The neocortical pseudostratified ventricular epithelium (PVE) of the embryonic mouse brain. **A** The neocortical PVE is the origin of most if not all neocortical neurons and is defined approximately as the dorsal pallial lining of the telencephalic ventricles. The sequence of neocortical histogenetic events is initiated at the rostrolateral margin of the neocortical PVE (*) and propagates caudomedially (large curved arrow, major axis of the transverse neuronogenetic gradient, TNG). **B** The cytological architecture of an oval cutout of the PVE is illustrated graphically. Cells of the PVE form a tight sheet (PVE sheet), attached to each other mechanically by adherens junctions and functionally by GAP junctions at the ventricular surface of the epithelium (small arrow)

neurons are formed and the total number which will be formed. We then proceed to a formulation of an algorithm by which these elementary parameters could set the schedule of overall operation of neuronogenetic process in a given species. Finally, we will consider plausible linkages between these workings of the proliferative process and their regulation, on the one hand, and those which provide for histogenetic specification, on the other. Here we refer to mechanisms of cell class specification and the specification of a regional protomap of the neocortex. Although these final themes are strongly theoretical and incompletely linked to experimental observations, it is our view that they point to favorable directions for future investigation, favorable directions which are largely within reach of current technology.

2
The Neocortical Pseudostratified Ventricular Epithelium

2.1
Cytologic and Architectonic Features of the PVE

The PVE is the dominant proliferative population of the ventricular zone (VZ) which lines the interior of the entire embryonic central nervous system (Takahashi et al. 1992; Takahashi et al. 1995). It is homeomorphic with the neural ectoderm of the embryonic neural plate which becomes interiorized in the course of neurulation (Sidman and Rakic 1982; Alvarez-Bolado and Swanson 1996). Visible regional demarcations emerge within the VZ in the course of early development of the neural tube, and to some extent these appear to define the borders between the principal subdivisions of the central nervous structures (Lumsden 1990; Guthrie et al. 1991; Ingham and Arias 1992; Martinez et al. 1992; Puelles and Rubenstein 1993; Shimamura et al. 1995). The region of this epithelium that gives rise to the neurons of the neocortex (neocortical PVE) is defined approximately as the dorsal pallial lining of the telencephalic vesicles (Fig. 1A). These evaginate in bilateral paired fashion shortly after closure of the anterior neuropore. The neocortical PVE within the cerebral vesicles is demarcated laterally and ventrally at the ventricular angle from PVE that gives rise to the basal ganglia and paleo-cortical structures and dorsally and medially from archicortical PVE (Fishell et al. 1993; Alvarez-Bolado and Swanson 1996; Bhide 1996). The telencephalic PVE is in continuity with but may be distinguished from the proliferative epithelium of the midline prosencephalic vesicle which will give rise to diencephalic and subdiencephalic structures. The expanse of PVE which serves as origin of the mesencephalic and rhombencephalic structures lies further caudally (Alvarez-Bolado and Swanson 1996).

Neuronogenesis within neocortical PVE stands apart in its scale and late expression in the overall course of embryonic development from that of other regions of the PVE (Jacobson 1978; Sidman and Rakic 1982). Thus, the neuronogenetic interval for the entire neocortical PVE, which is species characteristic in duration, is approximately 6 embryonic days in mouse (Caviness and Sidman 1973; Caviness 1982) but is as long as 60 days in monkey (Rakic 1974; Rakic 1995a) and over 100 days in humans (Sidman and Rakic 1982). Among the structures which arise from the general PVE of the neural tube, it is the last and has the most protracted neuronogenetic period. Only structures which arise from non PVE precursors, including the granular layer of the cerebellar cortex, the dentate gyrus of the hippocampal formation and the granular layer of the olfactory bulb, have a later and more protracted neuronogenetic period (Sidman and Rakic 1982).

The behavior and transformations of cells of the PVE as they go through the cell cycle are dynamic and complex. Although the cells of the epithelium are greatly variable in their configuration, globular to elongated, their shape is systematically dependent upon phase of the cell cycle (Fig. 1B) (Sauer 1936; Stensaas and Stensaas 1968; Takahashi et al. 1992; Takahashi et al. 1995). Through all phases the cell has an attachment to the ventricular surface. In mitosis the cell is globular and both the nucleus and the perikaryon are located immediately adjacent to the ventricular surface. During the G1 phase the nucleus moves away from the ventricular surface and the cell elongates progressively. At its maximum extent, which is somewhat variable from cell to cell, the cell initiates the DNA synthetic or S phase with soma and nucleus within the abventricular half of the epithelium. The tetraploid nucleus of the cell in G2 phase rapidly approaches the ventricular margin again to complete the cycle with mitosis. This up and down "elevator" motion of the nucleus in the course of the cell cycle has been called "interkinetic nuclear migration" (Sauer 1936; Boulder Committee 1970; Takahashi et al. 1995).

Little of the cytologic and regional diversity that is to arise from the neocortical and other regions of the PVE is revealed by cytologic and architectonic features of the epithelium which are homogeneous throughout. The integrity and regularity of the cellular arrangement of the epithelial sheet is assured by adherens junctions which bind cells together at the ventricular margin (Stensaas and Stensaas 1968; Hinds and Ruffett 1971) (Fig. 1B). Efficient direct cell to cell communication between large sets of proliferative cells is assured by gap junctions, again concentrated near the ventricular margin (Dermietzel et al. 1989; LoTurco and Kriegstein 1991; Bittman et al. 1997). More loosely linked modes of modulation of cellular behavior are afforded by an intercellular environment which allows heterogeneous classes of ligands to play upon the appropriate receptors at the cell surface (Kilpatrick and Bartlett 1993; Fulton 1995; Ghosh and Greenberg 1995; Temple and Qian 1995; Cavanagh et al. 1997; Lavdas et al. 1997).

The neocortical PVE belongs to a general class of proliferative epithelium which is widely represented within the developing central nervous system and somatic structures of vertebrates (Sauer 1936, 1937). Thus, it may be assumed that those mechanisms which regulate neocortical neuronogenesis and concurrently determine both the cytological and regional diversity of the neocortex are only special instances of developmental processes which are broadly distributed within the central nervous systems and somatic organs of vertebrates. Evolutionary selection pressures may have favored the pseudo-stratified architecture of the epithelium for its pervasive histogenetic roles in part, at least, because of the high cell packing efficiency of this type of epithelium. This packing efficiency is illustrated in mouse, for example, where the proliferative cells are 5–10 um in diameter at their maximum width (in prophase) with a mean cell packing density of some 2 cells/um^2 as projected upon the ventricular surface (Takahashi et al. 1995). Across species, it appears that there are strong constraints placed upon the thickness of the epithelium with a maximum thickness that is only 1.5–2.0 times that found at the outset of neuronogenesis (Caviness et al. 1995).

This relatively small increase in thickness is sufficient only to accommodate a doubling in epithelial volume (Takahashi et al. 1996a). However, the actual volumetric increase during the neuronogenetic interval is orders of magnitude greater, even in small mammals (Takahashi et al. 1996a; Takahashi et al. 1997). Since expansion in the radial dimension (i.e., thickness of the PVE, Fig. 1A) is limited, the large volumetric increase produces considerable expansion in the tangential dimensions. This tangential expansion is exactly what is required to produce the large "sheet" that is characteristic of the adult cortex. The process of expansion is allowed, presumably by the fact that both adherens and gap junctions are breakable and apparently do break at M phase (Goodall and Maro 1986; Dermietzel and Spray 1993; Fulton 1995; Bittman et al. 1997). This allows the relative position of early postmitotic cells to shift readily before cells reestablish their junctional linkages in G1 phase (Fishell and Hatten 1991; Walsh and Cepko 1993; Bittman et al. 1997).

3
Neocortex as Outcome of Neuronogenesis in the PVE

Neuronogenesis is the regulated formation of neurons and is the first step in the transformation of the hemispheric wall which contains only proliferating cells into the mature cortex which contains only mature neurons. Thus, neuronogenesis initiates the histogenetic sequence which leads to the formation of the neocortex and sets the stage for the events of neocortical pattern formation to follow. The cellular processes which follow neuronogenesis and continue the histogenetic sequence contribute directly to pattern formation. These processes are neuronal migration, repositioning after migration, growth, differentiation and histogenetic cell death (Fig. 2) (Sidman and Rakic 1973; Sidman and Rakic 1982; Finlay and Slattery 1983; Finlay and Pallas 1989). The histogenetic sequence, beginning with neuronogenesis and continuing through the cellular events leading to pattern formation, occurs during an extended and overlapping period. Each leaves its indispensable stamp upon the full three dimensional structure of the neocortex.

The histogenetic sequence which determines the structure of the neocortex in its radial dimension operates in ways that also provides for regional diversity of structure across the two tangential dimensions of the neocortex (Fig. 3). Thus, the laminar structure of neocortex is a general feature expressed across its radial dimension but there are regional differences in the details of cytology and cell arrangement within laminae which correspond to the neocortical cytoarchitectonic map (Brodmann 1909; Caviness 1975; Zilles 1990). Those histogenetic mechanisms which determine the character of the radial dimension are those which specify cell number, cell class and the orderly arrangement of cells by class into neocortical laminae, and which, thereby, assure its elementary neural processing functions (Hubel and Wiesel

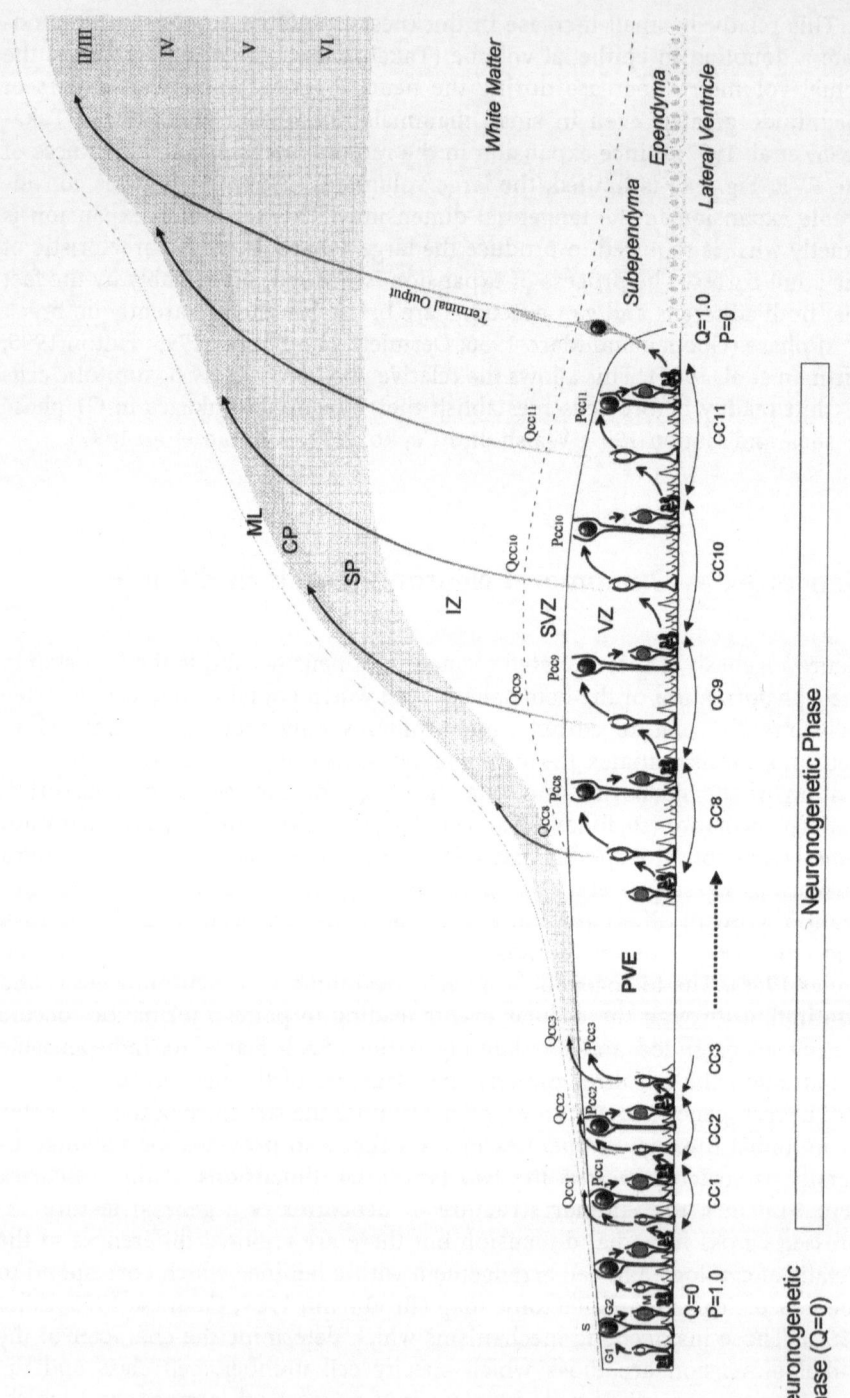

1965; Mountcastle 1978). The mechanisms which determine regional variation in this laminar structure of the cortex are those which lead to the production of regional architectonic variations and assure the assembly of architectonic units into distributed neural systems with regionally specific functions (Brodmann 1909; Rakic 1988; Felleman and Van Essen 1991).

3.1
The Radial Dimension of the Neocortex

The number of neurons in the cortex approaches 6 orders of magnitude in even the smallest mammalian species and varies upwards 3 further orders of magnitude across mammalian species of greatest size and behavioral complexity. This vast population of neurons is represented by two broad classes, neurons of long axon which are glutaminergic and excitatory and neurons of short axon which are GABAergic and inhibitory. (Cajal 1952, Rockel et al. 1980). The neurons of long axon, some 60%–75% of neocortical neurons depending upon species, are ordered by subclass into modular columnar processing streams which are aligned in the radial dimension of the cortex (Lorente de No 1938; Hubel and Wiesel 1965; Mountcastle 1978). These columnar arrangements of long axon neurons are arrayed in tangential register according to class and subclass, thereby forming the 6 laminae and their sublaminae which are the dominant architectonic feature of the neocortex (Fig. 3) (Brodman 1909; Krieg 1963; Caviness 1975; Zilles 1990). Thus, the polymorphic forms dominate layer VI, large pyramidal neurons layer V, granule cells, layer IV and medium and small pyramidal forms layers III and II, respectively. Layer I, at the surface, is principally a plexiform zone without dominant cellular forms. The long axon neurons appear to arise entirely from the PVE. The relatively smaller population of short axon neurons, perhaps

---

Fig. 2. The neocortical histogenetic sequence. Histogenesis of the neocortex is initiated with cell proliferation in the pseudostratified ventricular epithelium (PVE) which is approximately coextensive with the ventricular zone (VZ) at the margin of the lateral ventricle. The founder population proliferates exponentially. The onset of the neurogenetic phase corresponds to the initial cycle (CC_1) in which Q is greater than 0 and P correspondingly is less than 1.0. The cells of the PVE then execute a series of integer cycles (CC_1–CC_{11} in mouse) during which Q ascends to 1.0 (and P descends to 0.0). The PVE then becomes the cuboidal ependyma. Young postmitotic neurons of the Q fraction of each CC migrate across the intervening subventricular (SVZ) and intermediate zones (IZ) of the embryonic cerebral wall to attain the interface of cortical plate (CP) and marginal layer (ML) of the developing neocortex. The earliest formed are destined for the deepest level of the cortex while later formed cells will find positions at progressively more superficial levels. As postmigratory growth and differentiation proceed, somata of postmigratory cells become positioned progressively more deeply in the CP, and the earliest formed of these will inhabit the subplate (SP). In mouse, neuronal migration is completed over the first 2 days following birth. The CP at that stage will differentiate into cortical layers II/III and IV while the subplate at that stage will differentiate into layers V and VI. The IZ and SVZ become the white matter of the mature cerebrum

Radial Dimention

——▶ Cell Class Specification
within Regions

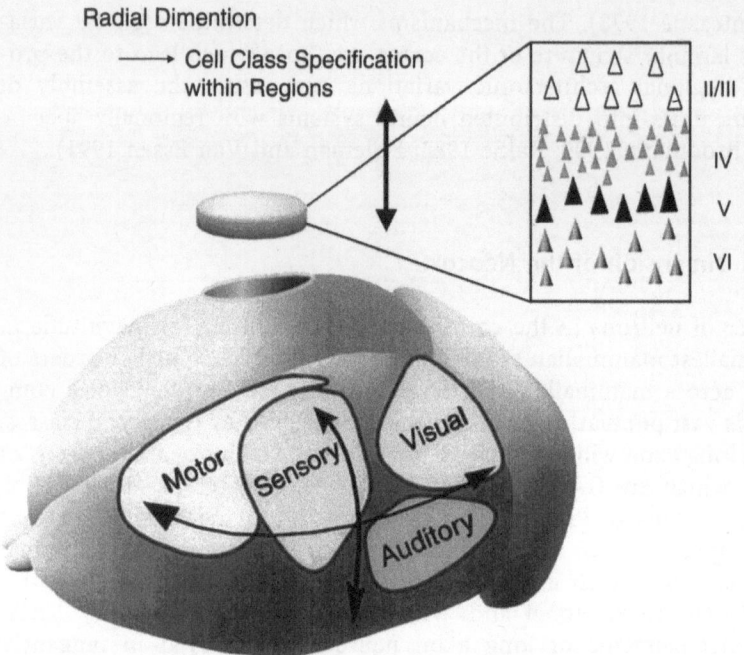

Tangential Dimention

——▶ Mapping of
Regional Specification

Fig. 3. Radial and tangential dimensions of the murine neocortex. The neocortex, in its radial dimension, has a general 6 layered architecture (above). The principal neuronal layers II/III–VI and their sublayers are each dominated by a layer-characteristic long axon neuronal class. Regional, or tangential, variations in the density, numbers and cytologic features of these basic neuronal forms correlate with regional mappings of specific neural systems representations: the motor representation frontally, the somatosensory representation parietally, the auditory representation temporally and the visual representation occipitally

25% of neocortical neurons in rodents, are distributed more diffusely through all layers of the neocortex but are somewhat more densely concentrated in its outer layers. They appear to arise only in part from the neocortical PVE with an uncertain and probably small complement arising from the ganglionic eminence of the basal forebrain (Anderson et al. 1997; Lavdas et al. 1998).

3.2
The Tangential Dimensions of the Neocortex

There are regional variations in the cytological character of neuronal classes and in the relative numbers and packing density across the vast tangential expanse of the cortex. These variations in radial structure serve as criteria for regional architectonic subdivision of the whole into a tangential mosaic or map of parcellation units (Brodmann 1909; Sanides 1972; Zilles 1990). There are approximately a dozen architectonic units in the smallest mammals and several dozens in larger species endowed with more elaborate or specialized behaviors and supported by complex patterns of neocortical information processing (Killackey et al. 1995; Krubitzer 1995; Northcutt and Kaas 1995). Although there is a substantial escalation in the total number of architectonic units with mammalian species of increasing size and behavioral complexity, the full assembly of units sorts into 4 sets of contiguous units with respect to neural systems and these sets share a common general topology of arrangement in the neocortex (Fig. 3) (Killackey et al. 1995; Krubitzer 1995; Northcutt and Kaas 1995). Thus, the set corresponding to executive/motor representations lie frontally (precentrally), somatosensory representations parietally (postcentrally), auditory temporally (laterally) and visual occipitally (posteriorly). Within each functional domain the multiple subsidiary mapping units are ordered hierarchically in terms of the exclusiveness of their affiliation with the primary modality on the one hand, their strength of shared association with 2 or more domains with those with increasing associative properties lying progressively near or at the interface of domain (Mesulam 1998). In addition, across species of ordered size and range of specialization of behavior there is a corresponding ordering of the size, number and complexity of interaction of the subsidiary mapping units and increasing associative interrelationships of those progressively near the periphery of each domain (Van Essen and Maunsell 1983; Felleman and Van Essen 1991; Mesulam 1998).

4
The Proliferative Process Within the Murine Neocortical PVE

The essence of neocortical neuronogenesis in mouse is that some 10^5 founder cells of the PVE give rise to approximately 10^7 neocortical neurons of multiple class. These are distributed in 6 cortical layers and their sublayers and partitioned across the tangential dimensions of the neocortex into nearly 2 dozen architectonic fields (Caviness 1975; Caviness et al. 1995; Takahashi et al. 1996a). The view of the workings of the proliferative process of the neocortical PVE which we present here arises from a series of analyses limited to the murine neocortical PVE. The perspective is constrained by the nature of the methods which were employed. Thus, the principal analyses were based upon population labeling with the S phase markers ^{3}H-thymidine (^{3}H-TdR) and/or bromodeoxyuridine (BUdR) (Takahashi et al. 1992; Takahashi et al. 1995; Takahashi et al. 1996a; Miyama et al. 1997). These methods allow a characterization of the kinetics of proliferation as mean values and variance of the mean for the entire population and are insensitive (and indeed "blind") to the behaviors of cells belonging to single lineages. A more limited set of studies, based upon retroviral insertion of the reporter *lacZ* gene, on the other hand, have allowed a glimpse of certain behaviors of restricted proliferative lineages (Cai et al. 1997a). These analyses have supported quantitatively rigorous, mathematical models and computer simulations of the proliferative behavior of the overall population of the PVE.

The investigations in mouse which form the core of this presentation draw heavily upon observations from now classical investigations into neocortical neuronogenesis which have been concordant in their principal findings across multiple species. These principles will be seen to form the cornerstones, not only of the quantitative model of neuron production formulated here for mouse, but also the theoretical basis which links the proliferative process to mechanisms of specification of cell class and region.

4.1
There Are Two Stages of Proliferative Activity in the PVE (Fig. 2)

Prior investigations have established that neuron production is initiated at a particular time and continues for a definable interval. During the period prior to the production of neurons, the cells of the PVE proliferate exponentially (Rakic 1995b). We refer to this initial stage of proliferative activity as the "preneuronogenetic phase" of proliferation. At the onset of neuron production, the second stage, which we have referred to as the "neuronogenetic interval" is initiated. At this time neurons arise from the PVE and initiate their migrations to the neocortex (Takahashi et al. 1996a; Caviness et al. 1997; Takahashi et al. 1999a). Over the course of the neuronogenetic interval the rate of neuronal production is NOT constant. Initially, the rate of neuron production is low, but it increases gradually to a peak and then declines rapidly as the neuronogenetic period ends (Hicks and D'Amato 1968; Rakic 1974; Bisconte and Marty 1975a; Bisconte and Marty 1975b; Luskin and Shatz 1985; Bayer and Altman 1991; Polleux et al. 1997). In addition, during the first part of the neuronogenetic interval the proliferative population increases, i.e., the PVE increases in volume and surface area. However, later in the neuronogenetic period the PVE regresses and eventually becomes the ependyma, non proliferative cuboidal epithelium (Fig. 2) (Takahashi et al. 1995). The disappearance of the PVE with this transformation corresponds to the end of the neuronogenetic process.

4.2
Neuron Production Advances in an Orderly Sequence

There is an approximate inside-out relationship between the sequence of neuron origin and eventual neuron position in the cortex (Fig. 2) (Hicks and

D'Amato 1968; Bisconte and Marty 1975a; McSherry 1984; Bayer and Altman 1991). The neurons of the deepest neocortical layer are the first to be formed while those of progressively more superficial layers are formed successively later. Thus, as the proliferative sequence advances in a region of the PVE the neurons arising in that region are destined for progressively more superficial layers.

4.3
The Proliferative State of PVE Varies Across the Surface of the Neocortex

In mouse and other rodents, a sampling of carnivores and primates, and therefore, presumably in all mammalian species, the neuronogenetic interval does not proceed in synchrony throughout the PVE. Specifically, at any given time during the neuronogenetic interval there is a regular pattern of progression of neuronogenesis across the PVE, referred to as the "transverse neuronogenetic gradient" or TNG (Fig. 1) (Hicks and D'Amato 1968; Rakic 1974; Bisconte and Marty 1975a; Rakic 1976; Rakic 1982; Smart and McSherry 1982; Smart and Smart 1982; McSherry 1984; Bayer and Altman 1991; Granger et al. 1995).

The TNG has its origin far rostrolaterally where the neocortical PVE is in continuity with the PVE of the basal ganglia which we will designate the origin of the TNG (* in Fig. 1) (Smart and McSherry 1982; Smart and Smart 1982; Bayer and Altman 1991). It is at this origin that the neuronogenetic interval of cortical histogenesis is initiated, that is, where postmitotic cells first exit the cell cycle. The neuronogenetic interval then propagates in a caudomedial direction so that at any given time the state of progression of the proliferative process is graded along the rostrolateral to caudomedial axis (major axis of the TNG) of the epithelium. According to this gradient regions most advanced with respect to proliferative sequence are positioned rostrolaterally and regions which are least advanced are positioned caudomedially. As a consequence of this gradient and the fact that layers are produced in an orderly sequence, the layers being produced rostrolaterally at any given moment during the neuronogenetic interval tend to be more superficial layers than those being produced caudomedially (Hicks and D'Amato 1968; Rakic 1974; Bisconte and Marty 1975a; Smart and McSherry 1982; Smart and Smart 1982; Bayer and Altman 1991).

4.4
The Cell Cycle in Histogenesis

The engine of the proliferative process is the cell cycle. Within the framework of organ histogenesis the cycle operates in two domains, the time domain and the output domain. By time domain, we refer to the time required for the cell to execute one complete cell cycle (cell cycle length or T_C) (Takahashi et al.

1999b). The significance of the time domain can be readily understood from the perspective of the number of cell cycles that can be executed in a particular interval. For example, if the cell cycle is 8 hours long, then 3 cell cycles can be executed per day; on the other hand, if the cell cycle is 16 hour long, then only 1 and 1/2 cell cycles can be executed per day.

Output domain refers to the fraction of postmitotic cells which exits the cycle, abbreviated here as "Q" for "quiescent" (Takahashi et al. 1996a) (Fig. 2). "P" an abbreviation for "proliferative" and the complement of Q, is the fraction of postmitotic cells which goes through G1 phase and reenters S phase. The neuronogenetic interval of histogenesis is initiated with the first cell cycle to give rise to postmitotic cells which leave the cycle to become young neurons. Before this initial cycle, the cells of the PVE proliferative exponentially, i.e., $Q = 0$ and $P = 1$. The neuronogenetic interval terminates in a region of the PVE with a cell cycle which gives rise to postmitotic neurons which all exit the cycle, i.e., when $Q = 1$ and $P = 0$. That is, in any given region of the PVE the interval of neuronogenesis corresponds to a series of cell cycles through which Q ascends from 0.0 to 1.0 (and P descends from 1.0 to 0.0). At the cell biological level the controls on Q and P are apparently active during the G1 phase of the cell cycle when the cell will "decide" whether to leave the cycle as part of the Q fraction or remain with the P fraction (Sherr 1993; Roberts et al. 1994; Sherr and Roberts 1995). Moreover, under physiological conditions of proliferation it is only during G1 that the proliferative cell is subject to modulation by cell external mechanisms (Murray and Hunt 1993; Sherr 1993; Roberts et al. 1994; Sherr and Roberts 1995; Caviness et al. 1999; Takahashi et al. 1999b).

4.5
A General Quantitative Model of Neuron Production

Neurons of the neocortex arise from a founder population through a discrete neuronogenetic interval, corresponding to a sequence of integer cell cycles (i.e., a neuron precursor cannot execute a partial cycle, see the section "*The Number of Integer Cycles*" for explanation) (Takahashi et al. 1995). With each passing cell cycle the size of the PVE will change as a function of P. The expansion of the PVE population will continue as long as P is greater than 0.5, but once P becomes less than 0.5, the proliferative PVE population becomes progressively smaller with each cell cycle (Takahashi et al. 1996a). The equation for calculation of the size of the PVE derived from a PVE founder cell population with unit size of 1 over the course of n cell cycles (PVE_n) is:

$$PVE_n = P_1 * \prod_{i=2}^{n} 2 * P_i \qquad (1)$$

where P_i is the P fraction at cell cycle i ($2 \leq i \leq 11$ in mouse) (Takahashi et al. 1996a).

With each passing cell cycle of the neuronogenetic interval neurons are produced as the PVE increases ($P > 0.5$) then decreases ($P < 0.5$) in size. The number of neurons formed during each cycle will be equal to twice the size of the PVE at that cell cycle (each cell will have two daughter cells) times Q or:

$$OUT_n = 2 * PVE_n * Q_n \tag{2}$$

where OUT_n is the output of the PVE at any given cell cycle n, PVE_n is the size of the PVE at that cell cycle (as given in equation 1) and Q_n is Q for that cell cycle.

Over the entire neuronogenetic interval the total neuron production (OUT_{TOTAL}) will correspond to the sum of neurons produced with each cycle or as derived from equation 2 above:

$$OUT_{TOTAL} = Q_1 + \sum_{n=2}^{TOTAL} 2 * PVE_{n-1} * Q_n \tag{3}$$

where TOTAL is the total number of cell cycles which constitute the neuronogenetic interval in a given spices Q reaches 1.0 at the completion of the final cycle; that is, the entire set of cells corresponding to $2 * PVE$ of the final cycle will exit the VZ as the terminal output of the proliferative process (Fig. 2).

The parameters that must be determined experimentally essential to satisfy this model for any given founder population are (1) the number of integer cycles which constitute the neuronogenetic interval and (2) the value of Q and P for each cycle. The experimental determination of these parameters, described in the following paragraphs, has been undertaken in the mouse.

4.6
Parameters of the Model: Experiments in Mouse

4.6.1
The Number of Integer Cycles

The number of cell cycles is calculated from T_C, as sampled experimentally across the neuronogenetic interval in a given region of the PVE (Takahashi et al. 1995). We undertook these experiments in mouse using cumulative exposure to BUdR. Proliferative cells of the PVE, continuously exposed to BUdR will incorporate this marker into DNA as they traverse S phase. Over time the proportion of cells labeled with the marker (LI: labeling index) increases until all proliferating cells are labeled. Saturation labeling occurs in an interval corresponding to T_C-T_S at an LI which corresponds to the growth fraction (GF: proportion of cells which are cycling). It is a critical property of the PVE that cells of the epithelium execute the cell cycle asynchronously (Takahashi et al. 1992; Takahashi et al. 1995). As a consequence, the proportion of cells of the PVE that are in any particular phase of the cell cycle (i.e., G1, S, G2 and M phases; the duration of each abbreviated as T_{G1}, T_S and T_{G2+M}, respectively) is equal to the duration of that phase as a fraction of T_C (Nowakowski et al. 1989). The Y-intercept and the slope of ascent of the LI correspond to $T_S/T_C * GF$, and GF/T_C, respectively. The interval required for all mitotic figures at the ventricular margin to become labeled corresponds to T_{G2+M}. Thus, the method of cumulative labeling with BUdR provides values for all parameters requisite to estimation of the Tc as well as the duration of the cell cycle phases and GF.

Such experimental determinations in the mouse embryo were executed initially in a dorsomedial region of the PVE, that is, a location which is relatively "down" the TNG. These experiments established that the GF in the dorsomedial region is essentially 1.0 throughout the neuronogenetic interval which extends from early E11–early E17. T_C in that region increases from just over 8 hr to approximately 20 hr (Fig. 4). There is no systematic change in T_{G2+M} or T_S. That is the doubling of T_C reflects entirely a near quadrupling of T_{G1}. In other words G1 phase is the only phase of the cycle whose duration is regulated in the course of murine neocortical neuronogenesis (Table 1).

The pattern of progression of T_C allows the calculation (Takahashi et al. 1992; Takahashi et al. 1995) that the neuronogenetic interval in mouse is constituted of 11 integer cycles (Fig. 2; CC_1 through CC_{11}). That is, the founder population of the dorsomedial murine neocortical PVE and its

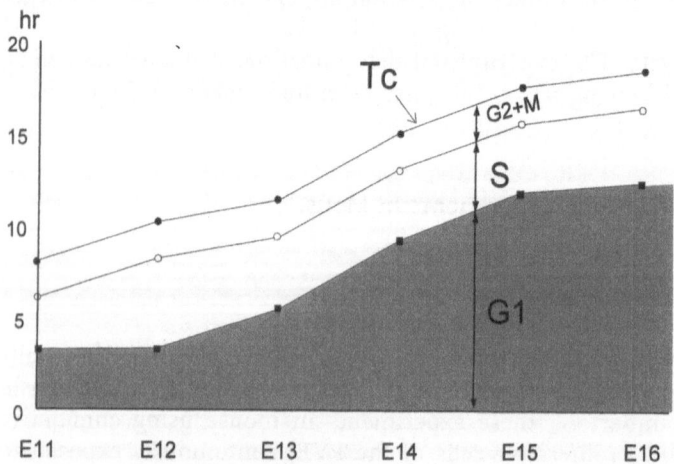

Fig. 4. Durations of cell cycle phases of the murine dorsomedial neocortical PVE. The neuronogenetic interval of the dorsomedial PVE continues 6 days independently of the region. In the dorsomedial region, for which the cell cycle parameters are presented in this figure, the neurogenetic interval is from embryonic day 11 into early embryonic day 17 (E11–E16). Over this interval the duration of the cell cycle (T_C, filled circles) increases from approximately 8 to 18 hr. The advance in T_C is attributable entirely to a near quadrupling of the duration of the G1 phase (T_{G1}, filled squares). There is no systematic variation in either the duration of S phase (T_S, area between filled squares and open circles) or the combined durations of the G2 and M phases (T_{G2+M}, area between filled and open circles)

Table 1. Summary of proliferative parameters of the murine dorsomedial PVE

Neuronogenetic interval	6 days	
T_C	Increases from ~8 to ~20 hr	Local Variations 5-7%*
T_{G2+M} and T_S	Invariant at 2 and approximately 6 hr	
Q	Ascends from 0.0 before CC_1 to 1.0 with CC_{11}	
Multiplier power of proliferative process	140 cells per founder cell	

* (Cai et al. 1997b).

progeny execute a total of 11 cell cycles over a 6 day period. Each cycle has a unique value of T_{G1} where the most rapid change in T_{G1} occurs over the E13–E14 interval. Virtually identical findings were obtained when the experiment was repeated in the lateral region of the PVE, i.e., in a location that is relatively "up" the TNG (Miyama et al. 1997). That is, in the lateral cortex there are also 11 cell cycles and the patterns of changes in T_C and T_{G1} are identical to dorsomedial cortex except that they are shifted in time by approximately 24 hours (Fig. 5A). Moreover, measurements with the percent labeled mitosis method indicate that the variation in the cell cycle length is small, i.e., about +/−5-7% in any small area of the PVE at any time (Cai et al. 1997b).

4.6.2
The Q and P Fractions

We have experimentally determined Q and P daily over the 6 day neuronogenetic interval of the dorsomedial neocortical PVE in mouse and then, by interpolation, reconstructed both Q and P values for the full set of 11 integer cycles (Takahashi et al. 1996a; Miyama et al. 1997) (Fig. 5B). The experimental basis for this determination has been a double labeling technique using two S-phase markers: an initial exposure to ^{3}H-TdR followed 2 hr later by exposure to BUdR to mark a "2 hr cohort" as ^{3}H-TdR only labeled cells. The initial labeling sequence is followed by continued exposure to BUdR for one set of animals (set 1) but no further exposure to the tracer for a second set (set 2). At an interval corresponding to T_C-T_S the number of cells labeled only with ^{3}H-TdR in set 1 corresponds to the number of cells of the Q fraction (N_Q) for the 2 hr cohort while that of the set 2 corresponds to the number of cells of the combined Q + P fractions (N_{Q+P}). Q is calculated as N_Q/N_{Q+P} and P as $1 - Q$.

Again the progression of Q and P in both dorsomedial and lateral regions are identical when plotted as a function of the number of elapsed cell cycles (Takahashi et al. 1996a; Miyama et al. 1997) (Fig. 5B). The increment in Q per cycle is low at first and the "pivot point" of the system where the number of postmitotic cells leaving the cycle is the same as the number returning to S phase (P = Q = 0.5) is achieved only during the 8th cycle, that is, only at a point nearly 75% of the way through the full set of 11 cycles. The continued

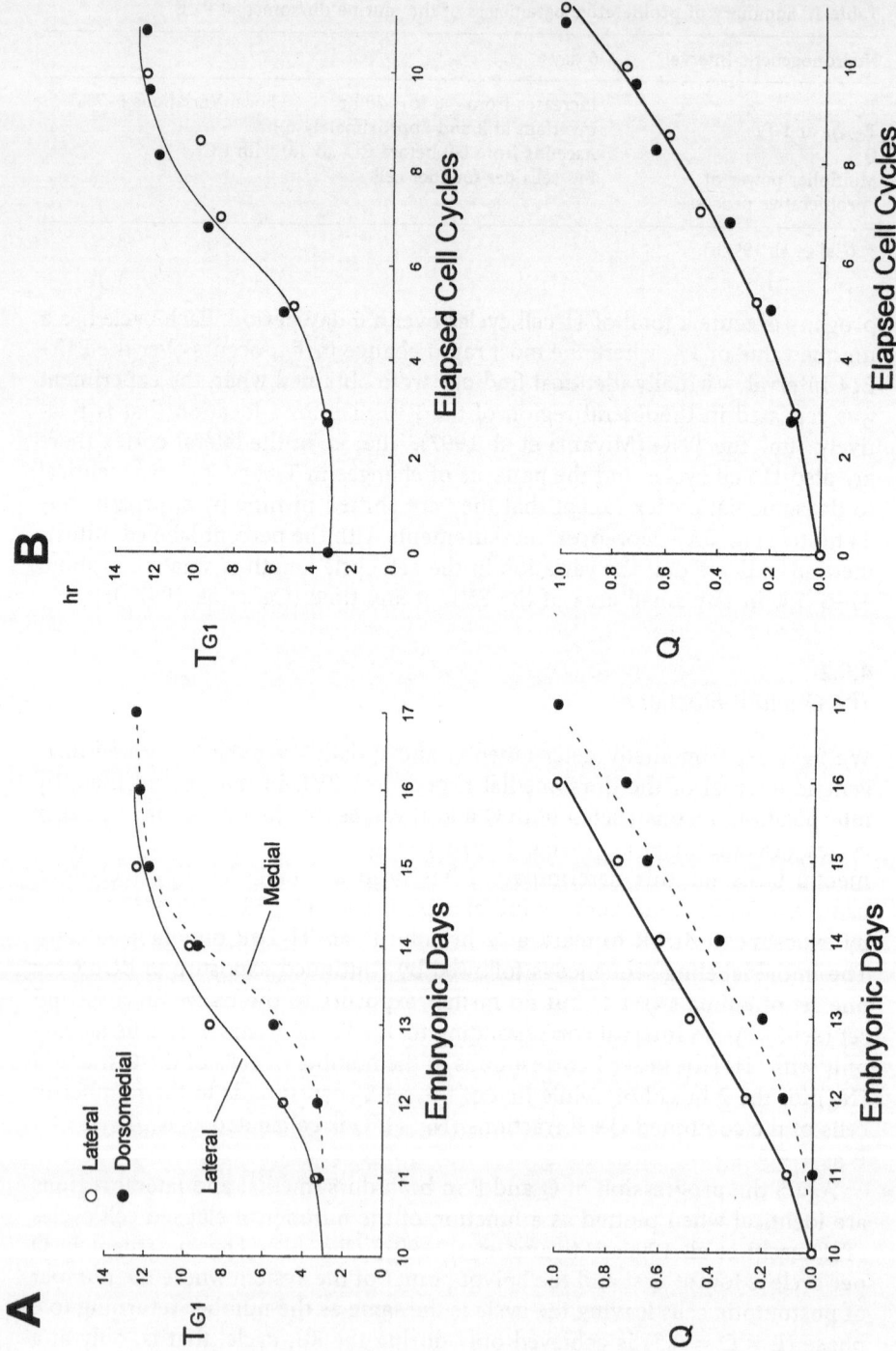

advance of the two parameters, Q to 1.0 and P to 0.0, is much more rapid over the terminal 3 cycles.

4.6.3
Neuron Production Model

The neuron production model, as it operates over the neuronogenetic interval, is expressed in equations (2) and (3), above. The output of each cell cycle is expressed as a function of Q and the size of the PVE for that cycle and the total output is the sum of output of the succession of individual cycles (Fig. 6). Consider, by illustration, the per cycle output of a single small area of PVE that occupies 1 unit of volume (e.g., height, length and width are all equal). Because the PVE increases approximately 50 fold in volume over the initial 8 cycles and only doubles in height, the PVE will increase approximately 25 times in area (5 times in each of its two tangential dimensions) between the beginning of the neuronogenetic interval (i.e., CC_1) and CC_8. As expected from the slow initial rise of Q, the initial rate of neuron production per cycle is low but it rises steadily to a peak around cell cycle 8. Afterwards the rate of cell production drops even though Q continues to increase because there is a marked reduction in the size of the PVE as Q is now greater than 0.5. With respect to the founder cell, i.e., a proliferating cell present at the outset of the 11 cell cycle neuronogenetic period, the average total cumulative output is approximately 140 cells. Of these, approximately 30% are formed through the first 7 cycles but 70% are formed at cell cycle 8 and after wards, i.e., after Q has become greater than 0.5.

These extrapolations from the experimental analysis of proliferative behavior may be compared to actual measurements based upon methods entirely different from those that led to the formulation of the present model. For example, the total volume and founder cell number of the murine PVE at the outset of CC_1, estimated from this model (maximum estimates of 0.5 mm^2 and 2.5×10^5 respectively) (Caviness et al. 1995), fall within a 2- to 4-fold approximation of measurements made by direct counting ((Rhee et al. 1998; Vaccarino et al. 1999); Vaccarino, personal communication). Further, if ES cells carrying the reporter gene, lacZ, are injected into the blastocyst of mouse, clones of labeled cells arising from PVE founder cells carrying the gene are distributed as a mosaic in the cortex of the adult mouse (Tan et al.

Fig. 5A,B. Proliferative parameters in dorsomedial and lateral neocortical PVE, that is, regions which are widely separated along the TNG. A The increase in the length of G1 (top) and in Q (bottom) during the progression of the neuronogenetic interval (abscissa) is given in terms of embryonic date. B The increase in the length of G1 (top) and in Q (bottom) during the progression of the neuronogenetic interval (abscissa) is given in terms of integer cell cycle number. Note that when expressed as a function of cell cycle number the ascent of the duration of the G1 phase (top) and of Q (bottom) is identical in the lateral and dorsomedial PVE

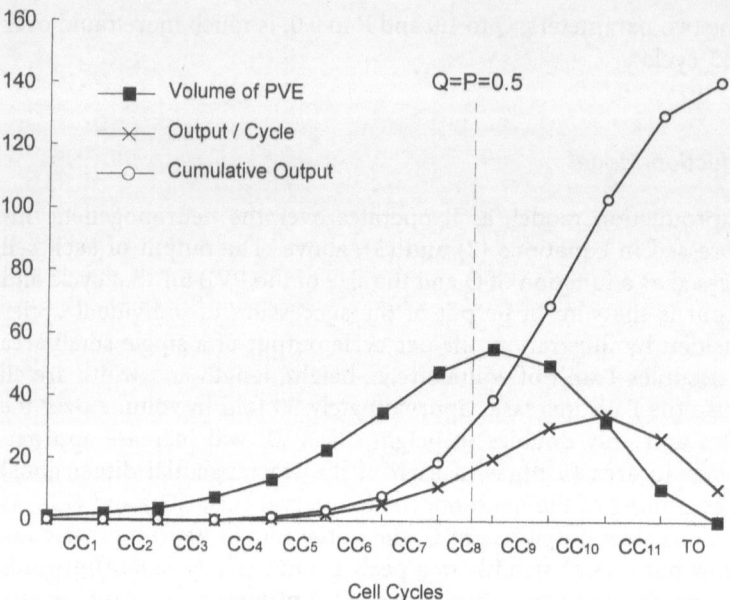

Fig. 6. Quantitative neuronogenesis expressed as function of cell cycle. Volume of the PVE, neuronal output per cycle and cumulative neuronal output are all expressed as a proportion of an original "unit" of PVE. Such a unit would expand to over 50 fold its original volume and then contract to zero. The cumulative output over the course of 11 cycles would be about 140 neurons for each founder cell in the original unit

1998). The number of cells in such clones has been determined to be consistent with a founder cell multiple of 150, matching closely the prediction from the present neuron output model. On both accounts, the accord between direct experimental observation and the predictions of the model is quite good, allowing for the unpredictable effects of a variety of experimental variables, possible strain differences in proliferative behavior and histogenetic cell death.

Estimates of the proportion of cell death based upon either the TUNEL method or counts of pycnotic nuclei, two methods which in many different laboratories have yielded estimates considered to be reliable and valid, suggest that the amount of cell death in the PVE is low, less than 1% per cell cycle at most (Thomaidou et al. 1997). Because these estimates are so low and also well within the standard error of the methods we have used to determine the proliferative parameters, we have not factored cell death into this output model.

In concluding this section, we draw attention to the fact that the essential parameters of the neuron production model for a founder population of defined size, are the number of cycles and the value of Q for each of the cycles over the neuronogenetic interval. The model is viewed as a general one, applicable across species with the values for each parameter characteristic of species. Note that those operative parameters are factors without time dimension, although as it has been seen, it is necessary to determine the temporal parameters of the proliferative process (i.e., Tc) in order to establish the number of integer cycles that constitute the neuronogenetic interval (Caviness et al. 1995). That is, with respect to cell production, time in *sensu strictu* is incidental to species and does not provide a parameter for the general regulation of output. This is not to suggest that operations which occur in time are not critical to the histogenetic process. On the contrary, it is probable that essential histogenetic mechanisms to which we will return later, unique in valence to species, are critically dependent upon the time dimension.

5
Higher Order Neuronogenetic Control

We have earlier set forth a view of the two cardinal operational properties of neuron production by the neocortical PVE. To recapitulate, the first of these is that the neuronogenetic interval of operation in any region of the PVE is initiated by the transition of Q from 0 to a value >0 and is defined as the interval required for progression of Q = 0 to Q = 1. Secondly, the ordering of neuronogenetic events does not occur in synchrony across the PVE; on the contrary, it is initiated at the rostrolateral margin of the epithelium and from there propagates caudomedially, corresponding to the major tangential axis of the epithelium, such that these ordered events are at any time progressively delayed with progression along this axis (Fig. 1).

The model of neuronal output, formulated in prior sections and based upon experiments in mouse, relates for any region of the PVE the variables critical to regulation of the output of neurons to the number of cell cycles in the neuronogenetic interval and the advance of Q per cell cycle. This is a formulation that we view as generalizable across mammalian species. For each species and for each point in the PVE there is a uniform enactment of the species characteristic sequence of cycle with advance of Q with each cycle. For each species, moreover, there is a characteristic rate and pattern of propagation of the neuronogenetic sequence across the PVE after its initiation at the most "upstream" portion of the TNG. This pattern of propagation depends not upon Q but upon the pattern of advance of Tc (actually determined by the regulated advance of T_{G1}). That is, the parameters which govern the general output functions of the PVE also serve at a higher regulatory level to determine for each species the overall cycle sequence and its pattern of propagation across the PVE. In the present section, we formulate what we propose to be a general algorithm for this higher order regulatory control as based upon the elemental proliferative parameters.

5.1
Number of Cell Cycles Regulated by Q

Of the two regulated parameters, i.e., T_{G1} and Q, it is Q whose regulation governs the operation of the overall proliferative process. As the elementary output model has been formulated, Q determines the number of cycles that will constitute the neuronogenetic interval in any given area of the PVE (Takahashi et al. 1997; Takahashi et al. 1999b). The path of advance of Q over the neuronogenetic interval continues from 0.0–1.0, corresponding to the mathematical definition of the neuronogenetic interval. In that the path is constrained to this range, it becomes obvious that the number of cycles which will constitute the neuronogenetic interval will be determined by the proportion of this path that is traversed by Q with each cycle. This relationship also governs the growth of and total output from a PVE founder population of defined number size (the founder multiplier) (Fig. 6). In mouse Q at cell cycle number CC is expressed as

$$Q = k * CC^{1.97} \tag{4}$$

where k is a constant corresponding to the proportionate advance of Q with each cycle. Independently of region of PVE, k is approximately .009 in mouse. This gives a total of 11 cell cycles through the neuronogenetic interval and the founder multiplier of 140. Over the species range mouse, rat, cat and monkey the duration of the neuronogenetic interval varies from 6 to 60 days, that is, a full order of magnitude. If progression of neuronogenesis in each of these species is expressed as percentile of elapsed neuronogenetic interval, it is observed that corresponding populations of neocortical neurons are formed during closely corresponding percentiles of the neuronogenetic interval in each species. This suggests that there is an essential commonality in the fundamental workings of the proliferative regulatory process across these species. To the extent that this is the case, values of k may also be estimated for other species where data sets relating to cell cycle lengths are available and the number of cell cycles in the neuronogenetic interval can be estimated (Takahashi et al. 1999b), i.e., for monkey (Kornack and Rakic 1998) and rat (Waechter and Jaensch 1972) at least to provide a species specific estimate of the change in Q as a function of CC.

5.2
Propagation of the Neuronogenetic Sequence Regulated by T_C

The nature of the molecular mechanism which effects the switch from Q = 0 to Q > 0 and the subsequent rise to Q = 1 is obscure. Suffice it to say at this point that one might imagine two plausible general mechanisms: (1) a process of induction of competence of proliferative cells to leave the cycle as Q fraction, initiated at the origin of the TNG in the rostrolateral neocortical

PVE and propagating from there or (2) a cell autonomous standing spatio-temporal gradient of sequence without dependence upon a propagated induction signal. Whichever mechanism is apt, once competence to advance in Q does arrive in a region, marking the initiation of neuronogenesis in that region, the overall proliferative process proceeds with predictable spatio-temporal precision exercised in the same sequence and with the same progression throughout the full extent of the PVE. We propose here a mechanism for this neuronogenetic sequence which is driven by the pattern of advance of T_C with cell cycle.

5.3
Propagation of Cell Cycle Domains

Cell cycle number is the unit of sequence underlying the events occurring during the neuronogenetic interval of cerebral histogenesis. As the progeny arising within each lineage advance from cycle to cycle, there is a step increase in T_{G1} and Q relative to their respective values in the prior cycle. In addition, because the proliferative process is initiated in and propagates from a rostrolateral origin in the neocortical PVE, the state of the proliferative process as expressed across the TNG shifts dynamically as neuronogenesis proceeds. This dynamism involves progressive shifts of cell cycle domains across the PVE (to be defined below, Fig. 7) and associated rapid tangential growth of the PVE.

That the PVE is a pseudostratified proliferative epithelium confers an additional complexity to this dynamic process in that the cells of the epithelium proliferate asynchronously. Thus, in all regions of the PVE there will be cells in all phases of the cell cycle (Sauer 1936; Sauer and Walker 1959; Takahashi et al. 1995). For reasons to become apparent in the discussion to follow, there is the additional complexity that at any given time and place in the PVE two cell cycles of the 11 cycle sequence are represented. Here we describe the elementary character of this dynamism from the perspective of what happens as cell cycle sequence advances in a single locale, in this instance the origin of the TNG, and also from the perspective of what happens with the propagation of cell cycle sequence along the major axis of the epithelium.

5.3.1
Initiation of Cycle at Origin

The neuronogenetic sequence is initiated at the origin of the TNG in the rostrolateral PVE (Figs. 1 and 7). The first cell cycle is designated CC_1, which is preceded by CC_0 and succeeded by CC_2. A small fraction of the cells that enter G1 of CC_1 become the first neocortical Q cells in that neocortical location sometime later during G1. The proportion of cells entering G1 of CC_1

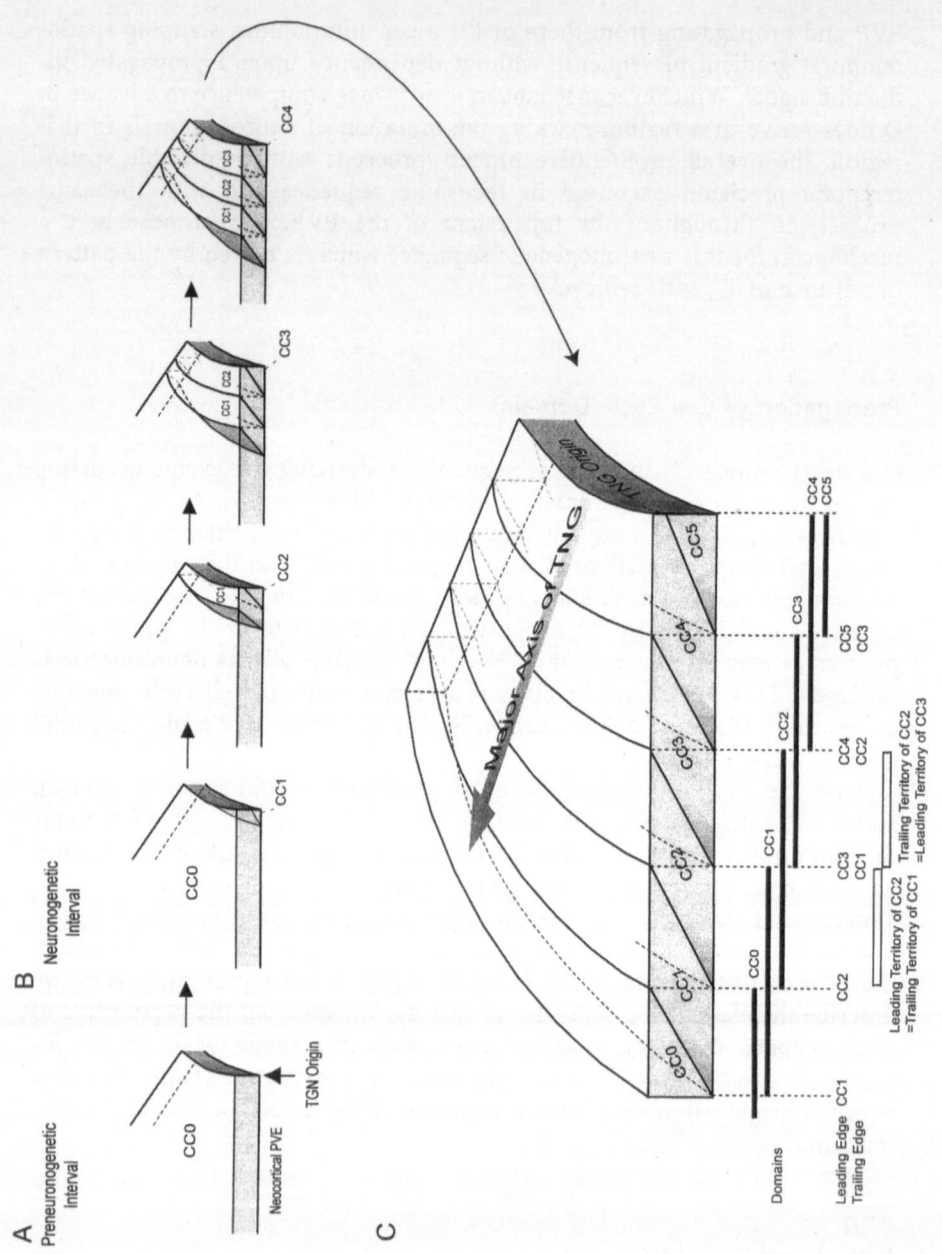

per hour (by exiting CC_0) is $1/T_{C\_CC0}$, where T_{C_CC0} is T_C for CC_0. Thus, after 1 hour there will be $1/T_{C\_CC0}$ cells in CC_1 in the PVE, after 2 hours there are $2/T_{C\_CC0}$, and so on with the rest of the cells still in CC_0. The first cells entering CC_1 are in the G1 phase of that cycle but with time all phases of CC_1 are represented. This series of events is repeated as CC_2 begins, i.e., after one hour $1/T_{C\_CC1}$ cells enter CC_2, etc. Note that at all times cells from two consecutive cell cycles are present, e.g., early the population consists of a mixture of CC_0 and CC_1 cells, then a mixture of CC_1 and CC_2 cells, etc. This mixture changes with the cells representing a single cell cycle being present for an extended period of time. For example, cells in CC_1 are continuously represented for a total time corresponding to T_{C_CC0} (time required for all progeny of CC_0 to exit that cycle and enter CC_1) + T_{C_CC1} (time required for the last progeny of CC_1 to exit that cycle and enter CC_2).

5.3.2
Propagation of Cycle Domains

The wave of initiation of CC_1 spreads beyond the origin down the major axis of the TNG (Fig. 7). We refer to this wave front that marks the first site of appearance of cells in a particular CC as the "leading edge" (LE). The leading edge (LE) of propagation of each successive CC will advance at a rate proportional to $1/Tc$ for the previous cell cycle. That is, the shorter the duration of a given cycle, the more rapid will be the pace of propagation of the LE of its successor (Fig. 7). Because T_C more than doubles in the course of the 11 cycle

---

Fig. 7A–C. Propagation of the neuronogenetic sequence along the major axis of the TNG. The abscissa, representing the tangential dimension of the PVE, corresponds to the projection of cycle domains along the TNG. A The preneuronogenetic interval (designated as CC_0). All cells of the neocortocal PVE belong to the P fraction. B The neuronogenetic interval is initiated when CC_1 emerges at the origin of the TNG (the first panel in B). The rest of the PVE is in the preneuronogenetic phase. As neuronogenesis proceeds (from left to right in B, horizontal arrows, continuing to panel C), cell cycle domains propagate across the expanse of the PVE. The propagation continues until CC_{11} sweeps through the whole expanse of the PVE, that is, from initiation through completion of the neuronogenetic interval. In this set of figures, domains only up to that of CC_5 are shown, an interval in excess of 40 hr after neuronogenesis is initiated in mouse. C The distribution of cell cycle domains across the murine neocortical PVE approximately 40 hr after the onset of neuronongenesis. Cycle domains are distributed in descending order, CC_{4-1}, from origin along the major axis of the transverse neuronogenetic gradient (TNG). The leading edge (LE) of a given domain corresponds to the trailing edge (TE) of the cycle that is 2 cycles ahead. Note that the leading half or "territory" of a domain of cells in a given cycle overlaps the trailing "territory" of the domain of the prior cycle (those for CC_2 are shown as open horizontal bars at the bottom). Thus, in this diagram each domain has the form of a parallelogram, reflecting the idea that at all positions throughout the expanse of the PVE, cells will be mixed systematically with respect to CC. The size of a given domain, corresponding to the width of each parallelogram, is the distance between the LE and TE (shown with horizontal thick lines for CC_1–CC_5 at the bottom of the figure), and the height is proportional to the relative prevalence of the cells of that CC

series, the proportionate rate progressively declines from cycle to cycle, eventually halving. The distance between each LE therefore lengthens.

This is analogous to the time dependent sequence of events occurring at origin described above except that these are changes which propagate through space with the passage of time (Fig. 7). Thus, at all locations cells from two consecutive cell cycles are present, and the LE for CC_N is also the "trailing edge" (TE) for CC_{N-2}. For example, early in neuronogenetic interval from caudomedial to rostrolateral across the surface of the PVE, i.e., moving up the TNG, successive domains will be encountered where the population consists of a CC_0 and CC_1 cells, then CC_1 and CC_2 cells, and so on (Fig. 7B). In other words, the mixture changes geographically with the cells representing a single cell cycle being present over an extended distance, i.e., the distance between the LE and the TE for CC_N. We refer to this region of the PVE included between LE and TE for CC_N as the domain of CC_N. Note that each CC domain contains a "leading territory" in which the proportions of cells of CC_N are rising and CC_{N-1} are falling and a "trailing territory" in which the proportions of cells of CC_N are falling and CC_{N+1} are rising. Moreover, each CC domain overlaps with both the preceding and the succeeding CC domain (Fig. 7B).

The presence of this wave-like sequence of progression of CC domains through the PVE is that the PVE is at any given time an overlapping mosaic pattern of cell cycle domains where the borders of each domain are continuously shifting. The number of domains active across the PVE at any given time has not been determined. The "distance" along the TNG between the lateral and dorsomedial PVE at the midcerebral coronal plane has been determined to be approximately 24 hr (Miyama et al. 1997) so that the full extent of the TNG must be longer than 24 hr. T_C's for the earliest cell cycles are around 8 hr. We estimate that in the course of 40 hrs after CC_1 is initiated at origin, the LE of CC_1 will have reached the end of the TNG at the caudomedial extreme of the PVE. At that time, that is, after a lapse of 40 hr, the overall PVE should map domains of CC_1–$CC_{4.3}$. That is, the domain of CC_5 should be approximately 30% complete at origin. Since this is when the T_C is shortest, this is the maximum number of domains that will be present at any one time in the course of the neuronogenetic interval. Since even at the earliest stage of the neuronogenetic interval T_C of successive cell cycles lengthens, successive cycle domains will occupy a proportionally larger span of the PVE so that the total number of mapped domains at any one time will progressively decline. With advance of the series of cycles toward CC_{11} and as T_C approaches 18 hr at the end of the neuronogenetic interval, approximately 2.2 cell cycles and thus only about 1.2 CC domains will be present. This corresponds to a reduction in the steepness of the gradients in T_C and T_{G1} present across the TNG (Miyama et al. 1997).

Accompanying this mosaic pattern is a distribution of the cell cycle differences such that at any given time the values of T_{G1} (and therefore T_C) and Q are distributed as a gradient along the principal axis of the TNG. In addition, the values of these parameters will advance as each successive cycle domain sweeps through a given region. An additional complexity arising from the wave-like progression of CC domains through the PVE is that during the first 7 cell cycles of the neuronogenetic interval when P > 0.5 the PVE is expanding (Fig. 6). Overall the PVE will enlarge some 25-fold in surface area, that is, about 5-fold in both axes of the tangential dimensions.

6
The Proliferative Process and Histogenetic Specification

The output model equations (1–3) and the general regulatory equation (4), define the proliferative parameters which govern the total number of neurons to arise from the founder population and the rate at which they arise with respect to cell cycle (Fig. 6). The parameters which drive output also regulate the overall operation of the PVE in terms of number of cycles and the program of propagation of CC domains across the epithelium. The cells which issue from the PVE are specified with respect to their class fates and within certain limits to their destinies with respect to architectonic field. These destinies are to some extent specified within the individual cell concurrently with its proliferative activity. It is thus likely that the mechanisms which govern these fundamental steps in histogenetic specification are linked to those which govern the proliferative process. We consider here, lines of evidence that begin to suggest the nature of such linkage.

6.1
Cell Number, Cell Class and Laminar Fate

There is a systematic inside-out relationship between sequences of neuron origin and laminar fate: The cells of the deepest layers are the first to be formed and those of the superficial layers are the last (Fig. 2). Is sequence of origin the determinant of laminar fate and by implication of cell class? Classic birthdating studies, based upon pulse labeling with ^{3}H-TdR, have indicated that cells arising concurrently actually have broad distributions through multiple cortical laminae and their sublaminae and that the inside-out relationship is only approximate (Hicks and D'Amato 1968; Rakic 1974; Bisconte and Marty 1975a; Bisconte and Marty 1975b; Luskin and Shatz 1985; Bayer and Altman 1991; Polleux et al. 1997). We have confirmed and extended this general finding in mouse by means of a double labeling method which specifies the cell cycle of origin and literally even the birth hour (born during any 2 hour period) within the cycle (Takahashi et al. 1996b; Takahashi et al. 1990a). The layers and sublayers of layers VI and V arise principally with cycles 1–7, those of layer IV principally from cycle 8 and those of layers II/III principally from cycles 9–11. This relationship is independent of whether the analysis is conducted in up gradient or down gradient locations with respect to the major axis of the PVE. However, the neurons born during any two hour period are distributed in a Gaussian fashion over a range of 20–50% of the full thickness of the neocortex.

These experiments indicate that sequence of neuron origin does not specify uniquely the class destiny and thereby, the laminar fate of a neuron, although the inside-out rule is regular and robust to specify uniquely the range of laminar positions and range of neuronal classes. Even with the aforementioned methodological refinement in the precision of specification of the sequence with which cells arise from the PVE, it is apparent that cells arising from the same cycle in the same region of the PVE include cells of multiple class, destined for multiple layers or sublayers. These findings, therefore, affirm the classic observation that sequence of origin alone is not sufficient to determine laminar fate or cell class. Instead, laminar fate of neurons is keyed to class and appears to be assured by mechanisms operating within the cortex after the completion of neuronal migration.

Although the set of neurons of *common origin*, that is arising simultaneously from the same region of the PVE, are heterogeneous with respect to class, these class fate options are constrained in two specific ways by sequence of origin. First of all a zone of distribution for the set of neurons of common origin is continuous in the radial dimension, comprehending all class options within that zone. Secondly the zone of distribution shifts systematically in an inside-outside fashion in the course of neuronogenesis. It is as though a narrow cell-class "band width filter" ascends a specified sequence of class options with progression of neuronogenesis. The width of the filter is always wider than the band corresponding to a single class but the multiple classes which are admitted must be classes that are within the sequence continuum.

Thus cell class sequences referenced to layers and arising momentarily but in 1, 2, 3 sequence from a region of the PVE of the following sort might be allowable:

1. VI-VI, V-V, IV
2. V, IV-IV-IV, II/III
3. VI, V, IV, II/III-II/III

Sequence referenced to layer such as the following would not be allowable:

1. VI-VI-VI, II/III, V
2. IV-IV-IV, II/III, VI

We suggest that the simplest lineage succession model consistent with these data and with little constraint and relatively few "programming resources" would be that all "long axon neuronal" lineages are equivalent in terms of class output potential. Those informational mechanisms specifying the range of class fate that are options at any moment in a region of the PVE would be close to threshold and the specific class assignment associated with a given mitosis might be determined stochastically. The critical constraint within this model would be that the range of options for a given region at a given moment would be closely specified, and that range of options would shift along the specified class sequence as neuronogenesis proceeds.

This model is, essence, a model of stochastic class determination within a homogeneous precursor lineage. We favor this model for its simplicity. We envision that there is a "transcriptional state" which is dependent on Q and/or T_{G1}. Thus, "transcriptional state" will advance as Q and/or T_{G1} advances systematically in the course of successive cell cycles. We suggest that the laminar fate of a cell which selects the Q fate is specified by the "transcriptional state" of the cell, and furthermore that the specification of the laminar fate is not absolute but merely a "predisposition" to a laminar destiny. Thus, the cells that leave the PVE at each cell cycle are in a particular "transcriptional state" that defines a window (or range) of class potentialities. For example, in the course of cycles 1–6 the window would be open to classes native to layer VI but closed to those of overlying layers. In the course of cycles 9–11 the transcriptional window would be open to classes native to layers II/III but closed to those of deeper layers. Within the class range viewed by the open window at a given cycle, the probability of class might have its unique threshold of expression among cells of the Q fraction for that cycle. Thus, the "transcription state" as specified by Q and/or Tg1 might provide a "memory" to each Q cell that directs the cell to its lamina through mechanisms that act within the cortex after the cell has completed its migrations. The donor/host transplant experiments of McConnell (McConnell 1989; McConnell and Kaznowski 1991) provide evidence that such a "memory" might exist and that it is erased upon re-entry into the cell cycle.

An alternative nonstochastic model of class specification would require that the PVE is heterogeneous with respect to long axon neuronal class lineages, and that, in effect, proliferative cells are "reserved" for specific laminae. The proliferative cells of such lineages dedicated to class at a given cycle would have to give rise to a specified number of cells after each division. That is, patterns of output with respect to the sequence of 11 cycles would need be severely constrained and rigidly controlled across the population. For example Q's for some fraction of the PVE cells would need to shift abruptly from 0 to very high values and then to 1.0 with lineage extinction in the course of a very few cycles. Retroviral experiments indicate such reserved populations do not exist and that the stochastic process is exactly what appears to operate with respect to the means by which postmitotic cells become assigned to the Q fraction (Cai et al. 1997a). Where lineages have been marked in this way by a reporter gene it is evident statistically that each daughter cell makes its Q vs P choice independently of its sister and that the distributions of choices made by large sets of lineages is best satisfied by a stochastic process of choice.

We emphasize that the probability of assignment to the Q fraction is not constant over the course of multiple divisions. Rather as we have seen earlier, it advances with each successive cycle. That is, mechanisms of broad effect act progressively with cycle to advance the probability that a postmitotic cells will be assigned to Q. The critical point is that this increase in Q acts independently upon each postmitotic cell. The stochastic model for assignment of cell to Q is readily modified to a stochastic model for assignment to cell class fate within a shifting window of fate options. That is, class fate option assignment would need operate only for the Q fraction complement and would operate stochastically within the range of options defined by the "transcriptional state" of the given cell cycle. Again, the donor/host transplant experiments of McConnell (McConnell 1989; McConnell and Kaznowski 1991) argue that the postmitotic cell "elects" a narrow range of class fates appropriate to the cell cycle which it has just completed.

6.2
Regional Specification Within the PVE

Circumstantial evidence of two general types has been accumulated to support the hypothesis that at or near the onset of neuronogenesis a protomap of the final neocortical architectonic map (Rakic 1988) is specified either within the PVE or shortly after the cells leave the PVE. That is, as neurons complete their terminal division and exit the cycle, their regional identity is already specified in terms of specific attributes the cell will require to operate within the systems exigencies of that region. The protomap appears to be specified with respect to the earliest formed neurons, that is those which will populate layers VI and, perhaps, V. The first line of evidence is that immediately upon completing their migrations, and perhaps even as they migrate, the neurons of layers VI issue axons which are organized in a way that observes a topology of organization which is essentially identical to that of the final cortical map (De Carlos and O'Leary 1992; Erzurumlu and Jhaveri 1992; Molnar and Blakemore 1995). This mapping order is established independently of prior contact with ascending thalamocortical axons which also appear to organize independently of corticothalamic axons (Woodward et al. 1990; Bicknese et al. 1994). The second line of evidence is that virtually immediately after migrations cells of layer V express cell class antigens which are region specific in ways that harmonize with the final neocortical map (Barbe and Levitt 1991; Cohen-Tannoudji et al. 1994). Cells of layers IV-II/III, by contrast, may not be specified with respect to region but rather may be recruited to region incidental to being assembled in radial register above the map assembled in VI and V (Miyama et al. 1997).

How could cell class be specified this early in neocortical development? What are the positional encoding determinants of the protomap in the PVE?

Even as early as the preneuronogenetic phase of proliferation, patterns of expression of various species of mRNA representing principal homeotic genes and other transcription factors, are distributed in a regionally characteristic fashion through the neocortical PVE (Puelles and Rubenstein 1993; Shimamura et al. 1995). Plausibly these patterns of mRNA expression play a role in directing the regional patterns of transcription requisite to formation of the protomap. Even if so, these patterns of mRNA expression are coarsely grained and insufficient to determine the fine grained regional diversity characteristic of the neocortex. Another or other complementary mechanisms of regional encoding must be acting to achieve the adult map, evolving and building upon the protomap in the PVE whose instantiation may be initiated during the preneuronogenetic phase.

We suggest that the operation of the proliferative process itself may in this way complement the transcription factor map in formation of the protomap within the PVE. This is tenable in that the parameters of the proliferative process, T_{G1}, Q and CC, are at all times throughout the neuronogenetic interval, distributed in a descending gradient fashion along the major axis of the TNG. That is, for any given locale in the PVE and at any moment in the neuronogenetic interval, the parameters of proliferation will have unique values (Miyama et al. 1997); cf. also, (Goodwin and Cohen 1969; Wolpert 1969). The values will be less than those of the locale located immediately rostral and lateral and greater than those of the locale located caudal and medial to the reference locale. In principle this gradation of proliferative parameters might be sufficient for encoding positional information. For such to be the case it would be necessary that the encoding mechanism be linked to transcriptional mechanisms necessary to implementation of the regional map. In other words, the same molecular controls that modify T_{G1} and Q progression could act also to control or regulate mapping specification.

Whereas such a general mechanism might be viewed at first glance as theoretically plausible (Goodwin and Cohen 1969; Wolpert 1969), upon closer examination of the needs of map specification in relation to the operation of the proliferative processes of the PVE, such an hypothesis can be but tentatively formulated. The dilemma for this theory, as for any gradient based theory of regional specification, is provision for a mechanism which will specify the boundaries which enclose populations of cells with the same mapping fate and partition them from adjacent populations of cells with unlike mapping fates. That is, some mechanism must operate to impose boundaries or discontinuities within the continuous gradient.

We suggest here, however, that a plausible positional encoding mechanism for boundary imposition could be provided by wave propagation of the CC domains but that this mechanism must be coupled to a signal triggered by the wave action of these domains. Specifically, mapping information appears to reside in neurons of layers VI and V (De Carlos and O'Leary 1992; Erzurumlu and Jhaveri 1992; Molnar and Blakemore 1995) and, we suggest is encoded as they arise with cycles 1–7. We envision the following operating properties of such a mechanism. As described above and shown in Fig. 7, at any given time there are multiple CC domains spanning the TNG in overlapping fashion. The critical mechanism complementary to domain propagation, we suggest, would be an encoding signal which emanates from origin as each TE passes through the origin and then is propagated rapidly (i.e., traveling across the PVE in a time much less than a single cell cycle) down the TNG. The borders of a mapping unit encoded by a given CC domain (CC_N) could correspond to its leading and trailing edges where they abut trailing and leading edges of domains of CC_{N-2} and CC_{N+2}, respectively (Fig. 7B). Alternately they might modulate the encoding process within the territory of overlap. The process would be repeated as such TE leaves the origin so that successive waves of parcellation would lead to a finely atomized mosaic of map units within the PVE. We suggest that this mechanism of parcellation would extend and be complementary to the relatively coarse grained parcellation implicit in the domains of distribution of transcription factors, already instantiated in the PVE by the outset of the neuronogenetic interval (Puelles and Rubenstein 1993; Shimamura et al. 1995).

It has been demonstrated that cells of the PVE are coupled by gap junctions from early in G1 phase but that coupling breaks in M phase (Bittman et al. 1997). It has also been found that the proliferative activity, and, presumably, the transcriptional profile, of the epithelium is strongly modulated by gap junction operation although the nature of the signals upon which this modulation depends is not known (Bittman et al. 1997; Goto et al. 1998). We suggest that upon entering G1 phase, cells would tend to become coupled again by gap junctions to other cells entering the same cycle, that is, to cells cycling in synchrony (cf., also (Goodwin and Cohen 1969; Wolpert 1969; Wolpert 1978). Thus, the encoding signal could be preferentially transmitted within each CC domain to cells with similar cell cycle parameters and, presumably, in a homogeneous transcriptional slate. To the extent that such a mechanism might serve specification of a unit of the protomap, it would be closely coordinate with the very molecular events which govern the numbers, and perhaps also the classes of neurons which will populate the laminae of the corresponding region of the neocortex.

7
The PVE: A Conserved Histogenetic Strategy

We have observed earlier, that the pseudostratified epithelial architecture of the neocortical PVE is only a special instance of a proliferative epithelial architecture which is general not only to the vertebrate CNS but also to a variety of somatic structures of epithelial nature. We suggested that evolutionary selection pressures may have favored the pseudostratified architecture of the epithelium for its pervasive histogenetic roles in part, at least, because of its high cell packing efficiency (per unit surface area). High cell packing efficiency may not be the only or even the most decisive histogenetic advantage which has favored evolutionary conservation of this epithelium, however. This review has highlighted additional properties of the PVE, in particular its elaborately integrated proliferative mechanisms, which support both the great heterogeneity and specificity of its histogenetic outcomes and these properties also may have favored the evolutionary conservative tenure of the pseudostratified proliferative epithelium across vertebrate species. Thus, the sequence of advance of the elementary proliferative parameters, Q and T_{G1}, governs at a higher regulatory level both the number of the cell cycle sequence enacted by the proliferative process (Q) and their pattern of propagation down the TNG gradient (T_{G1}, by regulating T_C). Mechanisms regulatory to these elementary parameters are also coordinate with the plausibly integral to those which govern histogenetic specification, that is specification of neuronal class and the specification and refinement of a protomap of the neocortex within the PVE.

From the three perspective provided in this review and given the unique structure and mode of operation of the epithelium, regulation of the overall process thus is seen to devolve upon the regulation of two elementary parameters: Q and T_{G1} as regulator of T_C. The entire process of enumeration is driven by those mechanism which control Q while the pace of propagation of cycle domains, and, therefore, the slope of the TNG, is determined by the pace of advance of T_C regulated by T_{G1}. How, then are Q and T_{G1} regulated and how is their regulation coordinated? These are largely unsounded depths of the overall process and a close examination of such leads that exist are beyond the scope of this review. We will mention only in conclusion that it is our sense that regulatory mechanisms operating around the G1 phase restriction point will prove to be the fulcrum of the overall regulatory linkage (Caviness et al. 1999; Takahashi et al. 1999b). It is through regulation of agents of molecular control, in particular the cycle facilitators CDK4/6 kinases, their regulatory subunit cyclin D and the inhibitor p27 that the progression of Q may be set with each cycle (Koff et al. 1993; Sherr 1993; Roberts et al. 1994; Sherr 1994; Massague and Polyak 1995). Moreover, it is at this point that cell external mechanisms including signals communicated via gap junctions appear to modulate the molecular events occurring at the restriction point (Kilpatrick and Bartlett 1993; Ghosh and Greenberg 1995; Lo Turco et al. 1995; Temple and Qian 1995; Bittman et al. 1997; Cavanagh et al. 1997; Goto et al. 1997; Goto et al. 1998; Caviness et al. 1999; Takahashi et al. 1999b). It is also at this point that mechanisms modulatory to cycle regulation are coupled to those which govern, at least in part, the transcriptional and translational profile of the cell (Touchette 1992; Weinberg 1995; Brown and Schreiber 1996; Gerhart and Kirschner 1997). Certain of these may be time dependent and this dependency is a plausible linkage between mechanisms of regulation of the cell cycle parameters T_{G1} to Q and those which determine cell class and provide for mapping (Shermoen and O'Farrell 1991; Ohsugi et al. 1997). Experimental systems competent to drive this search further must tease apart and identify in particular the linkage of transcription antecedent to histogenetic specification.

Acknowledgements. Supported by NIH grants NS12005 and NS33433, NASA grant NAG2-750 and a grant from Pharmacia-Upjohn Fund for Growth & Development Research. T.T. was supported by a fellowship of The Medical Foundation, Inc., Charles A. King Trust, Boston, MA.

References

Alvarez-Bolado G, Swanson L (1996) Developmental brain maps: structure of the embryonic rat brain. Elsevier, Amsterdam

Anderson S, Eisenstat D, Shi L, Rubenstein J (1997) Interneuron migration from basal forebrain to neocortex: dependence on Dlx genes. Science 278:474–476

Barbe MF, Levitt P (1991) The early commitment of fetal neurons to the limbic cortex. J Neurosci 5:519–533

Bayer SA, Altman J (1991) Neocortical development. Raven Press, New York

Bhide P (1996) Cell cycle kinetics in the embryonic mouse corpus striatum. J Comp Neurol 374:506–522

Bicknese A, Sheppard AM, O'Leary DD, Pearlman AL (1994) Thalamocortical axons extend along a chondroitin sulfate proteoglycan-enriched pathway coincident with the neocortical subplate and distinct from the efferent path. J Neurosci 14:3500–3510

Bisconte J-C, Marty R (1975a) Analyse chronoarchitectonique du cerveau de rat par radioautographie. I. Histogenese du telencephale. J Hirnforsch 16:55–74

Bisconte J-C, Marty R (1975b) Etude quantitative du marquage radioautographique dans le systeme nerveux du rat. II. Caracteristiques finales dans le cerveau de l'animal adulte. Exp Brain Res 22:37–56

Bittman K, Owens D, Kriegstein A, Lo Turco J (1997) Cell coupling and uncoupling in the ventricular zone of developing neocortx. J Neurosci 17:7037–7044

Boulder Committee (1970) Embryonic vertebrate nervous system: revised terminology: Anat Rec 166:257–262

Brodmann K (1909) Vergleichende Lokalisationslehre der Grosshirnrinde. Barth, Leipzig

Brown E, Schreiber S (1996) A signaling pathway to translational control. Cell 86:517–520

Cai L, Hayes N, Nowakowski R (1997a) Synchrony of clonal cell proliferation and contiguity of clonally related cells: production of mosaicism in the ventricular zone of developing mouse neocortex. J Neurosci 17:2088–2100

Cai L, Hayes N, Nowakowski R (1997b) Local homogeneity of cell cycle length in developing mouse cortex. J Neurosci 17:2079–2087

Cajal S, Ramon Y (1952) Histologie du Systeme Nerveux de l'Homme et des Vertebres. Consejo Superior de Investigaciones Cientificas, Madrid

Cavanagh J, Mione M, Pappas I, Parnavelas J (1997) Basic fibroblast growth factor prolongs the proliferation of rat cortical progenitor cells in vitro without altering their cell cycle parameters. Cereb Cortex 7:293–302

Caviness V (1975) Architectonic map of neocortex of the normal mouse. J Comp Neurol 164:247–263

Caviness V (1982) Neocortical histogenesis in normal and reeler mice: a developmental study based upon [^{3}H]thymidine autoradiography. Dev Brain Res 4:293–302

Caviness V, Sidman RL (1973) Time of origin of corresponding cell classes in the cerebral cortex of normal and reeler mutant mice: an autoradiographic analysis. J Comp Neurol 148:141–152

Caviness V, Takahashi T, Nowakowski R (1995) Numbers, time and neocortical neuronogenesis: a general developmental and evolutionary model. Trends Neurosci 18:379–383

Caviness V, Takahashi T, Nowakowski R (1997) Cell proliferation in cortical development. In: Galaburda A, Christen Y (eds), Normal and abnormal development of the cortex. Springer, Berlin, Heidelberg, New York, pp 1–24

Caviness V, Takahashi T, Nowakowski R (1999) The G1 restriction point as critical regulator of neocortical neuronogenesis. J Neurochem Res 24:497–506

Cohen-Tannoudji M, Babinet C, Wassef M (1994) Early determination of a mouse somatosensory cortex marker. Nature 368:460–463

De Carlos JA, O'Leary DM (1992) Growth and targeting of subplate axons and establishment of major cortical pathways. J Neurosci 12:1194–1211

Dermietzel R, Spray DC (1993) Gap junctions in the brain: where, what type, how many and why? Trends Neurosci 16:186–192

Dermietzel R, Traub O, Hwang TK, Bennett MVL, Spray DC, Willecke K (1989) Differential expression of three gap junction protein in developing and mature brain tissues. Proc Natl Acad Sci USA 86:10148–10152

Erzurumlu RS, Jhaveri S (1992) Emergence of connectivity in the embryonic rat parietal cortex. Cereb Cortex 2:336–352

Felleman DJ, Van Essen DC (1991) Distributed hierarchical processing in the primate cerebral cortex. Cereb Cortex 1:1–47

Finlay BL, Pallas SL (1989) Control of cell number in the developing mammalian visual system. Prog Neurobiol 32:207–234

Finlay BL, Slattery M (1983) Local differences in the amount of early cell death in neocortex predict adult local specializations. Science 219:1349–1351

Fishell G, Hatten M (1991) Astrotactin provides a receptor system for CNS neuronal migration. Development 113:755–765

Fishell G, Mason CA, Hatten ME (1993) Dispersion of neural progenitors within the germinal zones of the forebrain. Nature 362:636–638

Fulton BP (1995) Gap junctions in the developing nervous system. Persp Dev Neurobiol 2:327–334

Gerhart J, Kirschner M (1997) Cell, Embryos, and Evolution. Blackwell, London

Ghosh A, Greenberg ME (1995) Distinct roles for bFGF and NT-3 in the regulation of cortical neurogenesis. Neuron 15:89–103

Goodall H, Maro B (1986) Major loss of junctional coupling during mitosis in early chick embryos. J Cell Biol 100:568–575

Goodwin BC, Cohen MH (1969) A phase-shift model for the spatial and temporal organization of developing systems. J Theor Biol 25:49–107

Goto T, Takahashi T, Miyama S, Bhide P, Caviness V (1997) The effect of a gap junction uncoupling agent, 1-Octanol, on cell cycle in vitro in the neocortical proliferative epithelium. Soc Neurosci Abst 23:867

Goto T, Takahashi T, Miyama T, Bhide P, Caviness V (1998) Gap junctions exert a developmentally regulated mitogenic effect upon neocortical proliferative epithelium. Soc Neurosci Abst 24:280

Granger B, Tekaia F, Le Sourd A, Rakic P, Bourgeois J-P (1995) Tempo of neurogenesis and synaptogenesis in the primate cingulate mesocortex: comparison with the neocortex. J Comp Neurol 360:363–376

Guthrie S, Butcher M, Lumsden A (1991) Patterns of cell division and interkinetic nuclear migration in the chick embryo hindbrain. J Neurobiol 22:742–754

Hicks SP, D'Amato CJ (1968) Cell migration to the isocortex in the rat. Anat Rec 160:619–634

Hinds JW, Ruffett TL (1971) Cell proliferation in the neural tube: an electron microscopic and Golgi analysis in the mouse cerebral vesicle. Z Zellforsch 115:226-264

His W (1889) Die Neuroblasten und deren Entstehung im embryonalen Mark. Abh Math Phys Cl Kgl Saechs Ges Wiss 15:313-372

Hubel DH, Wiesel TN (1965) Receptive fields and functional architecture in two non-striate visual areas (18 and 19) of the cat. J Neurophysiol 28:229-289

Ingham P, Arias A (1992) Boundaries and fields in early embryos. Cell 68:221-235

Jacobson M (1978) Developmental Neurobiology. Plenum, New York

Killackey HP, Rhoades RW, Bennett-Clarke CA (1995) The formation of a cortical somatotropic map. Trends Neurosci 18:402-407

Kilpatrick TJ, Bartlett PF (1993) Cloning and growth of multipotential neural precursors: requirements for proliferation and differentiation. Neuron 10:255-265

Koff A, Ohtsuki M, Polyak K, Roberts JM, Massague J (1993) Negative regulation of G1 progression in mammalian cells; inhibition of cyclin E-dependent kinase by TGF-β. Science 260:536-539

Kornack D, Rakic P (1998) Changes in cell-cycle kinetics during the development and evolution of primate neocortex. Proc Natl Acad Sci (USA) 95:1242-1246

Krieg WJS (1963) Connections of the Cerebral Cortex. Brain Books, Chicago

Krubitzer L (1995) The organization of neocortex in mammals: are species differences really so different? Trends Neurosci 18:408-417

Lavdas A, Blue M, Lincoln J, Parnavelas J (1997) Serotonin promotes the differentiation of glutamate neurons in organotypic slice cultures of the developing cerebral cortex. J Neurosci 17:7872-7880

Lavdas A, Grigoriou M, Pachnis V, Parnavelas J (1998) The medial ganglionic eminence is a source of the early neurons of the developing cerebral cortex. Soc Neurosci Abst 24:282

Lorente de No R (1938) Cerebral cortex: architecture, intracortical connections, motor projections. In: Fulton JF (eds), Physiology of the Nervous System, Oxford University Press, London, pp 274-313

Lo Turco JJ, Kriegstein A (1991) Clusters of coupled neuroblasts in embryonic neocortex. Science 252:563-566

Lo Turco JJ, Owens DF, Heath MJS, Davis MBE, Kriegstein AR (1995) GABA and glutamate depolarize cortical progenitor cells and inhibit DNA synthesis. Neuron 15:1287-1298

Lumsden A (1990) The cellular basis of segmentation in the developing hindbrain. Trends in Neuroscience 3:329-335

Luskin MB, Shatz CJ (1985) Neurogenesis of the cat's primary visual cortex. J Comp Neurol 242:611-631

Martinez S, Geijo E, Sanchez-Vives M, Puelles L, Gallego R (1992) Reduced junctional permeability at interrhombomeric boundaries. Development 116:1069-1076

Massague J, Polyak K (1995) Mammalian antiproliferative signals and their targets. Curr Opin Gen Dev 5:91-96

McConnell SK (1989) The determination of neuronal fate in the cerebral cortex. Trends Neurosci 12:342-349

McConnell SK, Kaznowski CE (1991) Cell cycle dependence of laminar determination in developing neocortex. Science 254:282-285

McSherry GM (1984) Mapping of cortical histogenesis in the ferret. J Embryol Exp Morphol 81:239-252

Mesulam M-M (1998) From Sensation to cognition. Brain 121:1013-1052

Mione M, Cavanagh J, Harris B, Parnavelas J (1997) Cell fate specification and symmetrical/asymmetrical divisions in the developing cerebral cortex. J Neurosci 17:2018-2029

Mione MC, Danevic C, Boardman P, Harris B, Parnavelas JG (1994) Lineage analysis reveals neurotransmitter (GABA or glutamate) but not calcium-binding protein homogeneity in clonally related cortical neurons. J Neurosci 14:107-123

Miyama S, Takahashi T, Nowakowski R, Caviness V (1997) A Gradient in the duration of the G1 phase in the murine neucortical proliferative epithelium. Cereb Cortex 7:678–689

Molnar Z, Blakemore C (1995) How do thalamic axons find their way to the cortex? Trends Neurosci 18:389–397

Mountcastle VB (1978) The Mindful Brain: Part I. MIT Press, Cambridge, MA

Murray A, Hunt T (1993) The Cell Cycle. WH Freeman, New York

Northcutt RG, Kaas JH (1995) The emergence and evolution of mammalian neocortex. Trends Neurosci 18:373–379

Nowakowski R, Lewin SB, Miller MW (1989) Bromodeoxyuridine immunohistochemical determination of the lengths of the cell cycle and the DNA-synthetic phase for an anatomically defined population. J Neurocytol 18:311–318

Ohsugi K, Gardiner D, Bryant S (1997) Cell cycle length affects gene expression and pattern formation in limbs. Devel Biol 189:13–21

Parnavelas J, Barfield JA, Franke E, Luskin MB (1991) Separate progenitor cells give rise to pyramidal and non pyramidal neurons in the rat telencephalon. Cerebr Cortex 1: 463–468

Polleux F, Dehay C, Moraillon B, Kennedy H (1997) Regulation of neuroblast cell-cycle kinetics plays a crucial role in the generation of unique features of neocortical areas. J Neurosci 17:7763–7783

Puelles L, Rubenstein J (1993) Expression patterns of homeobox and other putative regulatory genes in the embryonic mouse forebrain suggest a neuromeric organization. Trends Neurosci 16:472–479

Rakic P (1974) Neurons in rhesus monkey visual cortex: systematic relation between time of origin and eventual disposition. Science 183:425–427

Rakic P (1976) Differences in the time of origin and in eventual distribution of neurons in areas 17 and 18 of visual cortex in Rhesus monkey. Exp Brain Res Suppl 1:244–248

Rakic P (1982) Early development events: cell lineages, acquisition of neuronal positions and areal and laminar development. Neuronsci Res Prog Bull 20:439–452

Rakic P (1988) Specification of cerebral cortical areas. Science 241:170–176

Rakic P (1995a) Corticogenesis in human and nonhuman primates. In: Gazzaniga MS (eds), The Cognitive Neurosciences, MIT Press, Cambridge, pp 127–145

Rakic P (1995b) A small step for the cell, a giant leap for mankind: a hypothesis of neocortical expansion during evolution. Trends Neurosci 18:383–388

Rhee J, Raballo R, Schwartz M, Vaccarino F (1998) Lineage and non-lineage specific effects of fibroblast growth factor (FGF2) on progenitor cells in the developing cerebral cortex. Soc Neurosci Abst 24:281

Roberts J, Koff A, Polyak K, Firpo E, Collins S, Ohtsubo M, Massague J (1994) Cyclins, cdks and cyclin kinase inhibitors. Cold Spring Harbor Symp Quant Biol 59:31–38

Rockel AJ, Horns RW, Powell TPS (1980) The basic uniformity of structure of the neocortex. Brain 103:221–244

Sanides F (1972) Representation in the cerebral cortex and its areal lamination patterns. In: Bourne GF (eds), The Structure and Function of Nervous Tissue V Structure III and Physiology III, Academic Press, New York, pp 330–453

Sauer FC (1935) Mitosis in the neural tube. J Comp Neurol 62:377–405

Sauer FC (1936) The interkinetic migration of embryonic epithelial nuclei. J Morphol 60:1–11

Sauer FC (1937) Some factors in the morphogenesis of vertebrate embryonic epithelia. J Morphol 61:563–579

Sauer ME, Walker BE (1959) Radioautographic study of interkinetic nuclear migration in the neural tube. Proc Soc Ext Biol NY. 101:557–560

Shermoen A, O'Farrell P (1991) Progression of the cell cycle through mitosis leads to a abortion of nascent transcripts. Cell 67:303–310

Sherr CJ (1993) Mammalian G1 cyclins. Cell 73:1059–1065

Sherr CJ (1994) G1 phase progression: cycling on cue. Cell 79:551–555

Sherr CJ, Roberts JM (1995) Inhibitors of mammalian G1 cyclin-dependent kinases. Genes Dev 9:1149–1163

Shimamura K, Hartigan D, Martinez S, Puelles L, Rubenstein J (1995) Longitudinal organization of the anterior neural plate and neural tube. Development 121:3923–3933

Sidman RL, Rakic P (1973) Neuronal migration, with special reference to developing human brain: a review. Brain Res 62:1–35

Sidman RL, Rakic P (1982) Development of the human central nervous system. In: Haymaker W, Adams RD (eds), Histology and histopathology of the neurons system, Charles C Thomas, Springfield, pp 3–145

Smart IHM, McSherry GM (1982) Growth patterns in the lateral wall of the mouse telencephalon. II. Histological changes during and subsequent to the period of isocortical neuron production. J Anat 131:415–442

Smart IHM, Smart M (1982) Growth patterns in the lateral wall of the mouse telencephalon. I. autoradiographic studies of the histogenesis of the iso-cortex and adjacent areas. J Anat 134:273–298

Stensaas LJ, Stensaas SS (1968) An electron microscope study of cells in the matrix and intermediate laminae of the cerebral hemisphere of the 45 mm rabbit embryo. Z Zellforsch 91:341–365

Takahashi T, Goto T, Miyama S, Nowakowski R, Caviness V (1990a) Sequence of neuron origin and neocortical laminar fate: relation to cell cycle of origin in the developing murine cerebral wall. J Neurosci 19:10357–10371

Takahashi T, Nowakowski R, Caviness V (1992) BUdR as an S-phase marker for quantitative studies of cytokinetic behaviour in the murine cerebral ventricular zone. J Neurocytol 21:185–197

Takahashi T, Nowakowski R, Caviness V (1995) The cell cycle of the pseudostratified ventricular epithelium of the murine cerebral wall. J Neurosci 15:6046–6057

Takahashi T, Nowakowski R, Caviness V (1996a) The leaving or Q fraction of the murine cerebral proliferative epithelium: a general computational model of neocortical neuronogenesis. J Neurosci 16:6183–6196

Takahashi T, Nowakowski R, Caviness V (1996b) Interkinetic and migratory behavior of a cohort of neocortical neurons arising in the early embryonic murine cerebral wall. J Neurosci 16:5762–5776

Takahashi T, Nowakowski R, Caviness V (1997) The mathematics of neocortical neuronogenesis. Dev Neurosci 19:17–22

Takahashi T, Nowakowski R, Caviness VS (1999b) Cell cycle as operational unit of neocortical neurogenesis. Neuroscientist 5:155–163

Tan S-S, Kalloniatis M, Sturm K, Tam P, Reese B (1998) Separate progenitors for radial and tangential cell dispersion during development of the cerebral cortex. Neuron 21:295–304

Temple S, Qian X (1995) bFGF, neurotrophins, and the control of cortical neurogenesis. Neuron 15:249–252

Thomaidou D, Mione M, Cavanagh J, Parnavelas J (1997) Apoptosis and its relation to the cell cycle in the developing cerebral cortex. J Neurosci 17:1075–1085

Touchette N (1992) pRb and the cell cycle: more than meets the eye: J NIH Res 4:56–59

Vaccarino F, Schwartz M, Raballo R, Nilsen J, Rhee J, Zhou M, Doetschman T, Coffin J, Wyland J, Yu-Ting E (1999) Changes in the size of the cerebral cortex are governed by fibroblast growth factor during embryogenesis. Nat Neurosci 2:246–253

Van Essen DC, Maunsell JHR (1983) Hierarchical organization and functional streams in the visual cortex. Trends in Neurosciences 6:370–375

Waechter RV, Jaensch B (1972) Generation times of the matrix cells during embryonic brain development: an autoradiographic study in rats. Brain Res 46:235–250

Walsh C, Cepko C (1993) Clonal dispersion in proliferative layers of developing cerebral cortex. Nature 362:632–635

Weinberg R (1995) The retinoblastoma protein and cell cycle control. Cell 81 : 323-330
Wolpert L (1969) Positional information and the spatial pattern of cellular differentiation. J Theor Biol 25 : 1-47
Wolpert L (1978) Gap junctions: channels for communication in development. In: Feldman J, Gilula NB, Pitts JD (eds), Intercellular junctions and synapses (Receptors and Recognition, Series B), Chapman and Hall, London, pp 83-96
Woodward WR, Chiaia N, Teyler TJ, Leong L, Coull BM (1990) Organization of cortical afferent and efferent pathways in the white matter of the rat visual system. Neuroscience 36 : 393-401
Zilles K (1990) Cortex. In: Paxinos G (eds), The human nervous system. Academic Press, Inc, New York, pp 757-802

Programmed Cell Death in Mouse Brain Development

Chia-Yi Kuan[1], Richard A. Flavell[2], and Pasko Rakic[1]

1
Introduction

Among the basic cellular events that shape the developing brain, programmed cell death (also called apoptosis) plays an essential role (Cowan et al. 1984; Oppenheim 1991). Although cell death during the development of the vertebrate nervous system was first described by anatomists of the nineteenth century, a mechanistic understanding of these events started only in the latter half of this century through two lines of research. First, the identification of nerve growth factor (NGF) by R. Levi-Montalcini and colleagues showed that cells died if deprived of trophic molecules and thus established the foundation for the "trophic theory" of neural development. Second, H. R. Horvitz pioneered genetic studies of programmed cell death in the nematode *Caenorhabditis elegans* and elucidated a genetic pathway involved in this process. Because the targeted disruption of specific genes in the mouse can now be performed, we can test directly whether the same cell death machinery is conserved in mammals. Moreover, the similarity of the organization and development of the central nervous system between the mouse and higher primates makes it an ideal experimental system for understanding the roles of programmed cell death in normal human brain development and congenital malformations. The present chapter focuses on the recent gene targeting studies elucidating the evolutionarily conserved cell death machinery in the context of mouse brain development.

2
Conceptual Framework of Programmed Cell Death

The discovery that cell death is an integral part of vertebrate embryonic development is largely attributable to E. Kallius and two of his pupils,

[1] Section of Neurobiology
[2] Section of Immunobiology, Howard Hughes Medical Institute, Yale University School of Medicine, 333 Cedar Street, New Haven, Connecticut 06510, USA

Results and Problems in Cell Differentiation, Vol. 30
Goffinet and Rakic (Eds.): Mouse Brain Development
© Springer-Verlag Berlin Heidelberg 2000

M. Ernst and A. Glucksmann (Hamburger 1992). In 1951, Glucksmann reviewed the various occurrences of cell death during normal vertebrate development and proposed a classification scheme consisting of phylogenetic, morphogenetic, and histiogenetic degenerations (Glucksmann 1951). Despite its limited mechanistic explanation, Glucksmann's scheme was very insightful and remains useful for appreciating the various developmental roles of cell death. Phylogenetic degeneration, such as the regression of the pronephros and mesonephros in higher vertebrates, is postulated to have a remote cause in evolution. Morphogenetic degeneration refers to the death of progenitor cells during early embryogenesis that is concerned with changes in the forms of organs and tissues. For example, cell death that occurred during the closure of the neural groove and evagination of optic vesicles was classified as morphogenetic degeneration. Finally, histiogenetic degeneration is thought to regulate the differentiation of tissues. The best example of histiogenetic degeneration is seen in limb extirpation and transplantation in chick embryos, which caused, respectively, an increase or decrease in the numbers of cells making up the spinal cord motor columns that normally innervate the limb (Hamburger 1934, 1939). Thus, the peripheral targets appear to produce specific chemicals that travel retrogradely in the axons to their respective nerve centers and somehow regulate the differentiation or survival of the centers. This hypothesis was later vindicated by the discovery of NGF, which led to the "trophic theory" of neural development (Levi-Montalcini 1987). In accordance with the trophic theory, cell death was generally viewed as a mechanism "for matching the size of each neuronal population to the magnitude of its target field" and "for eliminating erroneous projections" (Cowan et al. 1984).

In contrast to histiogenetic cell death, which has frequently been the subject of investigation, the concept of morphogenetic degeneration has become somewhat arcane, because its mechanism remained elusive until very recently. Nonetheless, cell death clearly occurs within the neuroepithelium before the axons of newly generated neurons reach their peripheral targets. This early cell death, classified as morphogenetic degeneration in Glucksmann's scheme and analogous to the lineage-dependent programmed cell death in *C. elegans*, cannot be easily explained by the trophic theory. We suggest that a morphogenetic and histiogenetic degeneration be distinguished in terms of "founder" versus "post-mitotic" neuronal cell death. By our definition, founder cells are the precursor cells that eventually give rise to neurons for a given brain structure (e.g. the ventricular zone before E12 in mice and E40 in monkeys). If the cell death of appropriate founders fails to occur, a large number of supernumerary progeny will result, which may cause malformations of the central nervous system. In contrast, the lack of post-mitotic cell death may modify the size of specific neuronal populations without affecting the global morphogenesis of the brain.

3
Mechanistic Framework of Programmed Cell Death

During the development of an adult *Caenorhabditis elegans* hermaphrodite, 1090 cells are generated, 131 of which undergo programmed cell death in a lineage-specific and cell-autonomous manner. Genetic analysis and molecular cloning have led to the elucidation of a cell death pathway in *C. elegans* (Fig. 1), which consists of three groups of genes (for a comprehensive review of this subject, see Metzstein et al. 1998).

The first group of genes, which includes ces-1 and ces-2 (ces, cell death specification), specifies the death of particular types of cells. The gain-of-function (*gf*) mutation of ces-1 and a partial loss-of-function (*lf*) mutation of ces-2 block the death of a specific group of neurons without causing other discernible cell death defects. Genetic studies indicated that ces-2 initiates cell death by inhibiting the function of ces-1, which in turn negatively regulates the cell-killing genes (Ellis and Horvitz 1991). Molecular cloning of ces-2 revealed that it encoded a member of the basic-leucine zipper (bZIP) family of transcription factors, suggesting that programmed cell death could be regulated by differential gene expression (Metzstein et al. 1996). The second group of genes affects most, if not all, of the 131 cells that undergo cell death and is thus involved in the execution phase of cell death. These global regulators of cell death include egl-1 (egl, egg-laying defective), ced-9 (ced, cell death abnormal), ced-4, and ced-3. Genetic studies show that ced-3 and ced-4 are required for all somatic programmed cell death. In contrast, the ced-9 (*gf*) mutation prevents most, if not all, programmed cell death. Since ced-3 (*lf*) and ced-4 (*lf*) mutations suppress the ectopic deaths resulting from the ced-9 (*lf*) mutation, ced-9 appears to be upstream to ced-3 and ced-4 in the apoptotic pathway (Hengartner et al. 1992). Consistent with this scheme, biochemical studies show that the CED-9 protein binds to CED-4, which in turn physically interacts with the CED-3 protein (Chinnaiyan et al. 1997; Spector

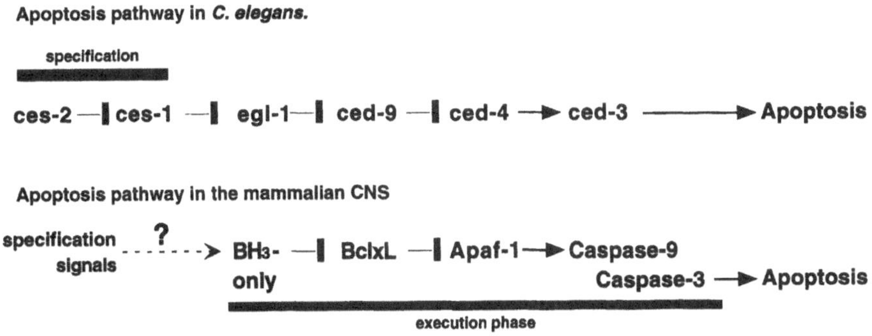

Apoptosis pathway in *C. elegans*.

specification

ces-2 —| ces-1 —| egl-1—| ced-9 —| ced-4 —▶ ced-3 ————————▶ Apoptosis

Apoptosis pathway in the mammalian CNS

specification   **?**
signals - - - - - -▶ BH3- —| BclxL —| Apaf-1—▶ Caspase-9
                      only   Caspase-3 —▶ Apoptosis

execution phase

Fig. 1. The conserved cell death pathway between the nematode *Caenorhabditis elegans* and mammals

et al. 1997). The gene egl-1 was first defined by its gain-of-function mutation, which caused the ectopic death of two hermaphrodite-specific neurons (HSNs), leading to an abnormal egg-laying phenotype. Later studies indicated that the loss-of-function mutation of egl-1 prevented most somatic pro- grammed cell death (Conradt and Horvitz 1998). Genetic analysis further indicated that egl-1 is upstream to ced-3 and ced-4, and biochemical studies showed that EGL-1 and CED-9 proteins interact physically. Thus, it appears that EGL-1 activates programmed cell death by binding to CED-9 and re- leasing the cell death activator CED-4 from a CED-9/CED-4 protein complex (Conradt and Horvitz 1998). The last group of cell death genes is involved in the process of phagocytosis which removes cell corpses and the degradation of DNA. To date, at least six genes have been isolated in this category, including ced-1, ced-6, ced-7, ced-2, ced-5, ced-10 and nuc-1 (nuc, nuclease abnormal).

Structural homologues of all the genes involved in the execution phase of cell death in *C. elegans* have been identified in mammals. Remarkably, these mammalian homologues appear to participate in the same cell death pathway as in the nematode (Fig. 1). The mammalian homologues of ced-3 comprise a family of cysteine-containing, aspartate-specific proteases called caspases (for review see Thornberry and Lazebnik 1998). The ced-4 homologue was identified as one of the apoptosis- activating factors (Apaf-1; Zou et al. 1997). The mammalian homologues of ced-9 belong to a growing family of Bcl-2 proteins, which share the Bcl-2 homology (BH) domain and are either pro- or anti-apoptotic (for review see Merry and Korsmeyer 1997; Adams and Cory 1998). The cloning of egl-1 indicates that it is similar to the BH3 domain- containing subfamily of Bcl-2 proteins (Conradt and Horvitz 1998).

4

Caspases-3 and -9 are Required
for Developmental Apoptosis of Neurons

In 1993, the *C. elegans* cell death gene ced-3 was cloned and shown to encode sequence homology to human interleukin-1β-converting enzyme (ICE) (Yuan et al. 1993). ICE belongs to the caspase family of proteases, which are syn- thesized as proenzymes in living cells and are activated by cleavage at key aspartate residues during apoptosis. Once activated, caspases cleave other caspases and also various cellular substrates including the DNA repair en- zyme poly (ADP-ribose) polymerase (PARP), leading to the ultrastructural changes that typify apoptosis (Kerr et al. 1972). To date, more than 14 members of the caspase family of proteases have been isolated; they have overlapped tissue distribution patterns and share similar cleavage specificity (Thornberry and Lazebnik 1998). The sheer size of this protease family, their overlapping tissue distribution, and their similar cleavage specificities lead to two fundamental questions regarding the biological function of caspases in vivo. First, are members of the caspase family functionally redundant? Second, is there a linear activation cascade among members of the caspase family?

Among members of the caspase family, caspase-3 (also called CPP32, Yama or apopain) exhibits the highest sequence homology to Ced-3 and shares similar substrate specificity (Xue et al. 1996). To study its biological function in vivo, we generated caspase-3 null mutant mice by gene targeting (Kuida et al. 1996). Most homozygous caspase-3 null mutant mice died perinatally with various brain malformations. Close examination of the embryonic tissues from these mice indicated that a general reduction in pyknotic cell deaths during early brain development (Fig. 2A,B). As a consequence of the reduced cell death in the embryonic central nervous system, supernumerary neurons were generated, which led to abnormalities such as multiple indentations of the cerebrum and periventricular ectopic cell masses (Fig. 2C,D). Interestingly, despite the severe brain phenotypes caused by the reduction in programmed cell death, the developmental apoptosis of thymocytes in the caspase-3 null mutants was preserved. Similarly, other lines of caspase-deficient mice (caspase-1, -2, -8 and -11) all exhibited preferential apoptosis defects rather than a global suppression of cell death, indicating that individual members of the caspase family play dominant and non-redundant roles in apoptosis in a tissue-selective or stimulus-dependent manner.

The mammalian homologue of the *C. elegans* cell death gene ced-4 was isolated in biochemical studies designed to identify the initiation factors of caspase-3 cleavage in human HeLa cells. These apoptotic protease-activating factors (Apafs) turned out to be cytochrome-c, which is usually sequestered in the mitochondria (Apaf-2; Liu et al. 1996), the human homologue of ced-4 (Apaf-1; Zou et al. 1997), and caspase-9 (Apaf-3; Li et al. 1997). The amino-terminal sequences of caspase-9 and Apaf-1 both contain a caspase-recruitment domain (CARD) that is indispensable for their interaction. In contrast, caspase-3 lacks the CARD motif and does not bind to Apaf-1 directly. Moreover, active-site mutations of caspase-9 function as dominant-negative inhibitors of caspase-3 in transfected cells (Li et al. 1997). Together, these data suggest a linear activation cascade between caspase-9 and caspase-3 in response to cytochrome-c released from the mitochondria during apoptosis (Fig. 3A). To test this hypothesis, we generated a caspase-9 null mutation by gene targeting (Kuida et al. 1998). Indeed, caspase-9-deficient mice showed a reduction in pyknotic cell deaths during early brain development, multiple indentations and ectopic cell masses in the cerebrum, and perinatal lethality, all reminiscent of the phenotypes of caspase-3 deficiency. The requirement of caspase-9 for normal caspase-3 processing during brain development was further confirmed by immunofluorescence staining, which showed the absence of activated caspase-3 in the embryonic brain tissue of caspase-9 null mutants. Furthermore, biochemical assays demonstrated that the cytochrome-c-mediated cleavage of pro-caspase-3 was defective in the cytosolic fractions of

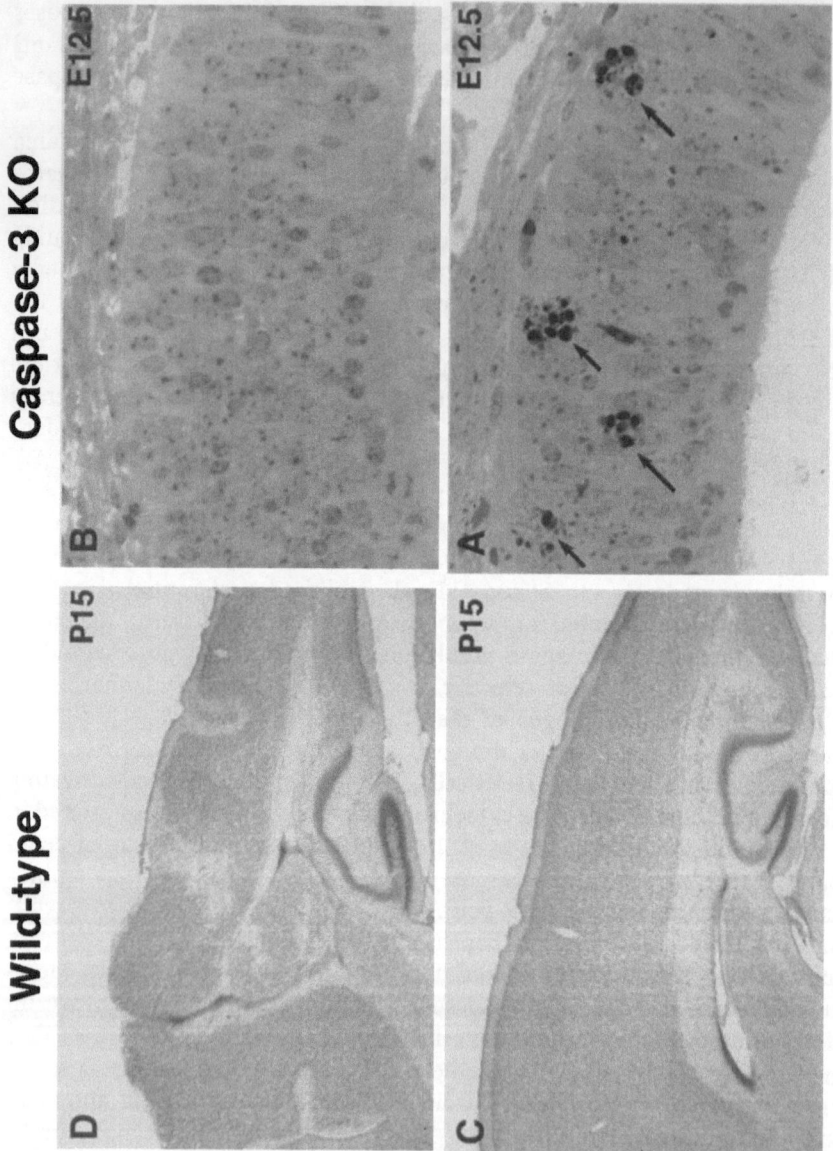

caspase-9 null mutants but was restored by adding in-vitro transcribed and translated caspase-9 (Fig. 3B,C). Together, these results establish a linear activation cascade from caspase-9 to caspase-3, which plays a critical role in programmed cell death during normal mouse brain development.

Targeted disruption of Apaf-1, the human homologue of CED-4, has also been reported, which resulted in similar brain abnormalities to those caused by a reduction in developmental programmed cell death (Cecconi et al. 1998; Yoshida et al. 1998). Moreover, the processing of caspase-3 was shown to be absent in the Apaf-1 null mutant embryonic brain (Cecconi et al. 1998). Taken together, it appears that Apaf-1, caspase-9, and caspase-3 are involved in an apoptotic pathway and have similar functions as their homologues in the nematode (Figs. 1 and 3C).

5
The Bcl-2 Protein Family Has Both Proapoptotic and Antiapoptotic Effects

In 1994, the *C. elegans* cell survival gene ced-9 was cloned and shown to encode a sequence homologous to the mammalian oncogene Bcl-2 (Hengartner and Horvitz 1994). Interestingly, Bcl-2 was first identified as a cell death suppressor gene in human follicular lymphoma. Moreover, overexpression of Bcl-2 inhibits programmed cell death in *C. elegans*, indicating an evolutionarily conserved function similar to ced-9 (Vaux et al. 1992). Subsequent studies revealed that Bcl-2 belongs to a family of cell death regulators that are either proapoptotic or antiapoptotic (for review see Merry and Korsmeyer 1997; Adams and Cory 1998). The prototypical cell survival proteins, Bcl-2 and BclxL, contain four distinct Bcl-2 homology domains, BH1, BH2, BH3, and BH4. Three proapoptotic (Bax-subfamily) proteins, Bax, Bak, and Bok, contain BH1, BH2, and BH3 motifs and resemble Bcl-2 fairly closely. Seven other Bcl-2 proteins, including Bik, Bid, and Bad, contain only the BH3 domain and are proapoptotic (BH3-only subfamily). Mutagenesis studies indicate that the BH1, BH2, and BH3 domains are crucial for dimerization. Pro- and antiapoptotic Bcl-2 proteins can form heterodimers and antagonize the function of each other, suggesting that the fate of death or survival may be determined by the relative concentration of these proteins within a cell (Oltvai et al. 1993).

Fig. 2A–D. Caspase-3 is required for programmed cell death in the developing mouse brain. **A,B** Targeted disruption of caspase-3 diminished pyknotic cell deaths in the embryonic mouse brain. **C,D** As a consequence of the reduced developmental apoptosis, mice with the caspase-3 null mutation exhibited multiple ectopic cell masses and indentations of the postnatal cerebrum. (Adapted from Kuida et al. 1996)

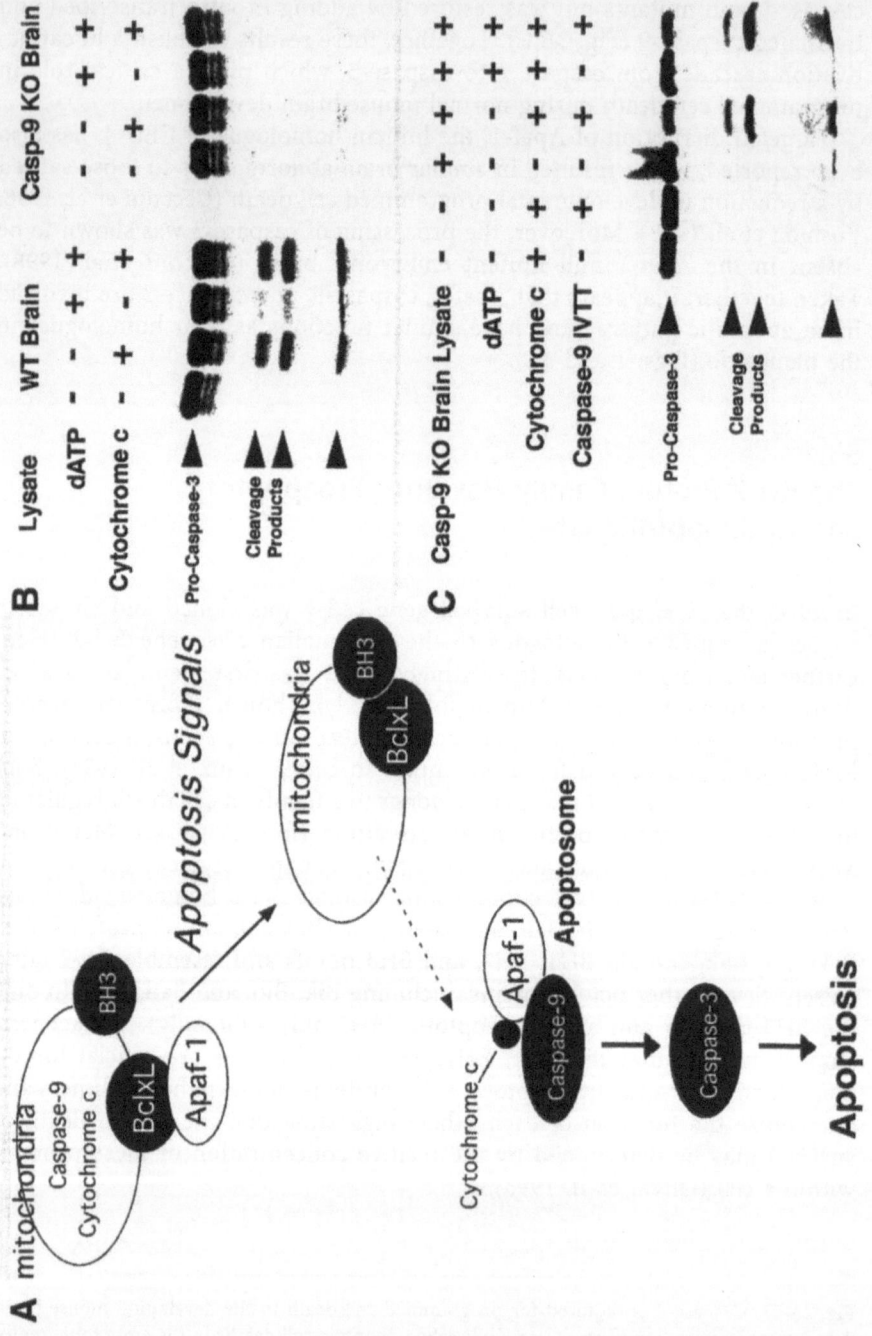

Biochemical and immunolocalization studies indicated that the Bcl-2 protein is an integral mitochondrial membrane protein (Hockenbery et al. 1990). Subsequent studies indicated that one major antiapoptotic effect of Bcl-2 is to prevent the release of cytochrome-c from mitochondria (Kluck et al. 1997; Yang et al. 1997). Furthermore, biochemical evidence indicated that BclxL, Caspase-9, and Apaf-1 form a ternary protein complex (Pan et al. 1998). Taken together, these observations suggest that the primary antiapoptotic mechanisms of BclxL are to sequester Apaf-1 and thereby prevent it from interacting with caspase-9 and to facilitate the retention of cytochrome-c, and possibly caspase-9, within the mitochondria (Fig. 3A; modified after Adams and Cory 1998). This hypothesis is even more appealing in light of the recent finding that the *C. elegans* cell death gene egl-1 encodes a sequence homologous to the proapoptotic BH3-only subfamily of Bcl-2 proteins and that EGL-1 physically interacts with CED-9 (Conradt and Horvitz 1998). Presumably upon activation by external stimuli, proapoptotic members of the Bcl-2 family, such as Bax, bind to the antiapoptotic protein BclxL and release Apaf-1, which can then interact with caspase-9 and cytochrome-c, leading to caspase activation.

After the cloning of the Bcl-2 protein family, transgenic and gene targeting approaches were then used to examine their biological functions in vivo. Transgenic mice overexpressing Bcl-2 exhibited somewhat enlarged brain (12% increase in weight) and a 40–50% increase in the cell number in the facial nucleus (Martinou et al. 1994). In contrast, Bcl-2 null mutations caused fulminant lymphoid apoptosis, polycystic kidneys, and hypopigmentation without any obvious cell death abnormality in the nervous system (Veis et al. 1993). The major cell death suppressor in the nervous system appears to be BclxL, as its null mutations caused extensive apoptosis of postmitotic neurons in the embryonic nervous system. In addition, massive apoptosis of the hematopoietic cells leading to embryonic lethality was also observed in the null mutation of BclxL (Motoyama et al. 1995). In contrast, naturally occurring neuronal death and apoptosis induced by the withdrawal of trophic factors were greatly reduced in the Bax-deficient mice (Deckwerth et al. 1996; White et al. 1998). Furthermore, the Bax deficiency was able to prevent the increased apoptosis of postmitotic neurons in BclxL null mutation, strongly indicating an intracellular balance between proapoptotic and antiapoptotic effects within the Bcl-2 protein family (Shindler et al. 1997).

The above-mentioned gene targeting studies of caspases, Apaf-1, and Bcl-2 family proteins together indicate that the cell death pathway is remarkably conserved between the nematode and mammals (Fig. 1). Another testimony

◄──

Fig. 3A–C. Activation cascade from caspase-9 to caspase-3 in the developing mouse brain. A Schematic diagram of current view of initiation of apoptotic events involving mitochondria. See text for explanation. B Cytochrome-c-dependent cleavage of pro-caspase-3 cleavage activity was absent in the cytosolic fractions of caspase-9 null mutants. C Pro-caspase-3 cleavage activity was restored by the addition of in-vitro transcribed-translated (IVTT) caspase-9. (Adapted from Kuida et al. 1998)

to the evolutionarily conserved mechanism is provided by the recent genetic analysis of the interaction between caspase-3 and BclxL in mammals. In *C. elegans*, the loss-of-function mutation of ced-3 (homologue of caspase-3) suppresses the ectopic cell deaths caused by the loss-of-function mutation of ced-9 (homologue of BclxL) (Hengartner et al. 1992). Therefore, we asked whether the null mutation of caspase-3 would decrease the ectopic neuronal apoptosis in BclxL-deficient mice as predicted by the epistatic interaction of these two genes. Indeed, the aberrant neuronal apoptosis due to the BclxL deficiency was abolished by the additional caspase-3 deficiency (Fig. 4C,D). In fact, the neuronal phenotype of the caspase-3 and BclxL double deficiency was literally indistinguishable from the caspase-3 single null mutation (Roth et al., in prep.).

6
Apoptotic Defects in Founders and Postmitotic Neurons Have Distinct Consequences

Based on the various phenotypes of the Bax, BclxL, caspase-3, and caspase-9 null mutations, we proposed a scheme to indicate the interactions between these cell death regulators during the development of the mammalian brain (Fig. 5). It appears that the normal function of BclxL is to inhibit the apoptotic effect of caspase-3 in the postmitotic neuronal population (steps 2 and 4; see Motoyama et al. 1995). The normal function of Bax is to inhibit the anti-apoptotic effects of BclxL (step 3; see Shindler et al. 1997). Intriguingly, current studies indicate that caspases-3 and -9 may have a preferential effect on early founder cells, which is not shared by Bax and BclxL (step 1; see Kuida et al. 1996, 1998).

The preferential effect of caspases on neuronal progenitor cells is best illustrated by cerebral malformations caused by null mutations of caspase-3 or -9 (Fig. 6A–D). The entire cell population of the vertebrate telencephalon is generated in the ventricular zones and migrates into distinct cerebral areas (Rakic 1988). Recent studies indicate a high incidence of apoptosis within the proliferative ventricular zones, which may regulate the size of the progenitor

---

Fig. 4A–D. Epistatic genetic analysis of BclxL and caspase-3 in developing mouse brain. A high density of pyknotic cells is found in specific locations during normal mouse brain development, such as in the lamina terminalis (*lam*) as shown in A, the absence of which is a hallmark of the caspase-3 null mutation phenotype (B). Except for these specific locations, very few sporadic pyknotic cell deaths were found in the neuroepithelial wall in normal mouse embryos. In contrast, targeted disruption of BclxL results in widespread pyknosis of postmitotic neurons located outside of the proliferative ventricular zone (*vz*) (C). Remarkably, the phenotype of increased apoptosis due to the BclxL deficiency is diminished by an additional caspase-3 null mutation, indicating an epistatic interaction between these two genes. (Adapted from Roth et al., in prep.)

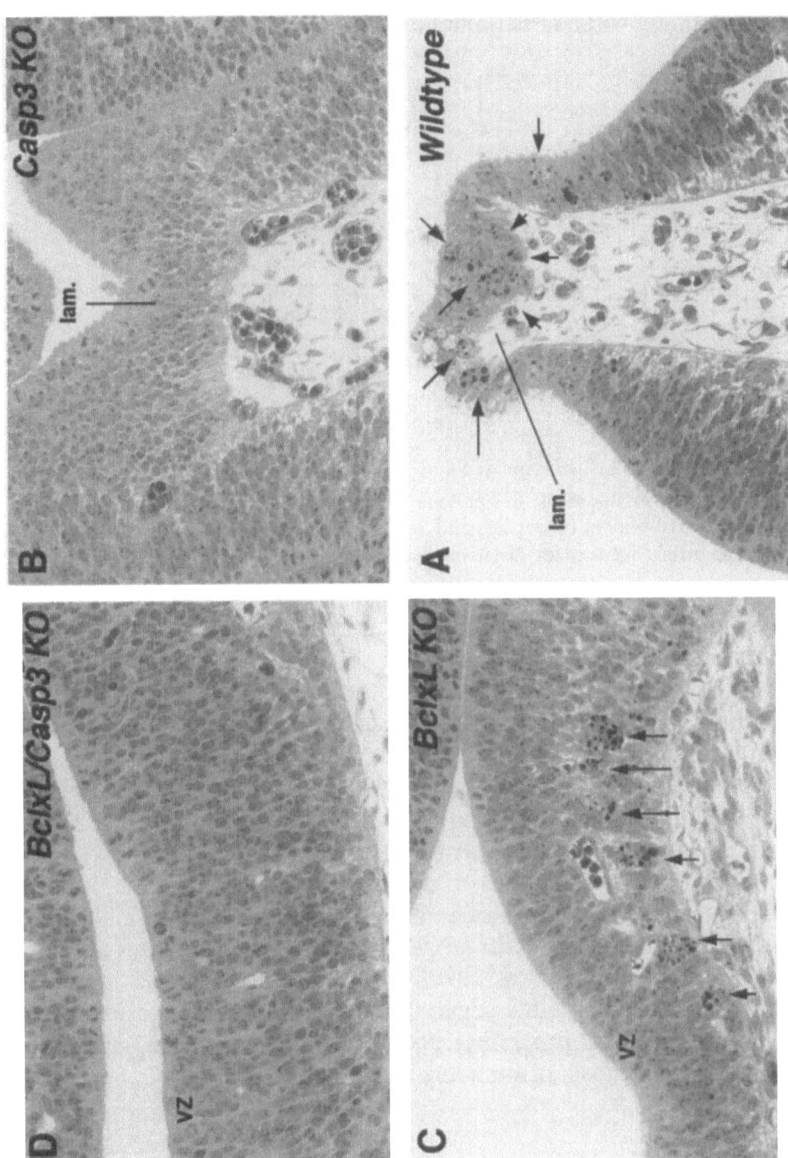

Apoptosis of Founder versus Post-mitotic Neurons

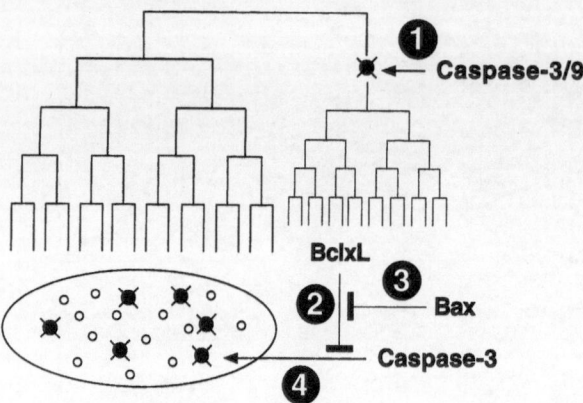

Fig. 5. Distinction between apoptosis of founders versus postmitotic neurons. Recent studies indicate differential involvement of caspases and Bcl-2-family proteins in apoptosis of founders and postmitotic neurons. See text for discussion of evidence

pool (Thomaidou et al. 1997). Conceivably, a reduction in the apoptosis of founder cells in the ventricular zone would produce a large supernumerary progeny, leading to a large cortical surface and thus multiple indentations and convolutions of the cerebrum, as typically observed in caspase-3 and caspase-9 null mutations (Fig. 6E,F). These cortical malformations are similar to the rat spontaneous mutation, telencephalic internal structural heterotopia (tish), and human periventricular heterotopia abnomaly (Eksioglu et al. 1996; Lee et al. 1997). In contrast, although the Bax and BclxL deficiencies cause either a reduction or an increase in cell death in several postmitotic neuronal populations, these mutations do not have a global effect on the early morphogenesis of the nervous system (Motoyama et al. 1995; Deckwerth et al. 1996; White et al. 1998). Taken together, the discrepancy of the phenotypes of these mutants suggests a unique role of caspases in regulating the neuroepithelial progenitor pool and indicates the importance of distinguishing the apoptosis of founders from that of postmitotic neurons.

7
c-Jun N-Terminal Kinases Regulate Brain Region-Specific Apoptosis

One important feature of programmed cell death is its occurrence at precise times and places during normal brain development (Kallen 1955). The temporal and spatial precision of cell death along the neuraxis was thought to be needed for the proper pruning of embryonic tissues (morphogenetic

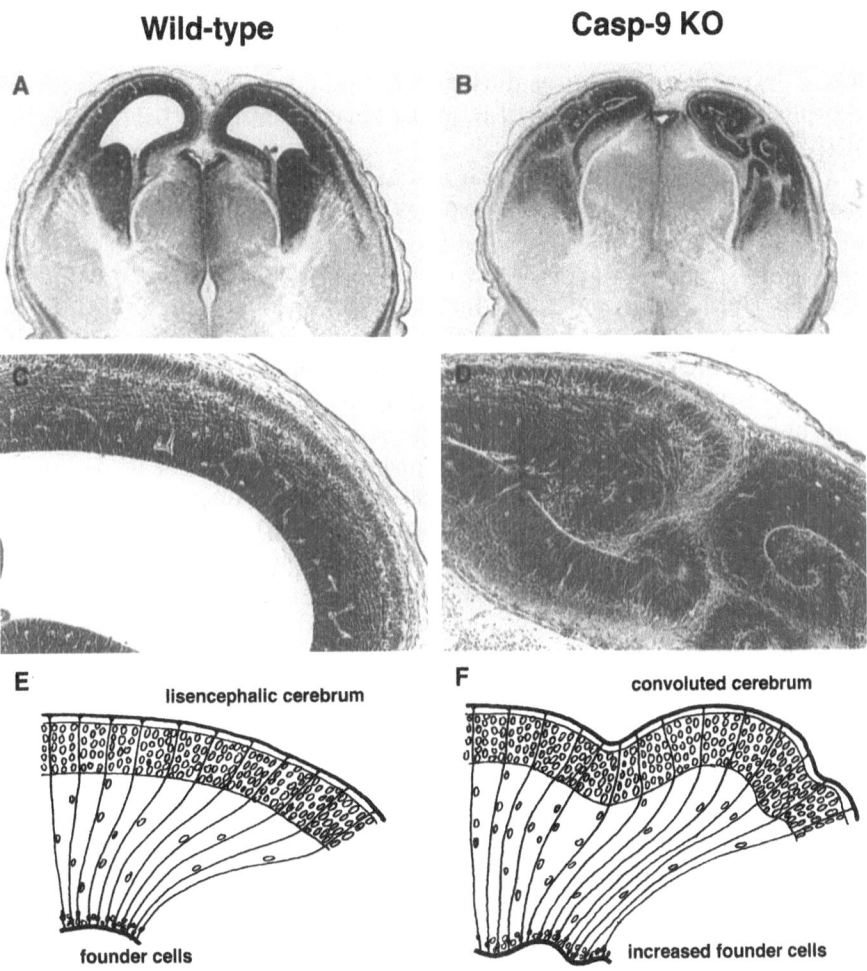

Wild-type   **Casp-9 KO**

Fig. 6A–F. Caspase-9 regulates size of neuroepithelial progenitor pool. Gene targeting of caspase-9 typically produces an expanded and convoluted cerebral cortex with an increased number of neurons (compare **A** and **C** wild-type to **B** and **D** mutants). **E,F** These abnormalities suggest that the caspase-9 null mutation primarily affects apoptosis of the founder cells in the ventricular zone. (Adapted from Kuida et al. 1998)

degeneration; Glucksmann 1951). Intriguingly, although inactive proenzymes of caspases are widely expressed in the nervous system, intense caspase activation only occurs at restricted locations. Thus, there exists either a neuroprotective mechanism to prevent excessive caspase activation and/or specific proapoptotic induction at restricted locations. This scenario is consistent with the cell death pathway in *C. elegans*, which has both specification and execution components; the former specifies the death of a particular lineage of cells whereas the latter is utilized in all forms of somatic cell death. The mechanism for specifying restricted programmed cell death during mouse brain development is at present much less well understood than the execution machinery of apoptosis. In this regard, a recent study showing severe dysregulated apoptosis during early brain development in mice lacking somatic isoforms of c-Jun N-terminal kinases (Jnk1 and Jnk2) is particularly relevant.

The Jnk family of protein kinases was first identified as a subfamily of the MAP kinases. Jnk mediates the phosphorylation of c-Jun at serines 63 and 73 and causes increased AP-1 transcription activity in response to external stress signals (for review see Ip and Davis 1998). The Jnk family has three different isoforms (Jnk1, Jnk2, and Jnk3) that have distinct substrate affinities and tissue distributions (Gupta et al. 1996). Jnk1 and Jnk2 are widely expressed in adult tissues whereas Jnk3 is mainly enriched in the nervous system (Martin et al. 1996). The exact roles of Jnk signaling in cell death are complicated, as both proapoptotic and antiapoptotic effects of Jnk activation have been reported in different cell types. It is likely that the net effect of Jnk activation on apoptosis may depend on the cell context and the simultaneously activated signal transductions within a cell (Ip and Davis 1998).

To address the biological functions of the Jnk protein kinase family, we generated mutant mice lacking each member of the family and derived compound mutants by crossing the Jnk single deficient mice (Yang et al. 1997; Dong et al. 1998; Yang et al. 1998). Although Jnk1, Jnk2, or Jnk3 single deficient and Jnk1/Jnk3 or Jnk2/Jnk3 double mutants all survived normally, Jnk1/Jnk2 dual-deficient mutants died on E11–12 (Kuan et al. 1999). Histological examination of these mutants revealed a great reduction in pyknotic cells at the lateral edges of the folding hindbrain leading to neural tube defects (Fig. 7A–D). A similar phenotype of hindbrain neural tube defect was also observed in caspase-3, caspase-9, and Apaf-1 null-mutant embryos, indicating that regional-specific cell death plays an important role in hindbrain neurulation (Cecconi et al. 1998; Hakem et al. 1998; Kuida et al. 1998; Yoshida et al. 1998). However, unlike the caspase null mutants, there was increased apoptosis and widespread caspase-3 activation within the forebrain of Jnk1 and Jnk2 dual-deficient embryos leading to a precocious degeneration (Kuan et al. 1999). These results suggest that Jnk1 and Jnk2 protein kinases may prevent excessive caspase activation in the forebrain and selectively induce apoptosis in the hindbrain (Fig. 7E,F). The identification of differential roles for Jnk signaling in apoptosis in a brain region-specific manner provides new avenues for understanding the specification phase of programmed cell death, during mouse brain development.

8
Concluding Remarks

From time to time, there arises an illusion that the basic principles of brain development are understood and what is left is to fill in the details. Although

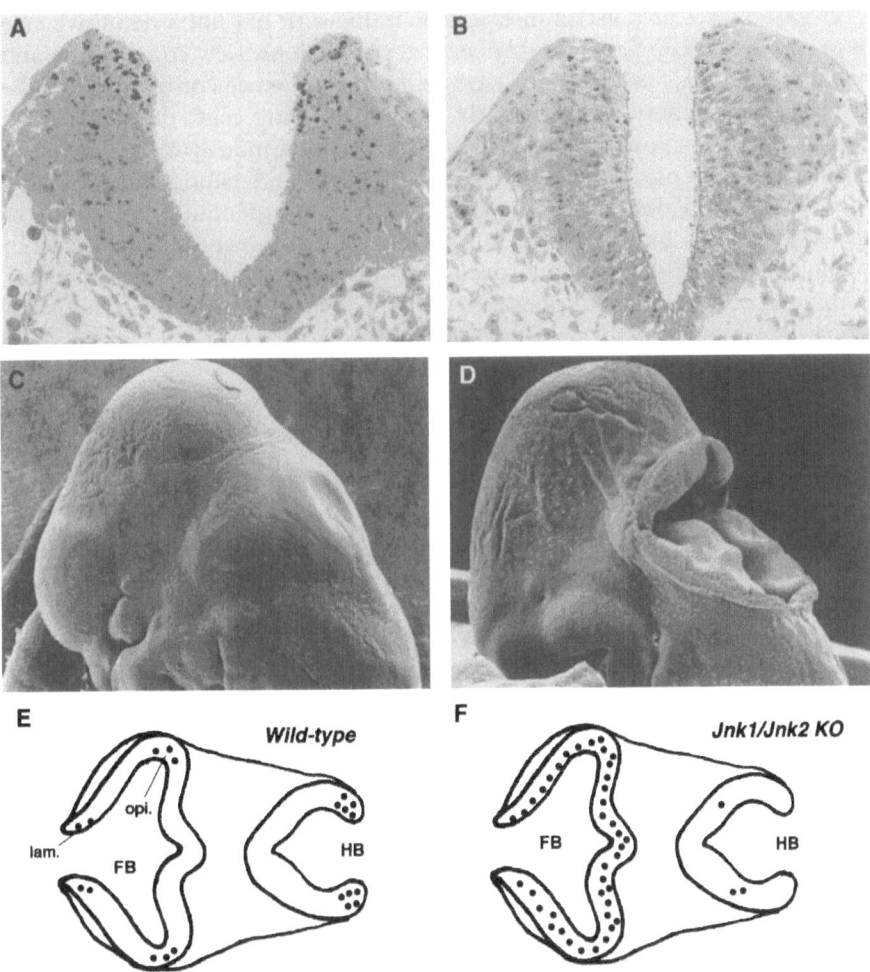

Fig. 7A–F. c-Jun N-terminal kinases Jnk1 and Jnk2 are required for brain region-specific programmed cell death during early brain development. Pyknotic cell deaths are typically located at lateral edges of the hindbrain prior to neural tube closure in wild-type embryo (**A**) but are greatly reduced in mouse mutants deficient in both Jnk1 and Jnk2 genes (**B**). As a consequence of this reduction in region-specific apoptosis, Jnk1 and Jnk2 dual-deficient embryos exhibited neural tube defects at the hindbrain (**C,D**). E,F In contrast, increased apoptosis and widespread caspase-3 activation were seen in embryonic forebrain of double mutants. (Adapted from Kuan et al. 1999). *FB* Forebrain; *HB* hindbrain; *lam.* Lamina terminalis; *opi.* optic invagination

in the past decade we have witnessed great progress in understanding the mechanism of programmed cell death during mammalian brain development, many important questions remain unanswered. Moreover, seemingly unifying mechanisms frequently turn out to be overly simplified or even misleading upon close inspection. The cell death pathway deduced in

C. elegans serves as a useful mechanistic framework but not a definitive rule of programmed cell death for all neuronal populations. New components and novel mechanisms are likely to emerge. Moreover, since components of the cell death machinery are constantly present in living cells, the mechanism that triggers the apoptotic process, either cell-autonomously or as a result of cell–cell interactions, is an important but poorly understood issue. Finally, there is good evidence that the mechanism of programmed cell death in development can be reactivated in later life, leading to neuronal apoptosis in pathological and degenerative disorders. Therefore, the understanding of programmed cell death in neural development may provide therapeutic insights into many human neurological diseases.

Acknowledgements. We thank our colleagues, especially K. Kuida, R. J. Davis, K. A. Roth, D. D. Yang, T. F. Haydar, and M. S.-S. Su, for many stimulating discussions leading to the ideas presented in this manuscript. This work was supported by NIH grants to P.R. and R.A.F. R.A.F. is an Investigator at the Howard Hughes Medical Institute.

References

Adams JM, Cory S (1998) The Bcl-2 protein family: arbiters of cell survival. Science 281:1322–1326

Cecconi F, Alvarez-Bolado G, Meyer BI, Roth KA, Gruss P (1998) Apaf1 (CED-4 homolog) regulates programmed cell death in mammalian development. Cell 94:727–737

Chinnaiyan AM, O'Rourke K, Lane BR, Dixit VM (1997) Interaction of CED-4 with CED-3 and CED-9: a molecular framework for cell death. Science 275:1122–1126

Conradt B, Horvitz HR (1998) The *C. elegans* protein EGL-1 is required for programmed cell death and interacts with the Bcl-2-like protein CED-9. Cell 93:519–529

Cowan WM, Fawcett JW, O'Leary DD, Stanfield BB (1984) Regressive events in neurogenesis. Science 225:1258–1265

Deckwerth TL, Elliott JL, Knudson CM, Johnson EM Jr, Snider WD, Korsmeyer SJ (1996) BAX is required for neuronal death after trophic factor deprivation and during development. Neuron 17:401–411

Dong C, Yang DD, Wysk M, Whitmarsh AJ, Davis RJ, Flavell RA (1998) Defective T cell differentiation in the absence of Jnk1. Science 282:2092–2095

Eksioglu YZ, Scheffer IE, Cardness P, Knoll J, Dimario F, Ramsby G, Berg M, Kamuro K, Berkovic SF, Duyk M, Parisi J, Huttenlocher PR, Walsh CA (1996) Periventricular heterotopia: an X-linked dominant epilepsy locus causing aberrant cerebral cortical development. Neuron 16:77–87

Ellis RE, Horvitz HR (1991) Two *C. elegans* genes control the programmed deaths of specific cells in the pharynx. Development 112:591–603

Glucksmann A (1951) Cell deaths in normal vertebrate ontogeny. Biol Rev 26:59–86

Gupta S, Barrett T, Whitmarsh AJ, Cavanagh J, Sluss HK, Derijard B, Davis RJ (1996) Selective interaction of JNK protein kinase isoforms with transcription factors. EMBO J 15:2760–2770

Hakem R, Hakem A, Duncan GS, Henderson JT, Woo M, Soengas MS, Elia A, de la Pompa JL, Kagi D, Khoo W, Potter J, Yoshida R, Kaufman SA, Lowe SW, Penninger JM, Mak TW (1998) Differential requirement for caspase 9 in apoptotic pathways in vivo. Cell 94:339–352

Hamburger V (1992) History of the discovery of neuronal death in embryos. J Neurobiol 23:1116–1123

Hamburger V (1934) The effects of wing bud extirpation on the development of the central nervous system in chick embryos. J Exp Zool 68:449-494

Hamburger V (1939) Motor and sensory hyperplasia following limb-bud transplantations in chick embryos. Physiol Zool 12:258-84

Hengartner MO, Horvitz HR (1992) C. elegans cell survival gene ced-9 encodes a functional homolog of the mammalian proto-oncogene bcl-2. Cell 76:665-676

Hengartner MO, Ellis RE, Horvitz HR (1992) Caenorhabditis elegans gene ced-9 protects cells from programmed cell death. Nature 356:494-499

Hockenbery D, Nunez G, Milliman C, Schreiber RD, Korsmeyer SJ (1990) Bcl-2 is an inner mitochondrial membrane protein that blocks programmed cell death. Nature 348:334-336

Ip YT, Davis RJ (1998) Signal transduction by the c-Jun N-terminal kinase (JNK)-from inflammation to development. Curr Opin Cell Biol 10:205-219

Kallen B (1955) Cell degeneration during normal ontogenesis of the rabbit brain. J Anat 89:153-161

Kerr JF, Wyllie AH, Currie AR (1972) Apoptosis: a basic biological phenomenon with wide ranging implications in tissue kinetics. Br J Cancer 26:239-257

Kluck RM, Bossy-Wetzel E, Green DR, Newmeyer DD (1997) The release of cytochrome c from mitochondria: a primary site for Bcl-2 regulation of apoptosis. Science 275:1132-36

Kuan CY, Yang DD, Samanta Roy DR, Davis RJ, Rakic P, Flavell RA (1999) The Jnk1 and Jnk2 protein kinases are required for regional specific apoptosis during early brain development. Neuron 22:667-676

Kuida K, Zheng TS, Na S, Kuan CY, Yang D, Karasuyama H, Rakic P, Flavell RA (1996) Decreased apoptosis in the brain and premature lethality in CPP32-deficient mice. Nature 384:368-372

Kuida K, Haydar TF, Kuan CY, Gu Y, Taya C, Karasuyama H, Su MSS, Rakic P, Flavell RA (1998) Reduced apoptosis and cytochrome c-mediated caspase activation in mice lacking caspase 9. Cell 94:325-337

Lee KS, Schottler F, Collins JL, Lanzino G, Couture D, Rao A, Hiramatsu KI, Goto Y, Hong SC, Caber H, Yamamoto H, Chen ZF, Bertram E, Berr S, Omary R, Scrable H, Jackson T, Goble J, Eisenman L (1997) A genetic animal model of human neocortical heterotopia associated with seizures. J Neurosci 17:6236-6242

Levi-Montalcini R (1987) The nerve growth factor 35 years later. Science 237:1154-1162

Li P, Nijhawan D, Budihardjo I, Srinivasula SM, Ahmad M, Almemri ES, Wang X (1997) Cytochrome c and dATP-dependent formation of Apaf-1/caspase-9 complex initiates an apoptotic protease cascade. Cell 91:479-489

Liu X, Kim CN, Yang J, Jemmerson R, Wang X (1996) Induction of apoptotic program in cell-extracts: requirement for dATP and cytochrome c. Cell 86:147-157

Martin JH, Mohit AA, Miller CA (1996) Developmental expression in the mouse nervous system of the p493F12 SAP kinase. Brain Res Mol Brain Res 35:47-57

Martinou JC, Dubois-Dauphin M, Staple JK, Rodriguez I, Frankowski H, Missotten M, Albertini P, Talabot D, Catsicas S, Pietra C, Huarte J (1994) Overexpression of BCL-2 in transgenic mice protects neurons from naturally occurring cell death and experimental ischemia. Neuron 13:1017-1030

Merry DE, Korsmeyer SJ (1997) Bcl-2 gene family in the nervous system. Annu Rev Neurosci 20:245-267

Metzstein MM, Hengartner MO, Tsung N, Ellis RE, Horvitz HR (1996) Transcriptional regulator of programmed cell death encoded by Caenorhabditis elegans gene ces-2. Nature 382:545-547

Metzstein MM, Stanfield GM, Horvitz HR (1998) Genetics of programmed cell death in C. elegans: past, present and future. Trends Genet 14:410-416

Motoyama N, Wang F, Roth KA, Sawa H, Nakayama KI, Nakayama K, Negishi I, Senju S, Zhang Q, Fuiji S, Loh DY (1995) Massive cell death of immature hematopoietic cells and neurons in Bcl-x-deficient mice. Science 267:1506-1510

Oltvai ZO, Milliman CL, Korsemeyer SJ (1993) Bcl-2 heterodimerize in vivo with a conserved homolog, Bax, that accelerates programmed cell death. Cell 74:609–19

Oppenheim RW (1991) Cell death during development of the nervous system. Annu Rev Neurosci 14:453–501

Pan G, O'Rouke K, Dixit VM (1998) Caspase-9, Bcl-xL, and Apaf-1 form a ternary complex. J Biol Chem 273:5841–5845

Rakic P (1988) Specification of cerebral cortical areas. Science 241:170–176

Shindler KS, Latham CB, Roth KA (1997) Bax deficiency prevents the increased cell death of immature neurons in bcl-x-deficient mice. J Neurosci 17:3112–3119

Spector MS, Desnoyers S, Hoeppner DJ, Hengartner MO (1997) Interaction between the *C. elegans* cell-death regulators CED-9 and ED-4. Nature 385:653–656

Thomaidou D, Mione MC, Cavanagh JF, Parnavelas JG (1997) Apoptosis and its relation to the cell cycle in the developing cerebral cortex. J Neurosci 17:1075–1085

Thornberry NA, Lazebnik Y (1998) Caspases: enemies within. Science 281:1312–1316

Vaux DL, Weissman IL, Kim SK (1992) Prevention of programmed cell death in *Caenorhabditis elegans* by human bcl-2. Science 258:1955–1957

Veis DJ, Sorenson CM, Shutter JR, Korsemeyer SJ (1993) Bcl-2-deficient mice demonstrate fulminant lymphoid apoptosis, polycystic kidneys, and hypopigmented hair. Cell 75:229–240

White FA, Keller-Peck CR, Knudson CM, Korsemeyer SJ, Snider WD (1998) Widespread elimination of naturally occurring neuronal death in Bax-deficient mice. J Neurosci 18:1428–1439

Xue D, Shaham S, Horvitz HR (1996) The *Caenorhabditis elegans* cell-death protein CED-3 is a cysteine protease with substrate specificities similar to those of the human CPP32 protease. Genes Dev 10:1073–1083

Yang DD, Kuan CY, Whitmarsh AJ, Rincón M, Zheng TS, Davis RJ, Rakic P, Flavell RA (1997) Absence of excitotoxicity-induced apoptosis in the hippocampus of mice lacking the Jnk3 gene. Nature 389:865–870

Yang J, Liu X, Bhalla K, Kim N, Ibrado AM, Cai J, Peng TI, Jones DP, Wang X (1997) Prevention of apoptosis by Bcl-2: Release of cytochrome c from mitochondria blocked. Science 275:1129–1132

Yang DD, Conze D, Whitmarsh AJ, Barrett T, Davis RJ, Rincón M, Flavell RA (1998) Differentiation of CD4$^+$ T cells to Th1 cells requires MAP kinase JNK2. Immunity 9:575–585

Yoshida H, Kong Y, Yoshida R, Elia AJ, Hakem A, Hakem R, Penninger JM, Mak TW (1998) Apaf1 is required for mitochondrial pathways of apoptosis and brain development. Cell 94:739–750

Yuan J, Shaham S, Ledoux S, Ellis HM, Horvitz HR (1993) The *C. elegans* cell death gene ced-3 encodes a protein similar to the mammalian interleukin-1 beta-converting enzyme. Cell 75:641–652

Zou H, Henzel WJ, Liu X, Lutschg A, Wang X (1997) Apaf-1, a human protein homologous to *C. elegans* CED-4, participates in cytochrome c-dependent activation of caspase-3. Cell 90:405–413

Neurotrophic Factors: Versatile Signals for Cell–Cell Communication in the Nervous System

Carlos F. Ibáñez[1]

1
Introduction

Cell–cell communication is the main business of the cellular components of the nervous system. This dialogue is primarily based on layered system of molecules released and received by the different cell types that form the nervous system and its targets. At one level, neurotransmitter and neuro-modulator substances dictate and propagate the activation states of neurons. In another layer, we find a group of regulatory molecules, typically poly-peptides of roughly a couple of hundred residues long, that control diverse aspects of the life of neurons, including survival, neuritic growth and dif-ferentiation states. Not surprisingly, many of these molecules have diverse functions also in tissues and organs outside the nervous system, where they are known as growth factors or cytokines. Although the term "neurotrophic factor" was originally associated with a survival-promoting activity, work on these molecules during the past ten years has considerably extended the functional spectrum of the term to comprise almost any growth factor or cytokine having some kind of effect on neurons. Important recent additions to the growing list of activities attributed to neurotrophic factors include short term effects on synaptic plasticity, growth cone steering and, quite paradoxically in retrospective, as active promoters of cell death. As much as cell–cell communication is of crucial importance for development, neuro-trophic factors play a central role during the assembly of the nervous system as regulators of cell number, axonal growth, target invasion and synaptic connectivity.

Departing from the original observations that led to the founding princi-ples of the field, encapsulated in the "neurotrophic hypothesis", this chapter moves on to summarize our current view of some basic aspects of the biology of neurotrophic factors. We will then revisit the predictions of the neuro-trophic hypothesis from a molecular perspective, using examples from recent

[1] Division of Molecular Neurobiology, Department of Neuroscience, Karolinska Institute, 17177 Stockholm, Sweden

Results and Problems in Cell Differentiation, Vol. 30
Goffinet and Rakic (Eds.): Mouse Brain Development
© Springer-Verlag Berlin Heidelberg 2000

experiments involving loss-of-function and gain-of-function genetic manipulations in transgenic mice.

2
The Neurotrophic Hypothesis

We can trace the roots of the neurotrophic factor field to the pioneer experiments made during the 1970's by Viktor Hamburger and his colleagues. In one of a series of experiments, these investigators counted the number of motor neurons in the spinal cord of chick embryos at different stages of development. They found a time window during which up to 40% of the number of neurons generated at earlier stages was eliminated (Hamburger 1975) (Fig. 1A). Because chick embryos come conveniently packed in eggs, small surgical manipulations can be made without disturbing the general capacity of the embryo to continue developing. When the same experiment was performed in developing embryos in which the target of innervation of the population of motor neurons examined – the developing limb bud, for example – had been removed, they saw again a decrease in cell number during the same period of natural cell elimination, except that this time many more neurons, almost the entire subpopulation, had disappeared (Fig. 1B). This observation suggested for the first time a role for the target of innervation in regulating the survival of innervating neurons. Predictably, when the counts were made in an embryo containing an additional, surgically transplanted limb bud, more neurons than normal survived the period of motor neuron cell death (Hollyday and Hamburger 1976) (Fig. 1C).

These experiments led to two fundamental concepts: first, during normal development, neurons are generated in excess numbers, the surplus of neurons being eliminated during a defined time window known as the period of "naturally occurring cell death", and second, the extent of cell death during this period is regulated by the target of innervation. How does the target regulate neuron survival? The answer to this question is encapsulated in the neurotrophic hypothesis (reviewed in (Korsching 1993; Levi-Montalcini 1987)), which predicates that during development, targets of innervation produce "neurotrophic" substances to which neurons become dependent for survival upon reaching the vicinity of the target (Fig. 2). Why do neurons die during normal development? The neurotrophic hypothesis goes on to postulate that target-derived survival substances are produced in limiting amounts, enough to maintain the survival of only a fraction of the innervating neurons, the rest being eliminated by programmed cell death. What could be the advantage of such scheme? It does provide a simple mechanism to automatically match the number of any developing population of neurons to the size of their corresponding target without relying in a predetermined hard-wired mechanism to control the size of each neuronal subpopulation.

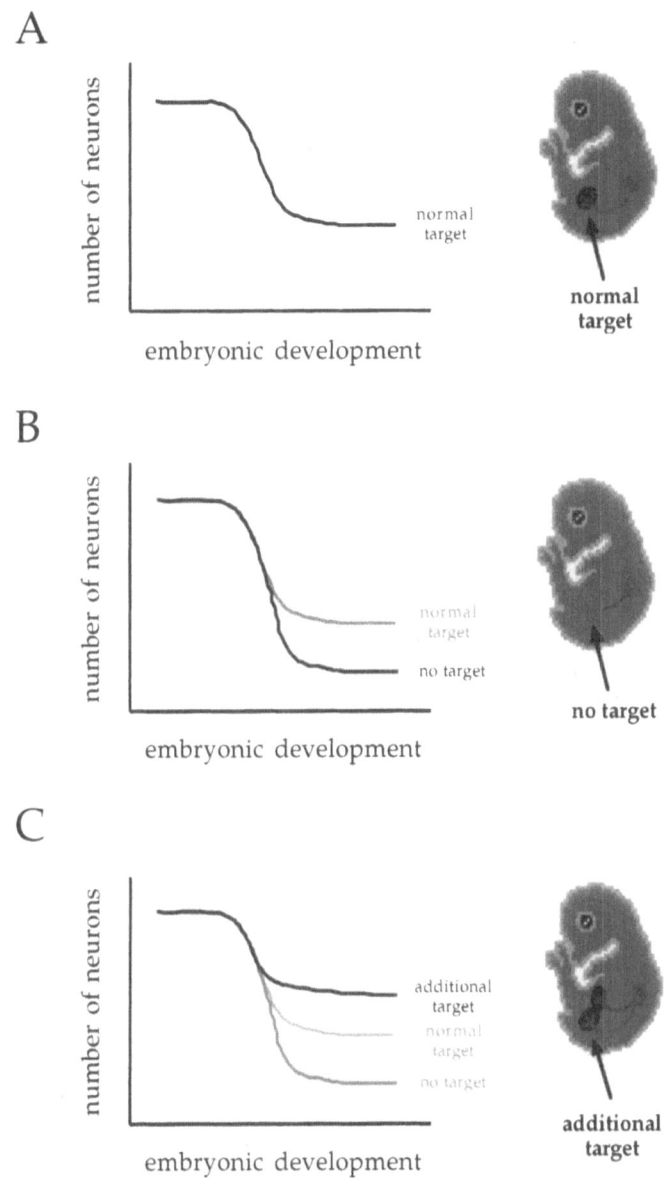

Fig. 1A–C. Programmed cell death of neurons during development of the nervous system is controlled by the target of innervation. (A) Up to 40% of spinal cord motor neurons degenerate between embryonic days 6.5 and 9.5 during the normal development of chick embryos. (B) Increased programmed cell death after target removal. (C) Rescue of motorneurons from programmed cell death by an enlarged target. Adapted from Hamburger (1975) and Hollyday and Hamburger (1976)

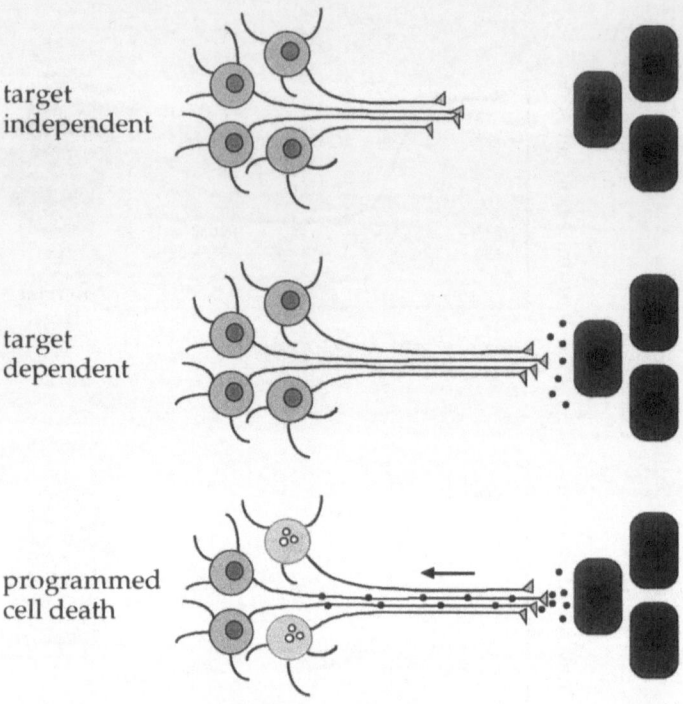

target
independent

target
dependent

programmed
cell death

Fig. 2. The neurotrophic hypothesis. Developing neurons growing axons towards their targets are initially independent of the target for survial. Upon target encounter, they become dependent on survival-promoting target-derived trophic factors. These are taken up by the axon terminal and retrogradely transported to the cell body. Target-derived survival substances are produced in limiting amounts, enough to maintain the survival of only a fraction of the innervating neurons, the rest being eliminated by programmed cell death. Adapted from Reichardt and Fariñas (1997)

Finally, is this what really happens? When it comes to neurons innervating other neurons, the answer to this question is much more complicated. However, in the case of populations of neurons innervating peripheral targets, such as motor, sensory or sympathetic neurons, experimental support for the neurotrophic hypothesis has come from elegant genetic manipulations of the dose of survival substances to which neurons are exposed during development. Perhaps in some of these cases, we could then elevate the neurotrophic *hypothesis* to the status of *theory*. Even there, however, the seemingly heretic notion that naturally occurring cell death can be brought about not only as a consequence of a short supply of survival substances but also by a dedicated set of killing molecules, a concept which immunologists have been all too familiar with, is rapidly gaining support among several developmental neurobiologists.

3
Neurotrophic Factors

Many of the concepts described above were established through work on the first growth factor to be discovered, nerve growth factor or NGF. In the peripheral nervous system, NGF functions as a target-derived survival signal for sympathetic neurons and subpopulations of, mainly heat- and pain-sensitive, sensory neurons. Subsequent to the discovery of NGF, many other growth factors were found to affect neuron survival in vitro and in vivo, and similar to most other gene products, these were found to belong to distinct families of structurally related proteins. Members of the same family are likely derived from a common ancestral molecule at some point during metazoan evolution. NGF belongs to a family of neurotrophic factors known as the neurotrophins, which in addition includes brain-derived neurotrophic factor (BDNF), neurotrophin-3 (NT3) and neurotrophin-4 (NT4). Table 1 shows some families of molecules with known neurotrophic activity. Although some of these were originally discovered as neurotrophic factors, such as the neurotrophins and the GDNF (glial cell line-derived neurotrophic

Table 1. Families of neurotrophic factors

Neurotrophic factor families	Family members	Signalling receptors	Accessory receptors
Neurotrophins	NGF, BDNF, NT3, NT4	TrkA, TrkB, TrkC (intrinsic tyr kinase)	p75 neurotrophin receptor* (contains death domain)
GDNF family	GDNF, NTN, PSP, ARTN	c-Ret (intrinsic tyr kinase)	GFRα1-4 (GPI-anchored)
Neurokine family	LIF, CNTF, CT-1	gp130, LIFRβ (associated tyr kinase)	CNTFRα (GDI-anchored)
Transforming growth factors	TGFβ-2, TGFα-3	TBRI, TBRII (intrinsic ser-thr kinase)	TBRIII
Insuline-like growth factors	IGF-1, IGF-2	IGFRs	Unknown
Fibroblast growth factors	FGF-1, FGF-2, FGF-5	FGFR	Proteoglycan
Hepatocyte growth factor	HGF	c-Met (instrinc tyr kinsase)	Unknown

* The p75 neurotrophin receptor has been shown to signal in the absence of Trk receptors (Dechant and Barde 1997). NGF, nerve growth factor; BDNF, brain-derived neurotrophic factor; NT, neurotrophin; Trk, tropomyosin receptor kinase; GDNF, glial cell line-derived neurotrophic factor; NTN, neurturin; PSP, persephin; ARTN, artemin; Ret, rearranged during transfection; GFRα, GDNF receptor family alpha; GPI, glysosyl phosphatyduylinositol; LIF, leukemia inhibitory factor; CNTF, ciliary neurotrophic factor; CT-1, cardiotrophin-1; TBR, TGFβ receptor; IGF, insuline-like growth factor; FGF, fibroblast growth factor; HGF, hepatocyte growth factor.

factor) ligand family, many other were first characterised as cytokines, such as LIF (leukemia inhibitory factor) and the TGFβs (transforming growth factor-βs), or as mitogenic growth factors, such as the FGFs (fibroblast growth factors) and HGF (hepatocyte growth factor).

Neurotrophic factors interact with multicomponent receptors formed by several subunits with distinct functions (reviewed in (Ibáñez 1998)). Typically, an accessory component allows or potentiates binding to a signalling component. The latter have intrinsic signalling capabilities, such as receptor tyrosine or serine-threonine kinases, or are associated with a cytoplasmic kinase. Accessory receptors may also signal in the absence of signalling components, although this has so far only been documented for the p75 neurotrophin receptor (Dechant and Barde 1997).

4
Beyond the Neurotrophic Hypothesis

Although the neurotrophic hypothesis helps to explain the role of the target in the control of naturally occurring cell death, it falls short of encompassing all the effects of neurotrophic factors. Work during the past ten years has established that these molecules can work in a variety of ways in addition to

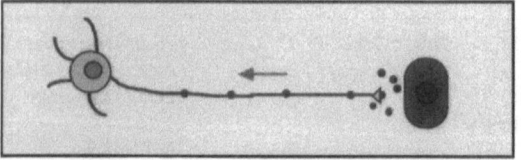

target–derived:
• sympathetic neurons get NGF from pineal and salivary gland
• Ia sensory neurons get NT-3 from muscle

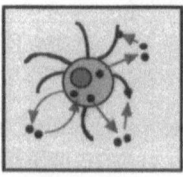

autocrine:
• adult sensory neurons are
  supported by their own BDNF

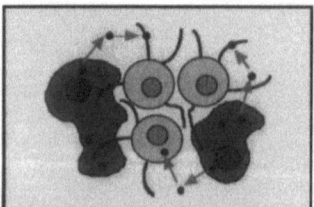

paracrine:
• early sensory neurons get NT-3 from
  surrounding cells
• central neurons get trophic factors
  from glial cells

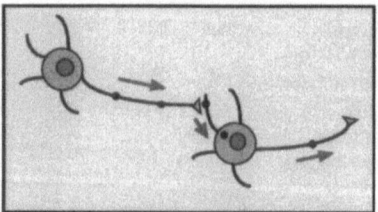

anterograde:
• DRG neurons transport BDNF to the spinal cord
• cortical neurons transport BDNF to the striatum

Fig. 3. Variable modes of action of neurotrophic factors: target derived (retrograde), autocrine, paracrine and anterograde

the classical target-derived mode of action (Fig. 3). Neurotrophic factors can act in a paracrine way within the tissue in which they are generated. Thus, for example, developing sensory neurons obtain NT3 during early development from surrounding cells before they are dependent on their target of innervation (Farinas et al. 1996), and several populations of central neurons obtain trophic support from surrounding glial cells. In several instances, neurons produce trophic factors for their own use in an autocrine way, as in the case of adult sensory neurons which are maintained by a BDNF autocrine loop (Acheson et al. 1995). Finally, neurotrophic factors are also transported anterogradely, down the axon, and are released and taken up by the innervating target (Altar and Disefano 1998), a mode of action in direct opposition to the one promoted by the neurotrophic hypothesis.

These alternative modes of delivery are echoed by the variety of biological functions attributed to neurotrophic factors (Fig. 4) which, in addition to cell survival, include neurite outgrowth, differentiation, and effects on synaptic plasticity. Interestingly, the neurotrophins BDNF and NT3 have been shown to enhance basal synaptic transmission as well as activity-dependent plasticity in hippocampal neurons through both pre- and post-synaptic mechanisms (Schuman 1999; Thoenen 1995). Moreover, some neurotrophins are also known to be released in an activity-dependent way (Blochl and Thoenen 1995; Goodman et al. 1996). The anterograde transport and activity-dependent release of neurotrophic factors, together with their effects on synaptic transmission, have in fact brought these molecules into the functional category of neuromodulators and neuropeptides.

The effects of neurotrophic factors on neurite extension has led to the investigation of their chemotropic activities, i.e. a direction-biased stimulation of neuritic growth. Figure 5A shows one of the "acid tests" used to investigate the presence of a diffusible chemotropic activity (Ebens et al. 1996). Although not all known neurotrophic factors would pass this test, the example shown illustrates the chemotropic action of HGF on axonal growth from developing motorneurons. Based on these and other results, Ebens et al. (1996) have in fact proposed that HGF produced by skeletal muscle may serve as a chemoattractant signal for ingrowing motor axons. A more dramatic example of the effects of neurotrophic factors on the direction of axonal growth is illustrated in Fig. 5B, which shows how a gradient of the neurotrophin BDNF triggers an attractive turning response of the growth cone of Xenopus spinal neurons in culture (Song et al. 1997). In the experiment shown, Song et al. (1997) also demonstrated how the same neurotrophin gradient induced repulsive turning of these growth cones in the presence of a competitive analogue of cAMP (Fig. 5B), suggesting that the same guidance cue may trigger opposite turning behaviours of the growth cone depending on its specific biochemical state.

In retrospective, it should not come as a surprise that potent and specific biochemical messengers such as these molecules have been recruited to control the complex behaviour of neurons. Semantic issues aside, however,

differentiation

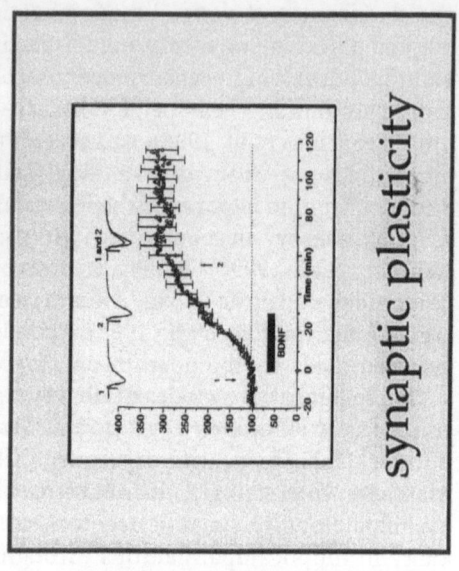

synaptic plasticity

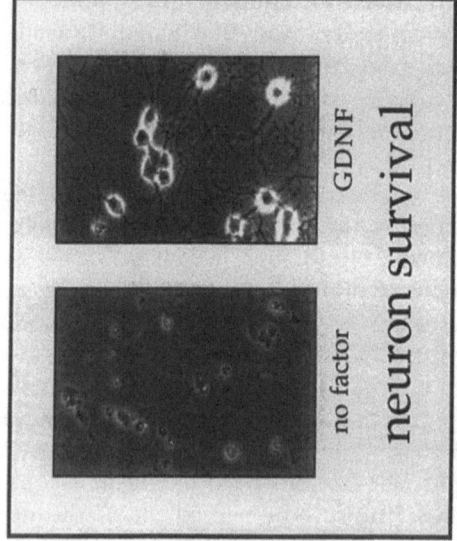

neuron survival

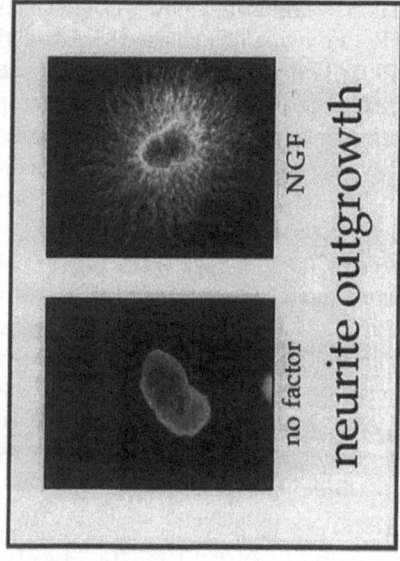

neurite outgrowth

the recent work on neurotrophic factors has clearly transformed these molecules into remarkably versatile signals for cell–cell communication in the nervous system.

5
Revisiting the Neurotrophic Hypothesis with Molecular Genetics

As we have seen, the neurotrophic hypothesis predicts massive developmental loss of neurons after target removal due to a shortage of target-derived neurotrophic substances. On the other hand, because neurotrophic factors are made in limiting amounts, the theory predicts increased survival of neurons following target enlargement, due to a surplus of neurotrophic support (Fig. 6). In recent years, these predictions have been very elegantly validated in a series of experiments using genetically modified mice. Instead of target removal or administration of blocking antibodies, homologous recombination techniques have been used to selectively eliminate genes coding for distinct neurotrophic molecules. In most cases, deletion of neurotrophic factor genes leads to selective neuronal losses. Typically, the deficits in these mice are more dramatically manifested among subpopulations of peripheral neurons (Reichardt and Fariñas 1997). Unfortunately, the limited life-span of these mice does not allow for a full evaluation of the survival of central neurons, which normally undergo programmed cell death during the first postnatal weeks. Nevertheless, the analyses made so far indicate that, at least during embryonic development, central neurons are not as severely affected as peripheral neurons by inactivation of individual neurotrophic factor genes. Thus, in comparison, peripheral neurons appear to depend upon a much more limited set of neurotrophic factors than central neurons. Alternatively, developmental cell death in the central nervous system may be regulated by different mechanisms, such as electrical activity, not directly involving target-derived neurotrophic factors.

The molecular genetics alternative to target enlargement has been implemented through overexpression of neurotrophic factor genes in transgenic mice. Here again, peripheral populations of neurons are the more conspicuously affected. Peripheral sensory and sympathetic ganglia become dra-

---

Fig. 4. Biological activities of neurotrophic factors. Clockwise from the top left: survival of embryonic chick paravertebral sympathetic neurons in vitro in the presence of GDNF (adapted from ref. (Trupp et al. 1995)), increased tyrosine hydroxylase expression and cell soma size induced by GDNF in central noradrenergic neurons in vivo (adapted from ref. (Arenas et al. 1995)), enhanced synaptic transmission stimulated by BDNF in hippocampal slices (adapted from ref. (Kang and Schuman 1995)), stimulation of neurite outgrowth by NGF in explants of embryonic chick paravertebral sympathetic ganglia (adapted from ref. (Trupp et al. 1995))

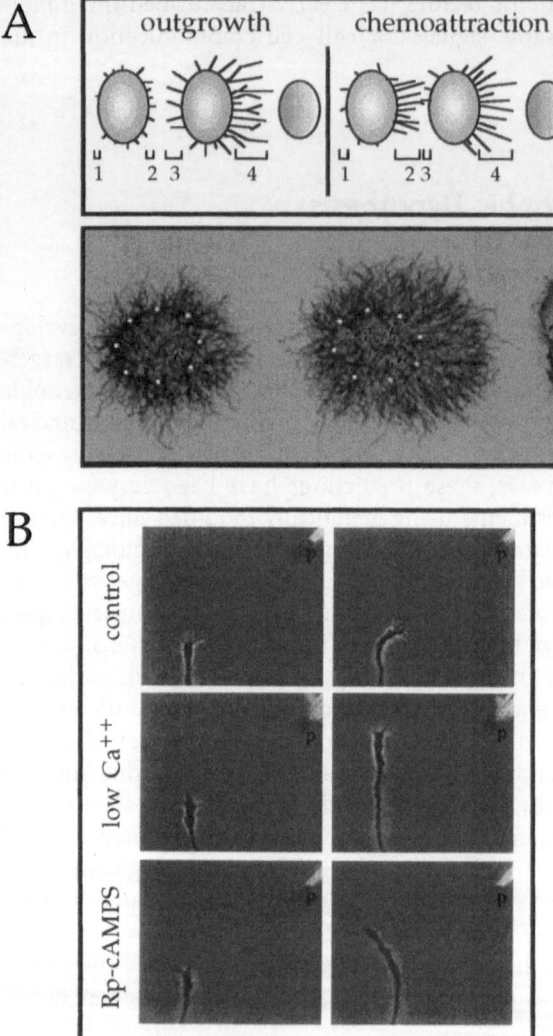

Fig. 5A, B. Chemotropic activities of neurotrophic factors. (**A**) "Acid test" of a soluble chemoattractant activity. An activity that promotes outgrowth in a concentration-dependent but direction-independent manner elicits a more profuse outgrowth from the far side of the proximal explant than from the near side of the distal explant (i.e., distance 3 > distance 2 in left part of the diagram). If the activity is chemoattractant, it stimulates a more profuse outgrowth from the near side of the distal explant than from the distal side of the proximal explant (i.e., distance 2 > distance 3 in right part of the diagram). The lower panel shows coculture of two ventral explants in tandem with forelimb mesenchyme (right) as source of HGF. Reproduced from ref. (Ebens et al. 1996). (**B**) A gradient of BDNF triggers an attractive turning response of the growth cone of Xenopus spinal neurons in culture (upper panels). This effect is dependent on extracellular calcium (middle panels) and can be transformed into a repulsive turning response in the presence of a competitive analogue of cAMP (bottom panels). Reproduced from ref. (Song et al. 1997)

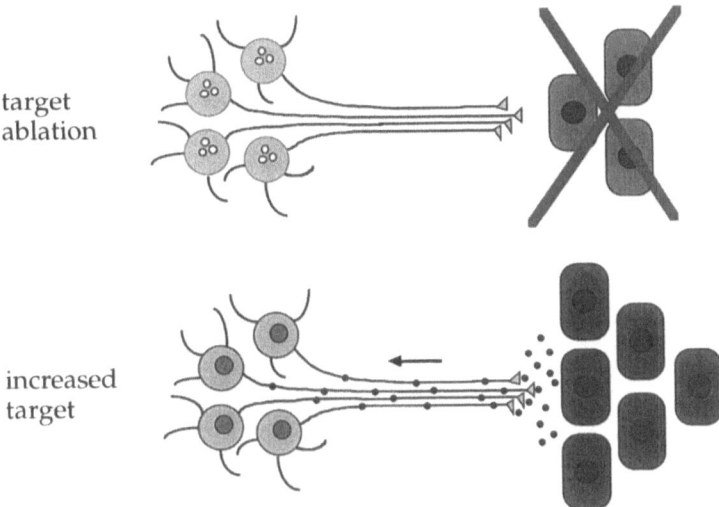

Fig. 6. Predictions of the neurotrophic hypothesis. Target removal will cause massive developmental loss of neurons due to a shortage of target-derived neurotrophic substances (upper diagram). Target enlargement will result in increased survival of neurons due to a surplus of neurotrophic support (bottom panel)

matically enlarged as a result of being exposed to abnormally elevated amounts of neurotrophic factors during the period of naturally occurring cell death, confirming that, at least in the periphery, neurotrophic substances are present in limited amounts as predicted by the neurotrophic hypothesis. Although gene overexpression might at first look like a more artificial way of manipulating the levels of a regulatory polypeptide, it is important to remember that in the case of a neurotrophic factor, the overexpressed protein has no other way to function than through endogenous receptors made by normal cells that would otherwise respond to normal levels of the factor. Gene overexpression can therefore be seen as a way to magnify an endogenous signalling pathway, and may be used to make manifest activities that might otherwise go unnoticed.

6
Selective Neuronal Loses and Maturation Deficits Following Inactivation of Genes Encoding Neurotrophic Factors or Their Receptors

Gene knock-out experiments have largely confirmed several of the predictions of the neurotrophic hypothesis, particularly among populations of neurons in the peripheral nervous system (Fig. 7): i) mice that develop in the

A

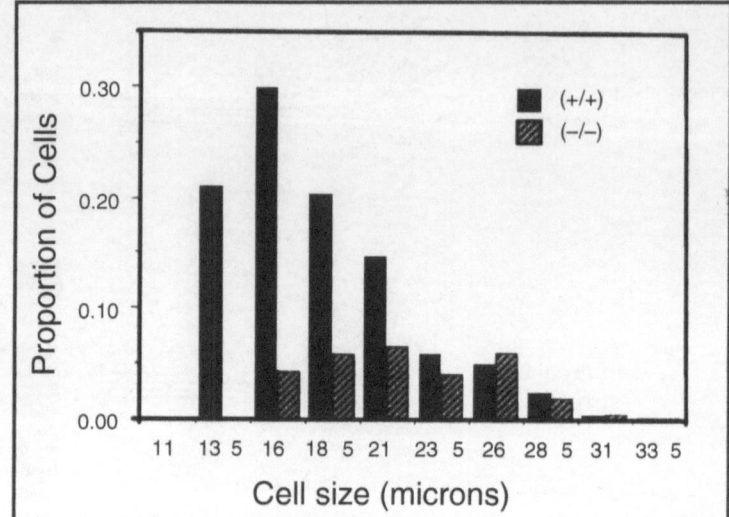

B

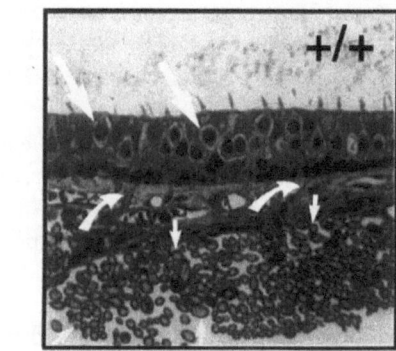
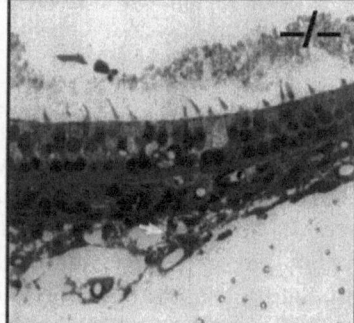

C

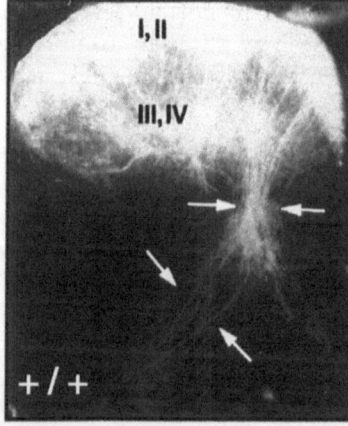
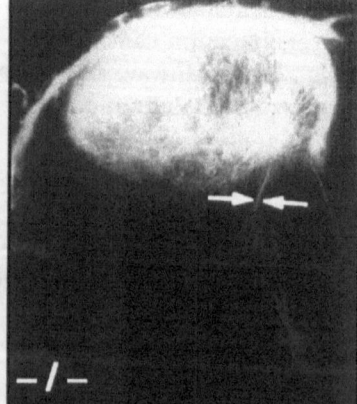

absence of NGF or its receptor TrkA almost completely lack sympathetic ganglia as well as 70% of their sensory neurons, predominantly small, heat- and pain-sensitive cells (Crowley et al. 1994; Smeyne et al. 1994); ii) inactivation of the BDNF gene, or its receptor TrkB, leads to selective losses in trigeminal, petrosal-nodose and vestibular ganglia (Ernfors et al. 1994; Jones et al. 1994; Klein et al. 1993); and iii) knock-out of the NT3 gene, or its receptor TrkC, results in specific absence of proprioceptive sensory neurons (Ernfors et al. 1994; Fariñas et al. 1994; Klein et al. 1994). These results are consistent with previous evidence indicating that these neurotophins serve as a target-derived trophic factors for those cells. Finally, inactivation of two members of the GDNF ligand family, GDNF (Moore et al. 1996; Pichel et al. 1996; Sánchez et al. 1996) and neurturin (Heuckeroth et al. 1999), or their specific accessory receptors GFRα1 (Cacalano et al. 1998; Enomoto et al. 1998) and GFRα2 (Rossi et al. 1999), has revealed deficiencies in enteric and parasympathetic neurons, respectively. These and other examples attest to the importance of target-derived neurotrophic factors for the development of the peripheral nervous system (for a more extensive overview of the effects of these and other gene knock-outs on the peripheral nervous system the reader is referred to the recent review by Reichardt and Fariñas 1997).

In contrast, the survival of many subpopulations of central neurons known to express receptors and to respond to exogenous neurotrophic factors in vitro and in vivo appears not to be compromised by disruption of neurotrophin genes. Initial studies on developing central neurons of neurotrophin knockout mice were hampered by the relatively short survival times of these animals, typically of a few days after birth. By extending the survival of these animals for up to 2 to 3 week, and combining several gene mutations, more recent studies have revealed that several neurotrophins, in particular BDNF and its receptor tyrosine kinase TrkB, are necessary for the survival of neurons in several regions of the postnatal brain. Mice lacking TrkB receptors show increased numbers of apoptotic neurons in several regions of the postnatal brain, including hippocampus, cerebral cortex, striatum and thalamus (Alcatara et al. 1997). The dentate gyrus region of the hippocampal formation appears to be the most affected. Interestingly, the peak in cell death in *trkB* knock-out mice occurs right after the period of programmed cell death (Alcantara et al. 1997), which in the rodent forebrain is thought to take place between the first and second postnatal weeks.

A much higher rate of cell death of postnatal dentate gyrus neurons was seen in double mutant mice with reduced expression in both the *trkB* and

---

Fig. 7A–C. Distinct losses of populations of peripheral neurons after inactivation of different neutrophin genes. (A) NGF knock-out mice have a selective loss of small diameter nociceptive neurons in dorsal root ganglia (DRG) (reproduced from ref. (Crowley et al. 1994)). (B) Massive loss of nerve terminals in the inner year of BDNF knock-out mice (reproduced from ref. (Ernfors et al. 1995)). (C) Loss of Ia spinal cord afferents in NT3−/− mice (reproduced from ref. (Ernfors et al. 1994))

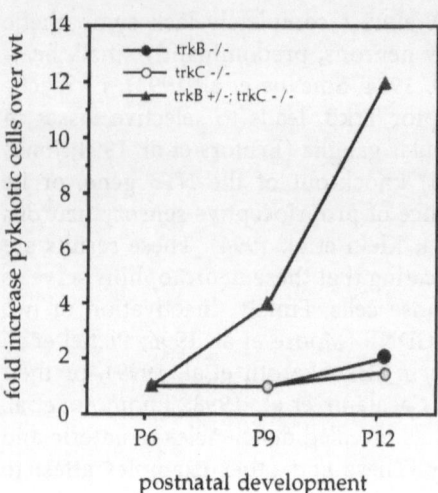

Fig. 8. Loss of neurons in the postnatal dentate gyrus of mice lacking TrkB and TrkC neurotrophin receptors. Adapted from Minichiello and Klein (1996)

trkC genes, indicating partially overlapping functions of these two neurotrophin receptors in the control of cell survival (Minichiello and Klein 1996). The death of neurons was developmentally regulated and followed the period of programmed cell death (Fig. 8). The dramatic cell death seen at P12 in *trkB+/−*; *trkC−/−* mice is remarkable, considering that these animals still carry a normal *trkB* allele. Mice lacking all expression of *trkB* and *trkC* genes died prematurely at P1 (Minichiello and Klein 1996). Because of the limited life span of these animals, it is not known whether all dentate gyrus neurons would undergo cell death in the complete absence of receptors for BDNF and NT3. Interestingly, no significant apoptosis was detected in the Ammon's horn, where hippocampal pyramidal neurons are located. The reason why dentate gyrus granule neurons are most vulnerable to the lack of neurotrophin signalling is not known. Intriguingly, another population of granule cells, that of the cerebellum, was also severely affected by combination of mutations in the *trkB* and *trkC* genes (Minichiello and Klein 1996). In this case, the deficits were more pronounced among the premigratory population in the external granule cell layer, with up to 7.5-fold increase in pyknotic cells compared to wild type. In agreement with these deficits, the cross-sectional area of excitatory, calbindin-positive mossy fibres was significantly reduced in *trkB*; *trkC* double knock-outs. Once fully matured and situated in the internal granule cell layer, cerebellar granule neurons become independent of neurotrophins, perhaps by upregulating receptors for other neurotrophic factors. In contrast, the total number of Purkinje cells, the other major cell type in the cerebellum, was unchanged in single or double mutants of the *trkB* and *trkC* genes (Minichiello and Klein 1996). However, a dramatic reduction in the thickness of the Purkinje cell dendritic tree of up to 50% was observed in double mutant mice, indicating a role for BDNF and NT3 in Purkinje cell dendritic differentiation. Considering the extensive degree of cell death seen among the cerebellar granule cell population, it is also possible that the reduction in dendritic arborization seen in Purkinjee cells in an indirect result of reduced interactions among these two neuronal populations. More recently, Minichiello et al. (1998) reported that mice with a mutation in the TrkB binding site to the adaptor protein Shc, responsible for one of the major outputs of the TrkB signal, have normal Purkinje cell arborizations (Minichiello et al. 1998), indicating that this event is largely independent of the Shc signalling pathway.

Using antibodies that specifically recognize activated Trk receptors, Schwartz et al. (1997) observed a dramatic reduction in Trk activation in cerebellar neurons of BDNF mutant mice (Schwartz et al. 1997). As in the case of trkB mutant mice, this was accompanied by a marked reduction in the arborization of Purkinje cells and a delayed development of the characteristic cerebellar layers. At P21, the external granule cell layer is still visible in *BDNF* −/− mice, while the molecular layer is remarkably thinner (Schwartz et al. 1997). Increased cell death was also seen among the granule cell population although not as pronounced as in double *trkB/trkC* mutants. Since BDNF is not synthesized by the granule cell targets, the Purkinje cells, but in granule cell afferents and by granule cells themselves, BDNF presumably acts in an autocrine or paracrine fashion to promote granule cell survival.

A role for TrkB and TrkC receptors in the growth and refinement of neural connections has also been proposed based on examination of hippocampal afferents in *trkB* and *trkC* mutant mice (Martínez et al. 1998). Commisural and entorhinal afferents were found to have a reduced number of axon collaterals and decreased densities of axonal varicosities in both *trkB*−/− and *trkC*−/− mice. Lower density of synaptic contacts and reduced expression of proteins responsible for synaptic vesicle exocytosis and neurotransmitter release (SNAREs) was also observed, particularly in mice lacking TrkB expression (Martínez et al. 1998). The ingrowth or layer-specific targeting of hippocampal connections was, however, unaltered in these mice. Together, these data indicate a role for TrkB and TrkC neurotrophin receptors in maturation and synaptogenesis of hippocampal connections.

In contrast to BDNF, NT3 and their respective Trk receptors, NGF and TrkA have a much more restricted expression in the developing and adult brain, primarily in cholinergic neurons of the basal forebrain and striatum. Mice lacking NGF or TrkA show normal number of basal forebrain cholinergic neurons but a reduced cholinergic fibre density in target areas, such as the hippocampus and cerebral cortex (Crowley et al. 1994; Smeyne et al. 1994). More recent analyses on *trkA* knock-out mice revealed a role for this receptor in the normal maturation of basal forebrain as well as striatal cholinergic neurons during development. Cholinergic neurons in the striatum of *trkA*−/− mice had a significant reduction in soma size at P7/8 and P20–25 (Fagan et al. 1997). At the later stages, the number of striatal cholinergic neurons was reduced in mutant mice by 25%, although no significant differences could be seen in the total number of neurons at this stage. Similar phenotypic defects were observed in cholinergic neurons of the basal forebrain. Knock-out cholinergic neurons were significantly smaller than wild type and had a consistently lighter staining for choline acetyltransferase (ChAT) one of the main markers of the cholinergic phenotype (Fagan et al. 1997). By P20–25, the number of ChAT-immunoreactive neurons in the medial septum of *trkA−/−* animals was reduced by 36% compared to control animals, and the remaining neurons were clearly atrophic. Importantly, a 2-fold increase in the number of TUNEL-positive cells was seen in the septum of *trkA* knock-out mice at P7, coincident with the period of programmed cell death in this region (Fagan et al. 1997). Together these results indicate that TrkA expression is not only required for the normal phenotypic maturation of septal cholinergic neurons, but also for their survival. Coincident with the observed atrophy and apparent loss of septal cholinergic neurons at P20–25, a profound deficit in cholinergic fibre density was seen in all regions of the hippocampus at this stage. Similar deficits were also noted in the cortex (Fagan et al. 1997), indicating abnormalities in cholinergic neurons of the nucleus basalis of Meynert.

Although the p75 neurotrophin receptor collaborate with Trk receptors to promote neurotrophin signalling, it is now clear that p75 has signalling capabilities on its own (Dechant and Barde 1997). In particular, recent studies have suggested that p75 can mediate cell death in certain cell types (Casaccia-Bonnefil et al. 1996; Frade et al. 1996; Rabizadeh et al. 1993). An increased number of basal forebrain cholinergic neurons has been reported by some investigators in mice lacking p75 expression, consistent with the idea that p75 may signal apoptosis in cholinergic neurons that express this receptor in the absence of TrkA (Van der Zee et al. 1996; Yeo et al. 1997). These data was later questioned by other investigators who reported decreased numbers of cholinergic cells in the basal forebrain of *p75−/−* mice (Peterson et al. 1997). Moreover, knock-out of the *trkA* gene leaves many neurons in the basal forebrain that are p75 positive and TrkA negative but that do not die over a 4-week period (Fagan et al. 1997), indicating that expression of p75 in the absence of TrkA is not sufficient per se to elicit apoptosis. The role of the p75 receptor in the control of neuron number has been difficult to establish with precision, as it may participate in both cell survival or death depending on the presence or absence of Trk receptors and neurotrophin ligands (Miller and Kaplan 1998). Together, these findings indicate that neurotrophins have critical roles in CNS development including and beyond neuronal survival.

7
Neurotrophic Factors Regulate Target Invasion

Despite the elegance and intellectual appeal of gene knock-out experiments, one limitation of this approach is that it prevents us from investigating any function that is downstream of the first deficit arising during the development of mutant animals. If survival is compromised by deletion of a neurotrophic factor gene, we are prevented from gaining any insights on the possible roles of that molecule in differentiation, outgrowth, target innervation, and so on, since, of course, these are attributes of "live" neurons only. Ectopic expression of neurotrophic factors is one way to circumvent this problem which has offered valuable insights into different aspects of the function of these molecules. Analyses of transgenic animals expressing neurotrophic factors ectopically have shown that these molecules are capable of affecting innervation patterns in vivo. Overexpression of NGF in pancreatic islets has been shown to induce dense sympathetic innervation (Edwards et al. 1989). In mice overexpressing NGF within sympathetic ganglia, these neurons still project to their normal target areas but fail to innervate them (Hoyle et al. 1993). These and other observations suggest that neurotrophic factors are not required for axons to reach the vicinity of their targets but are important for target invasion and connectivity.

At our laboratory, we have studied transgenic mice overexpressing different neurotrophic factors under the control of the *nestin* gene promoter, which targets expression to muscle and neural cell precursors (Zimmerman et al. 1994). The *nestin* gene promoter targets strong ectopic expression to the brain and spinal cord, particularly in the subventricular cell layer where neural precursor cells are located. Mice overexpressing BDNF or NT3 under the control of this promoter die prematurely around birth. At least in the case of the BDNF overexpressors, this appears to be due to cardiac malformations (Barbara L. Hempstead, personal communication).

Mice overexpressing NT3 had limb proprioceptive sensory deficits very similar to those displayed by NT3−/− mice but, unlike the knock-out mice, in the absence of neuronal loss (Ringstedt et al. 1997). In fact, elevated numbers of neurons were present in DRG of these transgenic mice, consistent with a boosting in cell survival produced by the ectopically expressed NT3. Despite the increased number of cells, parvalbumin expression, a marker of Ia proprioceptive neurons, was absent in DRG of transgenic mice, indicating a failure of Ia neurons to differentiate. This was likely due to the failure of prospective Ia neurons to form appropriate connections. In fact, NT3 overexpressing mice lacked muscle spindles in limb muscles, and the soleus nerve had a 50% reduction in the number of axons, indicating a lack of Ia muscle afferents. Moreover, tracing experiments revealed absence of central Ia afferents in the ventral motor columns of the spinal cord, despite a normal complement of axons in the dorsal roots entering the spinal cord. Careful examination revealed that the ectopic NT3 expression in the region around the central canal deflected Ia axons away from their normal ventral path towards the midline of the cord (Fig. 9). This misrouting would be sufficient to disrupt the monosynaptic reflex circuit and to bring about the proprioceptive sensory deficits seen in these mice. The absence of Ia afferents in muscle and the aborted Ia cell differentiation remain unexplained but could be a consequence of the failure of prospective Ia neurons to reach their central target.

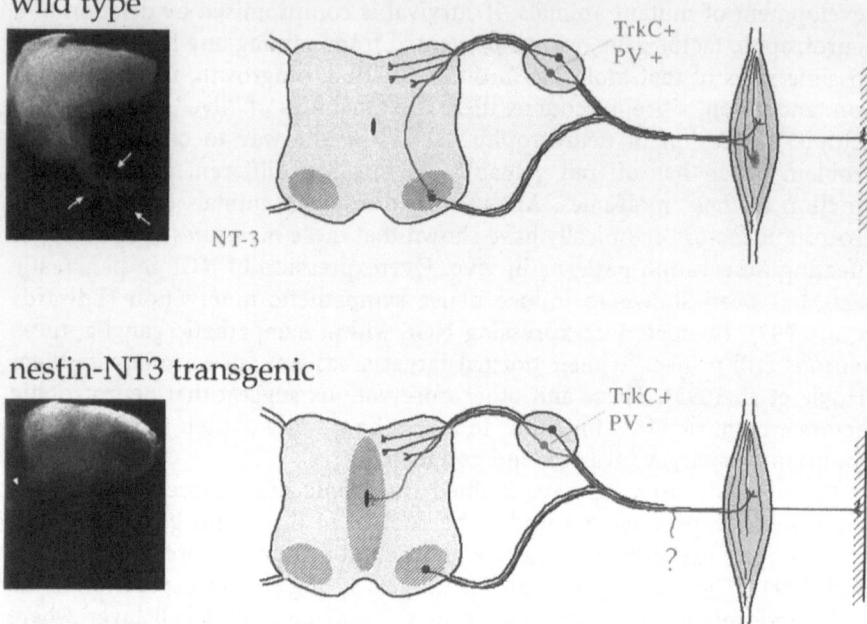

Fig. 9. Misrouting of Ia afferents in spinal cord of nestin-NT3 mice. In wild type mice (upper panels), Ia neurons (TrkC positive) express paravalbumin (PV), project to the muscle which induces the formation of muscle spindles, and enter the spinal cord projecting ventrally to contact NT3-expressing motorneurons. Nociceptive (TrkC negative) neurons innervate skin and their central axons terminate in the dorsal layers of the spinal cord. In nestin-NT3 mice (lower panels), prospective Ia neurons, as identified by expression of *trkCA* mRNA project to the dorsal spinal cord but turn away from their ventral path towards the midline, a region of high ectopic NT3 expression. Perhaps this failure to innervate their central target, aborts their differentiation program (PV negative). The fate of the peripheral projection of these neurons is unknown. The left panels show DiI tracing of Ia axons in spinal cord of wild type (top) and nestin-NT3 (bottom) mice (reproduced from ref. (Ringstedt et al. 1997))

Because ectopic expression of NT3 in developing muscle appears to be sufficient to rescue Ia innervation of motorneurons in NT3 knock-out mice, it has been argued that NT3, which normally is only expressed by developing motorneurons, plays no role in the guidance of Ia axons in the spinal cord (Wright et al. 1997). However, since overexpressed NT3 was likely to have diffused out from transgenic developing muscles, which in the mouse embryo are normally located in a ventrolateral position with respect of the spinal cord, it was unclear in that study whether ingrowing Ia axons might still have been exposed to a dorso-ventral gradient of NT3 as they traversed the spinal cord. Had NT3 nothing to do with guidance of Ia afferents, ectopic expression of this neurotrophin in the spinal cord should have no effect on the direction of growth of these axons. This prediction is at odds with our observations in nestin-NT3 transgenic mice, which show that NT3 is indeed capable of affecting the guidance of Ia axons in the spinal cord. Most likely, the growth of these axons in the spinal cord is controlled by an array of cues, of which NT3 is a component. Thus, this neurotrophin may act in conjunction with other molecules and guidance signals to modulate the direction of growth of the Ia axon once inside the spinal cord. Perhaps the ventral route is the default path for these axons imposed by repulsive or other type of cues, with NT3 and other molecules contributing a superimposed force field that modulates the direction of growth of the axon terminal.

Mice overexpressing BDNF under the *nestin* gene promoter had enlarged nodose/petrosal ganglia, again consistent with the rescue of neurons from programmed cell death by the elevated levels of BDNF expressed in these mice (Ringstedt et al. 1999). Despite the increased number of neurons, the number of taste buds in the gustatory epithelium of the tongue was drastically reduced. The number of fungiform papillae were significantly decreased, the papillae were smaller in size, and circumvallate papillae had a deranged morphology. Paradoxically, similar deficits had been seen in BDNF knockout mice, which, however, lack a substantial amount of neurons in the nodose/petrosal ganglia (Nosrat et al. 1997; Oakley et al. 1997). Although gustatory fibres reached the tongue in normal numbers in nestin-BDNF transgenic mice, the amount and density of nerve fibres in gustatory papillae were much lower than in wild type litermates. Instead of reaching towards their targets in the gustatory epithelium, which is the normal site of BDNF expression in the tongue, gustatory fibres appeared stalled among muscle fibres at the base of the tongue, a site of ectopic BDNF overexpression, where they formed abnormal branches and sprouts. It is likely that the malformations seen in gustatory papilla of BDNF overexpressing mice were a consequence of the failure of their BDNF-dependent afferents to reach their targets, probably because of influences from the ectopically expressed BDNF. These findings support the notion that mammalian taste buds and gustatory papillae require a normal complement of gustatory, BDNF-dependent innervation for appropriate development. Moreover, these data indicate that the correct spatial expression of BDNF in the tongue epithelium is crucial for the development of the normal pattern of gustatory innervation, and are in line with the general idea that neurotrophins play an important role in the regulation of target invasion and neuronal connectivity.

8
BDNF as a Maturation Factor for the Cerebral Cortex

The brains of transgenic mice overexpressing BDNF under the control of the *nestin* promoter showed a grossly aberrant architecture (Ringstedt et al. 1998). Microgyric sulcus formations consisting of invaginations of the marginal zone into the underlying cortical plate were seen in transgenic but not in wild type mice. In a few cases, heterotopic collections of cells were also observed in the marginal zone. Heterotopias had also been seen after exogenous administration of NT-4 (Brunstrom et al. 1997). Indicating that neurotrophins acting through the TrkB receptor are capable of affecting the migration or position of cells in the cerebral cortex. No cortical malformations were seen in nestin-NT-3 mice (C.F.I. and T. Ringstedt, unpublished observations), indicating that the effects seen with BDNF were not caused by non-specific overexpression of any neurotrophic factor. In contrast to control animals, calretinin labeling in the marginal zone of transgtenic mice was discontinuous, with stretches of marginal zone devoid of calretinin-positive cell bodies, and aggregations of Cajal-Retzius cells of abnormal morphology in microgyric sulci. In addition, the transgenic cerebral cortex showed a uniform distribution of labeling for microtubule associated protein-2 (MAP-2), a marker of differentiated neurons, without clearly delineated layers, indicating an aberrant lamination of the cortical plate. Support for this notion was obtained from BrdU-labeling experiments. These studies showed several early-born neurons reaching out all the way into the marginal zone, and many later-born neurons that did not migrate into the outer layers of the cortex and remained instead in deeper layers.

Strikingly, these abnormalities are similar to the phenotype of *reeler* mice, which are deficient in Reelin expression (Caviness Jr et al. 1988; Ogawa et al. 1995). Reelin, an early marker of Cajal-Retzius cells, is necessary for the formation of the inside-out layer organization of the cerebral cortex. In *reeler* mice, which lack expression of Reelin (D'Arcangelo et al. 1995; Hirosune et al. 1995), cortical layering is grossly abnormal: migrating neurons fail to split the preplate and instead line up below it, in the same order as they are born (Caviness Jr 1982; Caviness Jr et al. 1988; Goffinet 1984). The similarities observed between nestin-BDNF and *reeler* mice indicated that BDNF overexpression might have affected the levels of Reelin in transgenic mice. Analysis of mRNA and protein expression showed a profound and dose-dependent reduction in Reelin expression in the marginal zone of nestin-BDNF mice compared to wild type (Fig. 10A). Thus, many of the defects observed in these transgenic mice, including aberrant lamination and inverted layering of the cerebral cortex, could be the result of a reduction in Reelin levels produced by BDNF overexpression. BDNF was also shown to downregulate Reelin mRNA and protein expression after acute treatment of embryonic rat cortical cultures (Ringstedt et al. 1998), suggesting that the action of this neurotrophin on Reelin expression may be direct. These results demonstrate robust effects of BDNF on cortical Cajal-Retzius cells, and identify Reelin as a direct effector of this neurotrophin during brain development.

During normal development, very low BDNF expression can first be detected in the rodent brain at E13, with little change until P2, after which BDNF levels increase rapidly, with a peak at P14 (Friedman et al. 1991; Timmusk et al. 1994) (Fig. 10B) During postnatal brain development, and

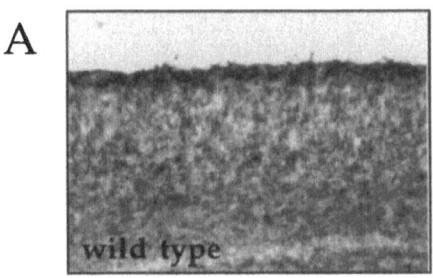
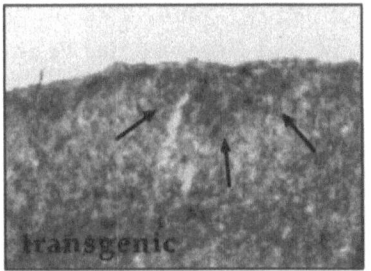

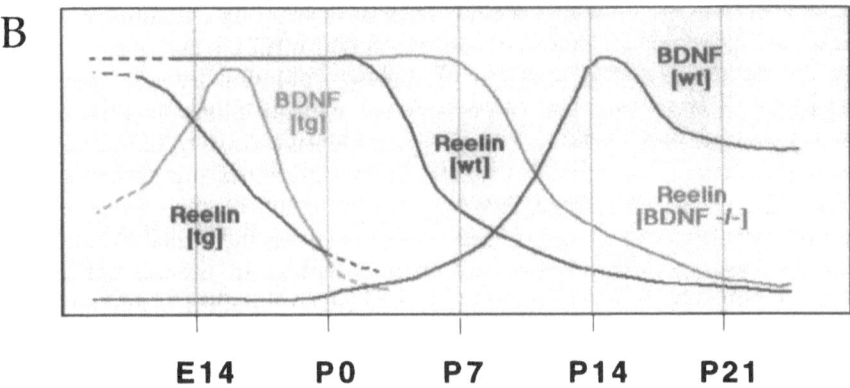

E14   P0   P7   P14   P21

Fig. 10A,B. BDNF is a negative regulator of Reelin expression during cortical development. (A) Reduced levels of Reelin expression in the marginal zone of nestin-BDNF transgenic mice. Arrows in left panel delineate an invagination of the marginal zone in the cortex of transgenic mice. The presence of marginal zone cells in this segment of transgenic cortex was confirmed independently by calretinin and cresyl violet staining. (B) Developmental patterns of expression of BDNF and Reelin mRNAs in the cerebral cortex of wild type (wt), nestin BDNF transgenic (tg) and BDNF knock-out (BDNF−/−) mice. Diagram based on data from Ringsted et al. (1998), Timmusk et al. (1994) and Schiffman et al. (1997)

concomitant with the completion of cortical lamination, Reelin expression is downregulated in cortical Cajal-Retzius cells (Schiffmann et al. 1997) (Fig. 10B). Downregulation of Reelin expression is followed by the disappearance of CR cells, which in the murine neocortex occurs during the second and third postnatal weeks (Del Rio et al. 1995; Derer and Derer 1990). BDNF upregulation coincides with the time-course of decline in Reelin levels and disappearance of CR cells that occurs during the normal development of postnatal rodent brain (Derer and Derer 1990; Schiffmann et al. 1997) (Fig. 10B), indicating the BDNF could be involved in several of these events. Interestingly, in mice lacking BDNF, a prolonged expression of Reelin has been seen in the marginal zone of the cerebral cortex (Ringstedt et al. 1998). At P7, BDNF−/− mice expressed almost 3-fold higher *reelin* mRNA levels compared to wild type mice (Fig. 10B), indicating that BDNF is required for the developmental downregulation of Reelin expression in cells of the marginal zone, and suggesting that this neurotrophin may acts as a cortical maturation factor.

9
Conclusions

Families of structurally and functionally related polypeptides control the differentiation, survival and maintenance of developing and adult vertebrate neurons. Targeted disruption of individual neurotrophic factor genes results in the nearly complete ablation of distinct subpopulations of peripheral neurons, underscoring the importance of neurotrophins as physiological target-derived survival factors in the peripheral nervous system. The physiological effects of neurotrophic factors in the central nervous system are only beginning to be understood through a combination of gene knock out and overexpression in vivo experiments. Taken together, the available data on the roles of neurotrophic factors and their receptors in the central nervous system indicates the existence of redundant survival pathways and additional roles for these molecules in central neurons, such as in the control of synaptic plasticity, gene expression, cell migration and cell fate.

Acknowledgements. Work at the author's laboratory is supported by grants from the Swedish Medical Research Council, the Swedish Cancer Society and the European Commission.

References

Acheson A, Conover JC, Fandl JP, Dechiara TM, Russell M, Thadani A, Squinto SP, Yancopulos GD, Lindsay RM (1995) A BDNF autocrine loop in adult sensory neurons prevents cell death. Nature 374:450–453

Alcantara S, Frisen J, Delrio JA, Soriano E, Barbacid M, Silossantiago I (1997) TrKB signaling is required for postnatal survival of cns neurons and protects hippocampal and motor neurons from axotomy-induced cell death. J of Neurosci 17:3623–3633

Altar CA, Distefano PS (1998) Neurotrophic trafficking by anterograde transport. Trends Neurosci 21:433–437

Arenas E, Trupp M, Åkerud P, Ibeáñez CF (1995) GDNF prevents degeneration and promotes the phenotype of brain noradrenergic neurons in vivo. Neuron 15:1465–1473

Blochl A, Thoenen H (1995) Characterization of nerve growth factor (NGF) release from hippocampal neurons: evidence for a constitutive and an unconventional sodium-dependent regulated pathway. Eur J Neurosci 7:1220–1228

Brunstrom JE, Grayswain MR, Osborne PA, Pearlman AL (1997) Neuronal heterotopias in the developing cerebral cortex produced by neurotrophin-4. Neuron 18:505–517

Cacalano G, Farinas I, Wang LC, Hagler K, Forgie A, Moore M, Armanini M, Phillips H, Ryan AM, Reichardt LF, Hynes M, Davies A, Rosenthal A (1998) GFRalpha-1 is an essential receptor component for gdnf in the developing nervous system and kidney. Neuron 21: 53–62

Casaccia-Bonnefil P, Carter BD, Dobrowsky RT, Chao MV (1996) Death of oligodendrocytes mediated by the interaction of nerve growth factor with its receptor p75. Nature 383:716–719

Caviness VS Jr (1982) Neocortical histogenesis in normal and reeler mice: a developmental study based upon [3H]thymidine autoradiography. Dev Brain Res 4:293–302

Caviness VS Jr, Crandall JE, Edwards MA (1988) The reeler malformation: implications for neocortical histogenesis. In: Peters A, Jones EG (eds) Development and maturation of cerebral cortex. Plenum, New York, pp 59–89

Crowley C, Spencer SD, Nishimura MC, Chen KS, Pittsmeek S, Armanini MP, Ling LH, Mcmahon SB, Shelton DL, Levinson AD, Phillips HS (1994) Mice lacking nerve growth factor display perinatal loss of sensory and sympathetic neurons yet develop basal forebrain cholinergic neurons. Cell 76:1001–1011

D'Arcangelo G, Miao GG, Chen SC, Soares HD, Morgan JI, Curran T (1995) A protein related to extracellular matrix proteins deleted in the mouse mutant reeler Nature 374:719–723

Dechant G, Barde Y-A (1997) Signaling through the neurotrophin receptor p75^{NTR} Curr Op Neurobiol 7:413–418

Del Rio JA, Martinez A, Fonseca M, Auladell C, Soriano E (1995) Glutamate-like immunoreactivity and fate of Cajal-Retzius cells in the murine cortex as identified with calretinin antibody. Cereb Cortex 5:13-21

Derer P, Derer M (1990) Cajal-Retzius cell ontogenesis and death in mouse brain visualized with horseradish peroxidase and electron microscopy. Neuroscience 36:839–856

Ebens A, Brose K, Leonardo ED, Hanson MG, Bladt F, Birchemeier C, Barres BA, Tessierlavigne M (1996) Hepatocyte growth factor scatter factor is an axonal chemoattractant and a neurotrophic factor for spinal motor neurons Neuron 17:1157–1172

Edwards RH, Rutter WJ, Hanahan D (1989) Direct expression of NGF to pancreatic beta-cells in transgenic mice leads to selective hyperinervation of the islets Cell 58:161–170

Enomoto H, Araki T, Jackman A, Heuckeroth RO, Snider WD, Johnson EM, Milbrandt J (1998) GFR-alpha-1-deficient mice have deficits in the enteric nervous system and kidneys. Neuron 21:317–324

Ernfors P, Lee K-F, Kucera J, Jaenisch R (1994) Lack of neurotrophin-3 leads to deficiences in the peripheral nervous system and loss of limb proprioceptive afferents Cell. 77: 503–512

Ernfors P, Lee KF, Jaenisch R (1994) Mice lacking brain-derived neurotrophic factor develop with sensory deficits. Nature 368:147–150

Eronfors P, Vandewater T, Loring J, Jaenisch R (1995) Complementary roles of BDNF and NT-3 in vestibular and auditory development. Neuron 14:1153–1164

Fagan AM, Garber M, Barbacid M, Silossantiago I, Holtzman DM (1997) A role for TrkA during maturation of striatal and basal forebrain cholinergic neurons in vivo. J Neurosci 17:7644–7654

Fariñas I, Jones KR, Backus C, Wang XY, Reichardt LF (1994) Severe sensory and sympathetic deficits in mice lacking neurotrophin-3. Nature 369:658–661

Farinas I, Yoshida CK, Backus C, Reichardt LF (1996) Lack of neurotrophin 3 results in death of spinal sensory neurons and premature differentiation of their precursors. Neuron 17: 1065–1078

Frade JM, Rodriguez-Tébar A, Barde YA (1996) Induction of cell death by endogenous nerve growth factor through its p75 receptor. Nature 383:166–168

Friedman W, Olson L, Persson H (1991) Cells that express brain-derived neurotrophic factor mRNA in the developing postnatal rat brain. Eur J Neurosci 3:688–697

Goffinet AM (1984) Events governing organization of postmigratory neurons. Studies on brain development in moral and reeler mice. Brain Res 319:261–296

Goodman LJ, Valverde J, Lim F, Geschwind MD, Federoff HJ, Geller AI, Hefti F (1996) Regulated release and polarized localization of brain-derived neurotrophic factor in hippocampal neurons. Mol Cellular Neurosci 7:222-238

Hamburger V (1975) Cell death in the development of the lateral motor column of the chick embryo. J Comp Neurol 160:535-546

Heuckeroth RO, Enomoto H, Grider JR, Golden JP, Hanke JA, Jackman A, Molliver DC, Bardgett ME, Snider WD, Johnson EM, Milbrandt J (1999) Gene targeting reveals a critical role for neurturin in the development and maintenance of enteric, sensory, and parasympathetic neurons. Neuron 22:253-263

Hirotsune S, Takahara T, Sasaki N, Hirose K, Yoshiki A, Ohashi T, Kusakabe M, Murakami Y, Muramatsu M, Watanabe S et al. (1995) The reeler gene encodes a protein with an EGF-like motif expressed by pioneer neurons. Nature Genetics 10:77-83

Hollyday M, Hamburger V (1976) Reduction of the naturally occurring motor neuron loss by enlargement of the periphery. J Comp Neurol 170:311-320

Hoyle GW, Mercer EH, Palmiter RD, Brinster RL (1993) Expression of NGF in sympathetic neurons leads to excessive axon outgrowth from ganglia but decreased terminal innervation within tissues. Neuron 10:1019-1034

Ibáñez CF (1998) Emerging themes in structural biology of neurotrophic factors. Trends Neurosci 21:438-444

Jones KR, Farinas I, Backus C, Reichardt LF (1994) Targeted disruption of the BDNF gene perturbs brain and sensory neuron development but not motor neuron development. Cell 76:989-999

Kang HJ, Schuman EM (1995) Long-lasting neurotrophin-induced enhancement of synaptic transmission in the adult hippocampus. Science 267:1658-1662

Klein R, Silossantiago I, Smeyne RJ, Lira SA, Brambilla R, Bryant S, Zhang L, Snider WD, Barbacid M (1994) Disruption of the neurotrophin-3 receptor gene Trkc eliminates Ia muscle afferents and results in abnormal movements. Nature 368:249-251

Klein R, Smeyne RJ, Wurst W, Long LK, Auerbach BA, Joyner AL, Barbacid M (1993) Targeted disruption of the trkB neurotrophin receptor gene results in nervous system lesions and neonatal death. Cell 75:113-122

Korsching S (1993) The neurotrophic factor concept: a reexamination. J Neurosci 13:2739-2748

Levi-Montalcini R (1987) The nerve growth factor 35 years later Science 237:1154-1162

Martínez A, Alcantara S, Borrell V, Del RJ, Blasi J, Otal R, Campos N, Boronat A, Barbacid M, Silos SI, Soriano E (1998) TrkB and TrkC signaling are required for maturation and synaptogenesis of hippocampal connections. J Neurosci 18:7336-7350

Miller FD, Kalpan DR (1998) Life and death decisions – a biological role for the p75 neurotrophin receptor. Cell Death Diff 5:343-345

Minichillo L, Casagranda F, Tatche RS, Stucky CL, Postigo A, Lewin GR, Davies AM, Klein R (1998) Point mutation in trkb causes loss of nt4-dependent neurons without major effects on diverse bdnf responses. Neuron 21:335-345

Minichiello L, Klein R (1996) TrkB and TrkC neurotrophin receptors cooperate in promoting survival of hippocampal and cerebellar granule neurons. Gene Dev 10:2849-2858

Moore MW, Klein RD, Farinas I, Sauer H, Armanini M, Phillips H, Reichardt LF, Ryan AM, Carvermoore K, Rosenthal A (1996) Renal and neuronal abnormalities in mice lacking gdnf. Nature 382:76-79

Nosrat CA, Blomlf J, Elshamy WM, Ernfors P, Olson L (1997) Lingual deficits in BDNF and NT3 mutant mice leading to gustatory and somatosensory disturbances, respectively. Development 124:1333-1342

Oakley RA, Lefcort FB, Clary DO, Reichardt LF, Prevedtte D, Oppenheim RW, Frank E (1997) Neurotrophin-3 promotes the differentiation of muscle spindle afferents in the absence of peripheral targets. J Neurosci 17:4262-4274

Ogawa M, Miyata T, Nakajima K, Yagyu K, Seike M, Ikenaka K, Yamamoto H, Mikoshiba K (1995) The reeler gene-associated antigen on Cajal-Retzius neurons is a crucial molecule for laminar organization of cortical neurons. Neuron 14:899–912

Peterson DA, Leppert JT, Lee KF, Gage FH (1997) Basal forebrain neuronal loss in mice lacking neurotrophin receptor p75. Science 277:837–838

Pichel JG, Shen LY, Sheng HZ, Granholm AC, Drago J, Grinberg A, Lee EJ, Huang SP, Saarma M, Hoffer BJ, Sariola H, Westphal H (1996) Defects in enteric innervation and kidney development in mice lacking gdnf. Nature 382:73–76

Rabizadeh S, Oh J, Zhong LT, Yang J, Bitler CM, Butcher LL, Bredesen DE (1993) Induction of apoptosis by the low-affinity NGF receptor. Science 261:345–348

Reichardt L, Fariñas I (1997) Neurotrophic factors and their receptors: roles in neuronal development and function. In: Cowan W, Jessell T and Zipursky S (eds) Molecular and Cellular Approaches ot Neural Development. New York: Oxford University Press

Ringstedt T, Ibáñez CF, Nosrat C (1999) Role of BDNF in target invasion in the gustatory system. J Neurosci 19:3507–3518

Ringstedt T, Kucera J, Lendahl U, Ernfors P, Ibáñez CF (1997) Limb proprioceptive deficits without neuronal loss in transgenic mice overexpressing neurotrophin-3 in the developing nervous system. Development 124:2603–2613

Ringstedt T, Linnarsson S, Wagner J, Lendahl U, Kokaia Z, Arenas E, Ernfors P, Ibáñez CF (1998) BDNF regulates reelin expression and cajal-retzius cell development in the cerebral cortex. Neuron 21:305–315

Rossi J, Luukko K, Poteryaev D, Laurikainen A, Sun YF, Laakso T, Erikainen S, Tuominen R, Lakso M, Rauvala H, Arumae U, Pasternack M, Saarma M, Airaksinen MS (1999) Retarded growth and deficits in the enteric and parasympathetic nervous system in mice lacking GFR alpha 2, a functional neurturin receptor. Neuron 22:243–252

Sánchez MP, Silossantiago I, Frisen J, He B, Lira SA, Barbacid M (1996) Renal agenesis and the absence of enteric neurons in mice lacking GDNF. Nature 382:70–73

Schiffmann SN, Bernier B, Goffinet AM (1997) Reelin mRNA expression during mouse brain development. Eur J Neurosci 9:1055–1071

Schuman EM (1999) Neurotrophin regulation of synaptic transmission. Curr Opin Neurobiol 9:105–109

Schwartz PM, Borghesani PR, Levy RL, Pomeroy SL, Segal RA (1997) Abnormal cerebellar development and foliation in bdnf−/− mice reveals a role for neurotrophins in cns patterning. Neuron 19:269–281

Smeyne RJ, Klein R, Schnapp A, Long LK, Bryant S, Lewin A, Lira SA, Barbacid M (1994) Severe sensory and sympathetic neuropathies in mice carrying a disrupted Trk/NGF receptor gene. Nature 368:246–249

Song HJ, Ming GL, Poo MM (1997) cAMP-induced switching in turning direction of nerve growth cones. Nature 388:275–279

Thonen H (1995) Neurotrophins and neuronal plasticity Science 270:593–598

Timmusk T, Belluardo N, Persson H, Metsis M (1994) Developmental regulation of brain-derived neurotrophic factor messenger RNAs transcribed from different promoters in the rat brain. Neuroscience 60:287–291

Trupp M, Rydén M, Jörnvall H, Timmusk T, Funakoshi H, Arenas E, Ibáñez CF (1995) Peripheral expression and biological activities of GDNF, a new neurotrophic factor for avian and mammalian peripheral neurons. J Cell Biol 130:137–148

Van der Zee C, Ross GM, Riopelle RJ, Hagg T (1996) Survival of cholinergic forebrain neurons in developing p75(ngfr)-deficient mice. Science 274:1729–1732

Wright DE, Zhou L, Kucera J, Snider WD (1997) Introduction of a neurotrophin-3 transgene into muscle selectively rescues proprioceptive neurons in mice lacking endogenous neurotrophin-3. Neuron 19:503–517

Yeo TT, Chuacouzens J, Butcher LL, Bredesen DE, Cooper JD, Valletta JS, Mobley WC, Longo FM (1997) Absence of p75 (NTR) causes increased basal forebrain cholinergic neuron size, choline acetyltransferase activity, and target innervation. J Neurosci 17:7594–7605

Zimmerman L, Lendahl U, Cunningham M, McKay R, Parr B, Gavin B, Mann J, Vassileva G, McMahon A (1994) Independent regulatory elements in the nestin gene direct transgene expression to neural stem cells or muscle precursors. Neuron 12:11–24

Growth Factor Influences on the Production and Migration of Cortical Neurons

Janice E. Brunstrom and Alan L. Pearlman[1]

1
Introduction

Production of neurons from progenitor cells is the first step in building the complex, six-layered cerebral cortex. The neurons that will populate the mature cortex are produced during development in an active proliferative neuroepithelium, the ventricular zone (VZ), adjacent to the cerebral ventricle (Rakic 1975; McConnell 1995; Caviness et al. 1996; Ross 1996). The first postmitotic neurons move out of the ventricular zone to form the preplate, just beneath the pia. Subsequent neuronal cohorts, generated in the neocortical ventricular zone, move into the preplate to form the cortical plate, which will eventually become layers 2 through 6 of cortex. At the earliest stages of cortical plate formation, preplate neurons are divided into two layers, the marginal zone, above the cortical plate, and the subplate, below it (Marin-Padilla 1971; Luskin and Shatz 1985; Allendoerfer and Shatz 1994). Many neurons move into the cortical plate under the guidance of the processes of radial glia (Hatten 1990; Rakic et al. 1994), but others, arising in distant proliferative zones, migrate tangentially into cortex, guided by cues that have not been defined (reviewed in Pearlman et al. 1998).

Although the cells of the pseudostratified neocortical VZ are relatively homogeneous in appearance, they are remarkably diverse in their capacity to produce offspring of different phenotypes. Progeny with distinct laminar fates are produced at different times in response to environmental cues, but the degree to which a progenitor can respond to these cues is regulated by programs intrinsic to the progenitor itself (McConnell and Kaznowski 1991; McConnell 1995; Frantz and McConnell 1996; Lillien 1998; Qian et al. 1998). The dynamics of neuronal production in the VZ and the analysis of molecular cues to migration and layer formation are extensively reviewed elsewhere in this volume. In this chapter we will focus on the role of three families of growth factors, the neurotrophins, fibroblast growth factors, and insulin-like growth factors in neuronal production, fate determination, and migration.

[1] Departments of Neurology and Cell Biology, Washington University School of Medicine, 660 South Euclid Avenue, St. Louis, Missouri 63110, USA

Results and Problems in Cell Differentiation, Vol. 30
Goffinet and Rakic (Eds.): Mouse Brain Development
© Springer-Verlag Berlin Heidelberg 2000

2
Trophic Factor Influences on Neurogenesis in the Ventricular Zone

2.1
Neurotrophins

The neurotrophin family includes the prototypic target-derived nerve growth factor, NGF (Levi-Montalcini 1976), along with several newer family members, BDNF (Barde et al. 1982), NT3 (Maisonpierre et al. 1990b), NT4 (Berkemeier et al. 1991; Hallbrook et al. 1991; Ip et al. 1992) and most recently NT6 (Gotz et al. 1994). The distribution of neurotrophins and their receptors suggests that neurotrophins play significant roles in neocortical development. Northern blot hybridization demonstrates that messenger RNA (mRNA) for NT3 is present early in neocortex at high levels during peak periods of neuroblast proliferation, and that its expression declines as neuronal maturation proceeds (Maisonpierre et al. 1990a). However, the cellular localization of this NT3 production has not been identified. In contrast to NT3, BDNF is initially present at low levels, then increases as proliferation ceases and cortical differentiation occurs (Maisonpierre et al. 1990a). Peak levels of messenger RNA for NT4 are present in the embryonic cerebral hemisphere by RNAse protection assay at early developmental stages (Timmusk et al. 1993). However, the precise location and timing of NT4 mRNA production has not been determined.

Neurotrophins exert their effects via a specific class of receptor tyrosine kinases (TrkA, B and C) (Ulrich and Schlessinger 1990; Barbacid 1994). These receptors have multiple splice forms, including truncated variants and forms with inserts in the catalytic domain that have reduced or absent tyrosine kinase activity (Klein et al. 1990a; Middlemas et al. 1991; Tsoulfas et al. 1993, 1996; Valenzuela et al. 1993; Eide et al. 1996). The low-affinity $p75^{NTR}$ receptor, a member of the tumor necrosis factor family (Meakin and Shooter 1992; Chao 1994) binds all members of the neurotrophin family and may further regulate their receptor specificity and function (Chao and Hempstead 1995; Ryden et al. 1995; Segal and Greenberg 1996).

Message for TrkC, like its ligand NT3, is expressed in the neocortical VZ at the earliest stages of development when the ventricular wall is composed entirely of progenitor cells (Lamballe et al. 1994). TrkB, the receptor for NT4 and BDNF, is present at high levels during corticogenesis (Klein et al. 1990b); receptor binding and phosphorylation studies indicate a developmental shift in TrkB from full length to truncated forms in the ventricular zone, intermediate zone and cortical plate that coincides with the end of neurogenesis and the formation of mature axonal connections (Allendoerfer et al. 1994; Escandon et al. 1994; Knusel et al. 1994). Immunolabeling and in situ hybridization studies demonstrate that TrkB is largely restricted to subsets of differentiating cortical neurons and their processes (Cabelli et al. 1996), and does not appear to be layer-specific (Pearlman et al. 1995).

Of the neurotrophins, NT3 holds the most promise as a regulator of the proliferation and fate of cortical progenitors. Abundant experimental evidence, in vitro and in vivo, indicates that NT3 functions as a powerful mitogen for neuronal and glial precursors in the peripheral nervous system, and as a survival factor for progenitors or very early neurons (Kalchiem et al. 1992; DiCicco-Bloom et al. 1993; Barres et al. 1994; Gaese et al. 1994; Elshamy and Ernfors 1996a,b; Verdi et al. 1996; Wilkinson et al. 1996). In the CNS, however, direct evidence of a similar mitogenic role for NT3 is lacking. Instead, NT3 has been shown to promote the neuronal differentiation of several CNS progenitors, including neural tube progenitors (Averbuch-Heller et al. 1994), cerebellar granule cells (Lindholm et al. 1993), and cortical progenitors (Ghosh and Greenberg 1995).

In cultured cortical progenitors, generated in the presence of basic FGF, NT3 appears to promote neuronal differentiation and antagonize precursor proliferation. Antibodies that block NT3 function cause a marked decrease in the number of differentiated neurons in these cultures, but do not affect precursor proliferation or the survival of post-mitotic neurons (Ghosh and Greenberg 1995). Thus, at moderately high doses (40–100 ng/ml), NT3 functions to regulate the exit of cortical progenitors from the cell cycle. However, NT3's effects on cortical progenitors could vary depending on dose. A noteworthy example of the dose dependent effects of NT3 is found in the sympathetic nervous system. Extremely low doses of NT3 (0.33–3 ng/ml) promote survival of cultured sympathetic neuroblasts while concentrations higher than 10 ng/ml induce mitotic arrest and neuronal differentiation (Verdi and Anderson 1994). Whether cortical progenitors have similarly dose-dependent sensitivity to NT3 has yet to be determined.

In summary, the timing and pattern of expression of neurotrophins and their receptors suggest distinct roles during neocortical development. In particular, NT3 and its receptor, TrkC, are highly expressed in the neocortical ventricular zone, and NT3 acts as a neuronal differentiation signal for cultured cortical progenitors. Additionally, NT3 has strong mitogenic and trophic actions on neuronal progenitors in the PNS that appear to be dose dependent, but a similar role for NT3 in neocortical development awaits further investigation.

2.2
Fibroblast Growth Factors

The fibroblast growth factors (FGFs) are a large family that currently consists of at least eighteen members, including FGFs 1–10 and 15–18, and four FGF homologous factors (FHFs 1–4; Smallwood et al. 1996) recently renamed as FGFs 11–14 (Coulier et al. 1997). Several FGFs exist as multiple isoforms of varying sizes; for example, FGF2 is expressed as an 18 kDa cytoplasmic- and cell surface-associated isoform, and as 22-, 23- and 24 kDa nuclear isoforms (reviewed in: Baird 1994; Eckenstein 1994; Mason 1994). The majority of FGFs contain a signaling sequence indicative of their role as secreted molecules (McWhirter et al. 1997; Greene et al. 1998; Ohbayashi et al. 1998). Furthermore, although FGFs 1, 2 and 9 are not secreted via the classical pathway, certain isoforms may be secreted by alternative pathways that are independent of the endoplasmic reticulum/Golgi apparatus (Florkiewicz et al. 1995; Mason 1994).

FGF1 and 2 are by far the most extensively studied family members. Both are expressed in the proliferative neuroepithelium of embryonic forebrain (Powell et al. 1991; Mason et al. 1994; Ozawa et al. 1996), with FGF2 expressed earliest in mouse neocortex at embryonic (E) day 9.5, followed by FGF1 at E11 (Nurcombe et al. 1993). Specific isoforms of FGF2 are present in a developmentally regulated fashion during peak periods of neurogenesis (Powell et al. 1991; Giordano et al. 1992; Weise et al. 1993). In the neocortex of the mouse at E14.5, the highest levels of FGF2 protein are detected in the ventricular zone (Dono et al. 1998).

Information regarding the expression of the newer FGF family members in developing brain is limited. PCR assays suggest that FGFs 3 and 5–9 are expressed in the embryonic mouse brain as early as E12 (Ozawa et al. 1996). Of these, FGF6 and 7 have been demonstrated in the embryonic VZ during restricted periods of neurogenesis (Mason et al. 1994; Ozawa et al. 1996). The FHFs (FGFs 11–14) appear very early in the embryonic telencephalon (Smallwood et al. 1996), but further analysis of their expression during embryonic cortical development is necessary. FGFs 15, 17, and 18 are expressed in the CNS during embryogenesis, but appear to be in regions outside the neocortex (McWhirter et al. 1997; Greene et al. 1998; Ohbayashi et al. 1998). The substantial sequence homology between the FGFs may lead to common detection of several family members, making expression studies of individual FGFs difficult to interpret (Eckenstein 1994). However, taken together, the above data indicate that one or more FGF ligands is present in the proliferative neuroepithelium of the early embryonic neocortex.

All of the FGFs, except 11–14 (the FHFs), act via four highly homologous transmembrane receptor tyrosine kinases (FGFRs) that exist as multiple splice variants (Johnson and Williams 1993; Eckenstein 1994; Greene et al. 1998). Isoforms of all four FGFRs are expressed in the neocortical VZ during early development (Peters et al. 1992, 1993; Yamaguchi et al. 1992; Orr-Urtreger et al. 1993; Dono et al. 1998), including the IIIc isoforms of FGFs 1–3 that are present in the VZ of the mouse as early as E10 through E14 (Qian et al. 1997). Although FGF ligands can cross-react with several FGFRs, each FGFR isoform displays unique ligand binding capabilities (Ornitz et al. 1996).

The activity of the FGFs are further regulated by heparin and heparan sulfate proteoglycans (HSPGs; Ornitz et al. 1992), which are abundantly expressed in the developing cortex (Stipp et al. 1994). HSPGs appear to play a role in FGF-ligand presentation, activation and signaling (Ornitz et al. 1992; Eckenstein 1994; Mason 1994; Bartlett et al. 1998). For example, heparan sulfates regulate the ability of FGF2 to bind to receptors on neural progenitors (Brickman et al. 1995), and very early murine neuroepithelial cells secrete a unique HSPG, a perlecan variant (Joseph et al. 1996), that differentially binds to FGF1 or FGF2, depending on its glycosylation state (Nurcombe et al. 1993).

Functional studies of FGFs in the CNS have thus far been carried out primarily with FGF1 and FGF2, because they are widely available and activate multiple FGF receptors. In dissociated primary cultures or cell lines both FGF1 and FGF2 induce neuronal and glial precursor proliferation, cell differentiation and survival (reviewed in Baird 1994; Eckenstein 1994; Bartlett et al. 1998). The mitogenic potential of several other FGFs (4, 7, 8 and 9) has also been demonstrated for neurons (DeHamer et al. 1994; Lee et al. 1997) and glia (Naruo et al. 1993). For the purposes of this chapter, we will focus on the functional roles of FGF2 (basic FGF) during corticogenesis since it has been studied most extensively and has recently been implicated in cortical neurogenesis in vivo (Dono et al. 1998; Ortega et al. 1998).

FGF2 is mitogenic for early embryonic cortical progenitors in culture (Kilpatrick and Bartlett 1993; Ghosh and Greenberg 1995; Qian et al. 1997), including those that give rise to glutamatergic neurons (Vaccarino et al. 1995), and FGF2 induces isolated ventricular zone cells to divide, even in the absence of cell-contact or serum (Qian et al. 1997). Furthermore, FGF2's influence on single progenitors appears to be dose dependent, since at low concentrations (0.1 ng/ml) FGF2 is mitogenic for neuronal progenitors, but at higher concentrations (10 ng/ml) FGF2 redirects the fate of these progenitors to become multipotent cells that generate glial and neuronal phenotypes (Qian et al. 1997). Additionally, it appears that only 10 % of early progenitors (from E10 mouse) are FGF2-responsive (Qian et al. 1997), and that these are true stem cells capable of self-renewal (Kilpatrick and Bartlett 1993; reviewed in Bartlett et al. 1998).

FGF2 has also been shown to be a survival factor for both embryonic cortical and hippocampal progenitors in high density cultures; in its absence, they stop proliferating and die (Murphy et al. 1990; Vicario-Abejon et al. 1995). In contrast, FGF2 does not appear to affect the survival of progenitors grown in isolation (Qian et al. 1997).

FGFs may also regulate cell cycle kinetics during corticogenesis via its influence on intercellular coupling between VZ cells (Nadarajah et al. 1998). FGF2 increases the expression of the gap-junction-associated protein connexin-43 when added to cortical cultures; this effect is mediated by FGF receptor tyrosine kinases and is associated with an increase in intercellular coupling (Nadarajah et al. 1998). The degree of intercellular coupling in the VZ correlates with connexin expression (Nadarajah et al. 1997), phase of the cell cycle, the stage of neurogenesis, and the availability of functional gap junctions (Bittman et al. 1997). Intriguingly, intercellular coupling also appears to determine the percentage of VZ progenitors that enter into S-phase (Bittman et al. 1997).

The influence of FGF2 on intercellular coupling has important implications for the laminar fate determination of cortical neurons. Transplantation studies show that cells make laminar fate choices during S-phase of the cell cycle (McConnell and Kaznowski 1991), and that an early progenitor's ability to choose an appropriate deeper-layer fate depends on cell–cell contact (Bohner et al. 1997). Furthermore, VZ progenitors from late neurogenesis are restricted to an upper cell fate (Frantz and McConnell 1996), coincident with declining connexin levels (Nadarajah et al. 1997) and reduced cell coupling (Bittman et al. 1997). Thus, a progenitor's ability to respond to FGF2 by increasing its communication with other VZ cells may determine its ability to respond to laminar fate-specific cues; this could explain the observations that early neocortical VZ cells are less restricted in their fate decisions than are VZ progenitors from later stages of neurogenesis (McConnell and Kaznowski 1991; Frantz and McConnell 1996).

Finally, recent analyses of the brains of mutant mice lacking FGF2 show a reduction in neuronal density in the neocortex, particularly in layer 5, suggesting a role for FGF2 during corticogenesis in vivo (Dono et al. 1998; Ortega et al. 1998).

In summary, the expression of FGFs and their receptors in the embryonic VZ suggests that FGFs function as trophic factors for cortical precursors. FGF2 is expressed in the embryonic neocortical ventricular zone and is necessary for normal cortical neurogenesis. Additionally, FGF2 is a powerful tool for investigating the effects of FGFs on cortical progenitors, because of its ability to cross-react with several FGF receptors, including receptor isoforms expressed in the early telencephalic VZ. However, it is likely that several FGF family members contribute to the complex environmental milieu that regulates the timely proliferation and differentiation of cortical neurons.

2.3
Insulin-Like Growth Factors

Substantial evidence supports a role for insulin-like growth factors (IGF1 and IGF2) in development of the CNS (reviewed in D'Ercole et al. 1996). Named for their structural homology to insulin, these two similar peptides activate a cell-surface receptor tyrosine kinase, the type 1 IGF receptor (IGFR1), which shares homology with the insulin receptor. Although a second very different IGF receptor exists (IGFR2), all of the growth-promoting effects of IGFs are thought to be mediated by IGFR1. Activation of IGFR1 initiates multiple signaling cascades, including the Ras/MAP kinase pathway common to many growth factors. The actions of IGFs are also regulated both positively and negatively by a family of six high-affinity extracellular binding proteins (IGFBP1-6).

A role for IGFs during embryonic cortical development is supported by in situ hybridization studies that demonstrate high expression of IGFR1 throughout the telencephalon in the embryonic (E14) rat (Bondy et al. 1990); levels decrease dramatically postnatally (Werner et al. 1989). Although IGF1 mRNA is not detectable in the early embryonic telencephalon by in situ hybridization (Bondy et al. 1990), RNAse protection assays demonstrate message for both IGF1 and IGF2 in the embryonic brain at the earliest ages examined (E14; Lund et al. 1986; Rotwein et al. 1988), while in situ hybridization indicates that IGF2 is strongly expressed in the embryonic choroid plexus (Stylianopoulou et al. 1988; Bondy et al. 1990). Thus, IGF2 delivered via the CSF may serve as a ligand for the IGFR1-expressing cells in the neocortical proliferative zones (Bondy et al. 1990).

IGFs and high doses of insulin (thought to act via IGFR1) regulate cell proliferation and neuroblast survival in the PNS and CNS, in vitro and in vivo, as demonstrated for sympathetic neuroblasts (DiCicco-Bloom and Black 1988; Zackenfels et al. 1995) and cerebellar granule cells (Galli et al. 1995; Tanaka et al. 1995; Ye et al. 1996). In the neocortex, precursors that express IGFR1 proliferate when cultured in the presence of IGF ligands (Nielsen et al. 1991). IGF1 ligand is secreted by early cortical neuroepithelial precursors (E10 mouse) and acts as an autocrine/paracrine factor necessary for their survival (Drago et al. 1991).

As in neurotrophin knock-out animals, targeted deletions of either IGF1 or IGFR1 have yielded surprisingly few brain abnormalities to date. However, these mice do have retarded growth and small brains, possibly secondary to a defect in oligodendrocyte development and myelination (Liu et al. 1993; Beck et al. 1995). Although the gross morphology of the brains is otherwise normal, detailed examination of the neocortex has not been carried out. In addition, these results must be interpreted with caution, given the absence of phenotypic effects in many knock-out mice that is generally attributed to compensatory or redundant mechanisms.

In summary, the IGF receptor, IGFR1, is highly expressed in the early telencephalic VZ, and ligands for IGFR1 are also present in the brain during these proliferative periods. IGFs promote the survival and proliferation of several neuronal precursors, including very early neuroepithelial progenitors of the cerebral cortex.

2.4
Trophic Collaborations

It is important to stress that, although experimental studies tend to isolate the effects of various trophic factors in order to better understand their roles, many of these molecules coexist in the neocortical VZ and it is highly improbable that any of them act independently. In fact, several lines of evidence suggest that growth factors collaborate to promote the survival of embryonic neurons and neuronal precursors. (Cattaneo and McKay 1990; Drago et al. 1991; Frodin and Gammeltoft 1994; Meyer-Franke et al. 1995; Lindholm et al. 1996; Hanson et al. 1998). Studies in knockout mice remind us that growth factors can compensate for one another, as in the case of IGF rescuing neurons that have been deprived of BDNF (Lindholm et al. 1996). Members of different growth factor families can also regulate one another's activity (Rajah et al. 1996; Verdi et al. 1996; Ferhat et al. 1997).

We also emphasize that the neocortical proliferative zones express many other trophic molecules that influence progenitor behavior that we have not reviewed because of space limitations. These include cytokines, e.g. EGF/TGF-alpha or CNTF (Lillien 1998), that preferentially regulate glial cell production, interleukins (Michaelson et al. 1996), PDGF (Williams et al. 1997), and PACAP (DiCicco-Bloom et al. 1998). Even the neurotransmitters, viewed in a different light, are 'trophins' capable of regulating the cell cycle and promoting the survival of cortical progenitors (reviewed in Lauder 1993), see (Lo Turco et al. 1995; Antonopoulos et al. 1997; Pabbathi et al. 1997).

3
Trophic Factor Influences on Glial-Guided Radial Migration

During cortical neurogenesis, the majority of neurons migrate out from the ventricular zone towards the pial surface along a transient population of specialized cells, the radial glia. (Rakic 1972; Levitt and Rakic 1980; Misson et al. 1988; Hatten 1990). Once neuronal migration is complete, radial glia transform into astrocytes (Voigt 1989; Culican et al. 1990). The ability of an astroglial cell to maintain its radial morphology, rather than assume a more mature glial form, appears to be dependent on soluble factors present in the early embryonic forebrain (Hunter and Hatten 1995; Anton et al. 1997; Rio et al. 1997).

Although a number of neuronal and glial receptor systems have been implicated in glial-guided neuronal migration (for reviews see Rakic et al. 1994; Hatten and Heintz 1998; Pearlman et al. 1998), very little is known about the trophic or tropic molecules that stimulate neurons to migrate along radial glia or direct them to their destination. In vitro, microchemotaxis assays suggest that neurotrophins, platelet-derived growth factors, or GABA may be involved, since they all induce cortical or spinal neuroblasts to move in a directed fashion across a microporous filter (Behar et al. 1994, 1995, 1996; Forsberg-Nilsson et al. 1998). Radial glia express the neurotrophin receptor TrkB (Cabelli et al. 1996), as well as BMP6, a cytokine of the TGF-beta superfamily (Schluesener and Meyermann 1994). To date, however, the neuregulins (NRGs) are the only growth factor family that has been directly linked to glial-guided migration in the cerebral cortex (Anton et al. 1997).

Neuregulins (glial growth factor, GGFs) are produced by migrating neurons and their precursors; they promote neuronal migration and are also necessary for maintenance of the radial glial phenotype (Anton et al. 1997; Rio et al. 1997). An interdependence between immature CNS neurons or their precursors and radial glia was identified several years ago (Hatten 1985, 1987; Gasser and Hatten 1990). It is reminiscent of a similar relationship in the PNS, where NRG-secreting neuroblasts indirectly promote their own survival and differentiation by stimulating the production of NT3 by neighboring Schwann cells (Verdi et al. 1996). Drawing on this analogy, it is tempting to speculate that NRG-stimulated radial glia might also be a source of NT3 to promote the survival and/or differentiation of cortical neurons and progenitors.

The neurotrophin BDNF has an indirect effect on the formation of cortical layers at the final stages of radial migration. Transgenic mice overexpressing BDNF have cortical lamination defects similar to those in the reeler mutant mouse (Ringstedt et al. 1998; see Ibáñez, this volume), which lacks the extracellular matrix glycoprotein reelin that is normally produced by marginal zone neurons. Mice overexpressing BDNF downregulate reelin production by marginal zone cells during the embryonic period when layers are forming. BDNF null mutants have prolonged postnatal production of reelin, indicating that BDNF is a negative regulator of reelin.

4
Trophic Factor Influences on Tangential Migration

There is mounting evidence that a significant proportion of cortical neurons reach their destination via non-radial routes (O'Rourke et al. 1995, 1997). Tangential migration results in the wide dispersion of clonally related cells (Reid et al. 1997), and appears to be a favored route for neurons fated to become GABAergic interneurons (Tan et al. 1998). Furthermore, many of these GABAergic neurons travel exceptionally long distances from regions outside the neocortex (Anderson et al. 1997; Tamamaki et al. 1997; Lavdas et al. 1999), contrary to the long-standing view that the neocortical ventricular zone is the sole source of neocortical neurons (Bayer et al. 1991).

The embryonic marginal zone (MZ; future layer 1) is a complex network of neurons, neuronal and radial glial processes, and extracellular matrix (Marin-Padilla 1971; Pearlman and Sheppard 1996) that is essential for the orderly formation of the neocortex (Caviness 1982). Substantial evidence now indicates that a tangential in-migration of neurons also plays an important role in the normal development of the MZ and that neurotrophins may regulate the formation of this layer (Brunstrom et al. 1997). Our work has shown that one member of the neurotrophin family, NT4, causes excess neurons to migrate into the MZ and thus suggests that NT4 may play a role in determining the number of neurons present in this layer during normal and pathological cortical development (Brunstrom et al. 1997).

4.1
NT4 Produces Heterotopic Accumulations of Neurons in the MZ in vitro

In organotypic slice cultures prepared from neocortex in mouse, NT4 produces collections of cells (heterotopias; Fig. 1) in the MZ that are identifiable as neurons by immunolabeling for a neuron-specific isoform of β-tubulin (TuJ1; Lee et al. 1990; Fig. 1D), and the microtubule-associated protein (MAP2; not shown). Heterotopias in the marginal zone are evident in cortical slices 20–24 h after exposure to NT4 (100 ng/ml). Both the total number of heterotopias and the size of each heterotopia (measured as the number of neurons per heterotopia) doubles over 2–3 days in culture Fig. 1H, I), and the largest heterotopias are found in lateral neocortex. Additionally, a greater percentage of neurons in the MZ of NT4-treated slices express an immature phenotype (TUJ1+; MAP2−) compared with control slices. Thus, application of NT4 to early embryonic neocortex in vitro causes a continuous addition of immature neurons to the MZ that is greatest in the lateral aspect of the neocortex.

4.2
NT4, but Not BDNF, Produces Heterotopias
in a TrkB-Mediated Response

NT4 and BDNF are both preferred ligands for the high-affinity receptor tyrosine kinase TrkB (Klein et al. 1989, 1991). We applied varying doses of each of these TrkB ligands to cortical slices (E14 mouse) and found that MZ heterotopias are produced by doses of NT4 as low as 20 ng/ml. In contrast, no heterotopias are produced with ten-fold higher doses of BDNF (200 ng/ml; Fig. 1C).

To demonstrate that the induction of abnormal collections of neurons by NT4 is mediated by the TrkB receptor, we applied NT4 to cortical slices from mice lacking the TrkB receptor (Klein et al. 1993; Snider 1994). Heterotopic

---→

Fig. 1A–H. Heterotopic collections of neurons are produced by NT4 in organotypic slices. **A–C, E,F** Hematoxylin-stained sections from cortical slices (E14 mouse). **A** After treatment with NT4 (100 ng/ml; 48 h), collections of cells (heterotopias, *arrowheads*) in the marginal zone (*MZ*) disrupt normal cortical architecture. **B** Heterotopias are not present in control slices (48 h) or C after BDNF treatment (200 ng/ml; 72 h). **D** Laser confocal micrograph of heterotopic cells immunolabeled with an antibody (TuJ1) to a neuron-specific form of β-tubulin. **G** Outline of a slice hemisphere. *Boxed area* represents region shown in A; *V* ventricle, *L* lateral, *D* dorsal. **E** Heterotopia (*arrows*) in an NT4-treated slice from a *trkB+/+* mouse. **F** Heterotopias are not present in NT4-treated slices from *trkB−/−* littermates. **H** Mean (±SEM) number of heterotopias per slice-hemisphere increases from 24 h to 48 h of NT4 treatment, but does not change significantly between 48 and 72 h. **I** Mean (±SEM) number of cells/heterotopia increases steadily from 24 h to 72 h. *$P < 0.002$; **$P < 0.005$; $\Delta P < 0.05$; ANOVA, unpaired t-test comparing means at 48 and 72 h with 24 h. *Bars* in B,C and F 100 μm; D 25 μm. *CP* cortical plate; *VZ* ventricular zone; *PIA* pia-arachnoid. (Reprinted with permission from Brunstrom et al. 1997a)

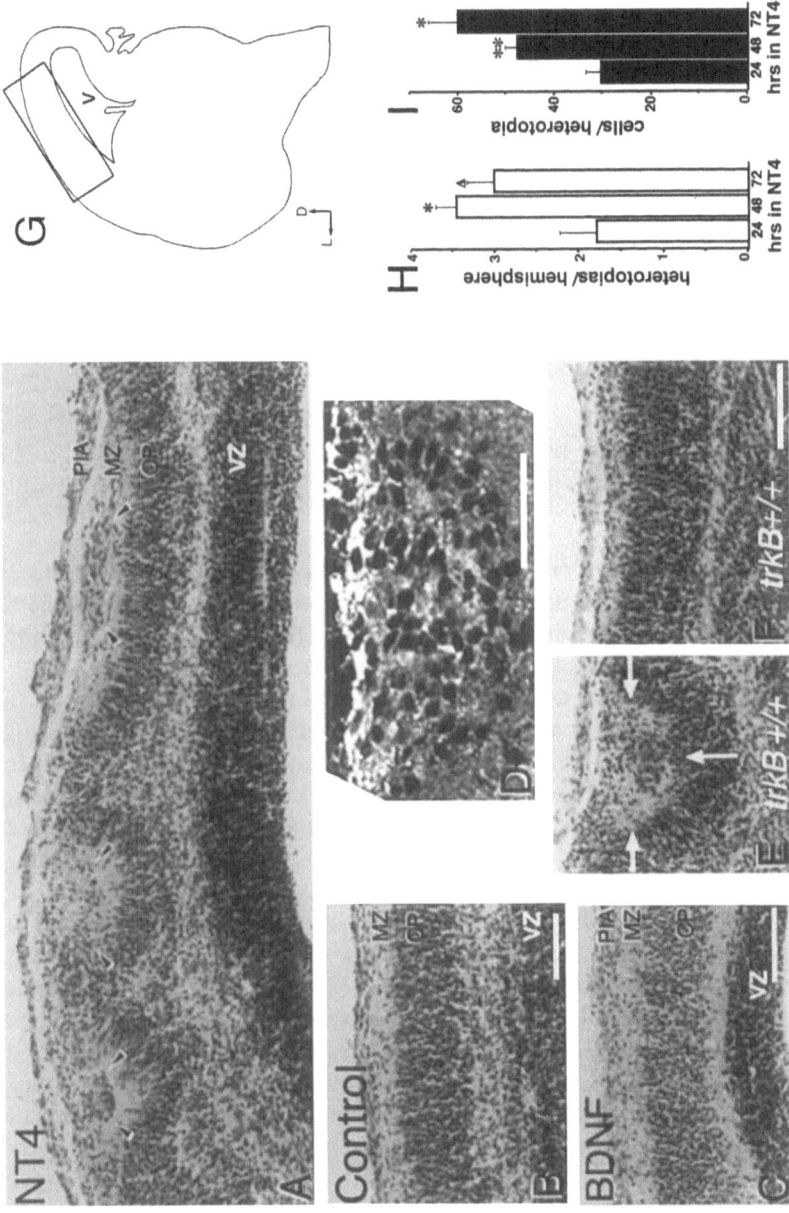

collections of neurons do not form in cortical slices from trkB−/− animals (Fig. 1F), but are present in slice cultures from heterozygous and wild-type littermates (Fig. 1E).

Taken together, these results indicate that TrkB-expressing neurons can distinguish between the two TrkB-ligands. The differential responsiveness of early cortical neurons to BDNF and NT4 could result from variations in the structure of TrkB. For example, there are several splice variants of TrkB, including variants of the extracellular domain (Garner et al. 1996; Strohmaier et al. 1996) and a truncated form that lacks tyrosine kinase activity (Klein et al. 1990a). However, intriguing new studies in mice that have a mutation of the Shc binding site of TrkB provide evidence that NT4 and BDNF may differentially activate distinct TrkB-mediated signaling pathways (Minichiello et al. 1998).

4.3
NT4 also Produces Heterotopic Neuronal Collections in vivo

We delivered neurotrophins to embryonic neocortex in vivo (Fig. 2) by administering two consecutive intraventricular injections of NT4, BDNF, NT3, or PBS at E16 and E18 in the rat. (At E16 in rat, cortical development is comparable to E14 in mouse.) Brains were examined at E19.

All of the brains treated with NT4 contain striking abnormalities. The most prevalent finding is a dramatic increase in the number of cells throughout the entire MZ. These cells form a dense band over the hemispheres that is thickest in the anterior (not shown) and lateral aspects of the neocortex (Fig. 2A,E) and extends rostrally to just behind the olfactory bulb and posteriorly to the occipital poles. Elliptical collections of cells (Fig. 2F) are present in the MZ in the posterior dorsomedial neocortex. Round collections of cells (Fig. 2B) that distort the underlying cortical plate are found in the MZ in posteromedial (retrosplenial) cortex.

Subtle, isolated MZ abnormalities are evident in extreme anterolateral cortex, cingulate cortex or retrosplenial cortex in a few of the brains injected with BDNF. However, in the vast majority of BDNF-treated brains, as well as in those injected with NT3 or control solutions, no abnormalities were noted (Fig. 2D) except those due to injury from the injection itself (Fig. 2C).

5
Pathogenesis of NT4-Induced Heterotopias

We used both the organotypic slice preparation and intraventricular injection methods to investigate the pathogenesis of these NT4-induced malformations (Brunstrom et al. 1997).

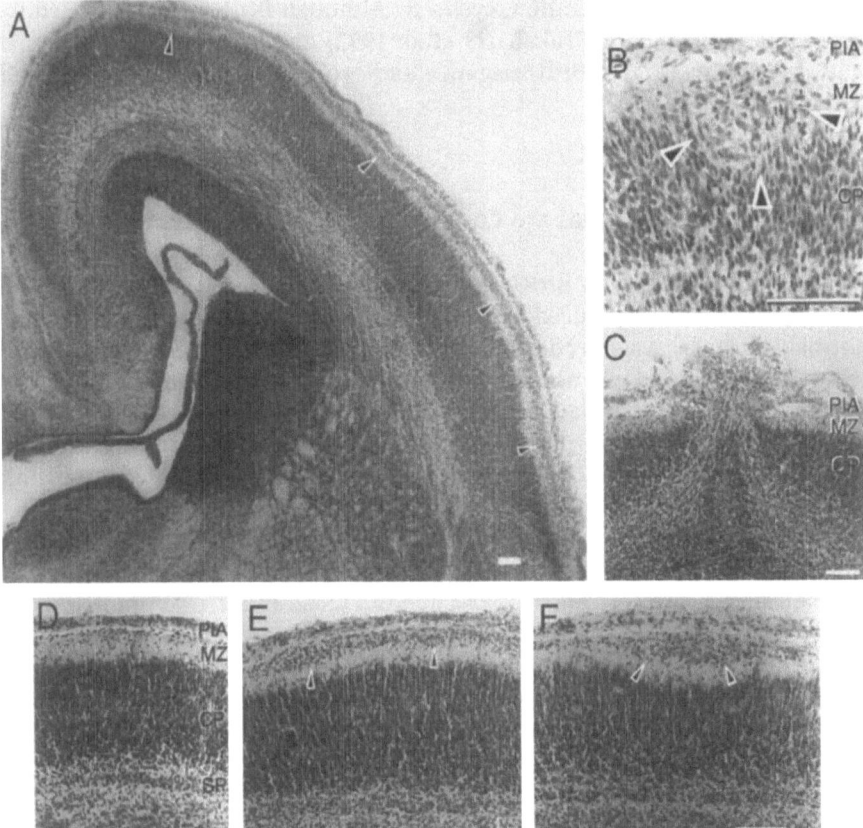

Fig. 2A–F. Heterotopic collections of neurons are produced by NT4 in vivo. Hematoxylin-stained coronal sections (18 μm) from embryos (E19 rat) injected intraventricularly at E16 and E18 with BDNF (**D**) or NT4 (**A,B,E,F**). **A,E** Increased cells in the marginal zone (*MZ*) after NT4: These cells form a dense band (*arrowheads*) that is thickest laterally (to the *right*) and anteriorly (not shown). **F** Elliptical collections are evident posteriorly in dorsal cortex. **B** Round collections of cells that indent the underlying cortical plate (*CP*) are found in extreme posteromedial cortex. These are quite different from the puncture-wound ectopias (**C**) that are found at the injection site in 97% of injected brains. *PIA* pia-arachnoid; *VZ* ventricular zone; *SP* subplate. *Bars* 100 μM. (Reprinted with permission from Brunstrom et al. 1997a)

5.1
NT4 Does Not Induce Cell Proliferation in the Marginal Zone

We exposed cortical slices to NT4 or PBS-containing media for 6–24 hrs, then pulse-labeled them with bromodeoxyuridine (BrdU) to label cells synthesizing DNA prior to and during the formation of the earliest heterotopic neuronal collections. Additional slices were cultured with BrdU from the outset to label all cells entering S-phase within the first 10–24 hrs of exposure to NT4. Sections from these slices were immunolabeled for BrdU and for another marker of proliferation, cyclin A. Although BrdU is incorporated by dividing cells in the VZ (Takahashi et al. 1992) and by capillary endothelial cells, cells in the MZ are BrdU-negative and cyclin A-negative, both with and without added NT4.

5.2
NT4-Induced Heterotopias are Composed of Marginal Zone Neurons

We labeled MZ cells with BrdU at their time of origin to determine whether cells within the NT4-induced collections are generated at the same time as normal MZ cells. An injection of BrdU given to the pregnant mouse quite early (E11–E11.5) labels normal MZ cells as well as cells within the NT4-induced heterotopias in later cortical slice cultures (E14). Similarly, the excess MZ neurons (TUJ1-positive) produced in vivo after intraventricular administration of NT4 in the embryonic rat are labeled in S-phase at the same time as preplate neurons (E13). The number of early generated, BrdU-labeled neurons in the marginal zone is increased more than two-fold by subsequent administration of intraventricular NT4 in vivo. This NT4-induced increase in MZ neurons is greatest anteriorly. Thus, NT4 induces an accumulation of excess MZ neurons that are born at the same time as normal MZ neurons. Neurons within these NT4-induced accumulations express the calcium-binding protein calretinin, and are therefore phenotypically similar to a major subtype of MZ cells, the Cajal-Retzius neurons (Ramon y Cajal 1890; Ogawa et al. 1995).

5.3
NT4-Induced Accumulation of Neurons is not at the Expense of the Subplate

Since both the MZ and subplate layers are derived from the preplate (Marin-Padilla 1971; Luskin and Shatz 1985), the excess MZ neurons produced by NT4 could potentially be misplaced subplate neurons. We demonstrated that this is not the case, however, since there is no change in the number of early born (E13 rat) BrdU-labeled cells in the subplate after in vivo exposure to NT4 (Brunstrom et al. 1997).

5.4
Heterotopic Neurons are not Misplaced Cortical Plate Cells

We used cell birthdating to label neurons of cortical layers V and VI as they are being generated in the ventricular zone (Caviness 1982; Bayer and Altman 1991). In our slice cultures, BrdU-labeled cortical plate cells migrate to the top of the cortical plate normally in the presence or absence of exogenous

NT4, but are not present within the heterotopic collections. However, in areas where the normal architecture of layer I has been severely disrupted by heterotopia formation, cortical plate neurons appear to move out between the heterotopias, filling the space between cortical plate and pia which is vacated when MZ neurons cluster into abnormal collections.

To determine whether cortical plate neurons contribute substantially to the increased numbers of neurons in the MZ after NT4 administration in vivo, we counted the BrdU-labeled cells present in the MZ after intraventricular injections of NT4 or PBS. In both conditions, a single injection of BrdU at E15 (rat) heavily labels CP cells as well as a cohort of MZ cells, consistent with the overlap in birth dates between cortical layers (Bayer and Altman 1991). However, the number of BrdU-labeled cells in the MZ is not significantly different between the two conditions, indicating that NT4 does not cause cortical plate neurons to move into the MZ.

5.5
Heterotopias do not Result from the Trauma of Intraventricular Injection

The morphology of cortical ectopias produced by needle punctures (Rosen et al. 1992) is very different from the accumulations produced by NT4 (see Fig. 2). Furthermore, we found the same dramatic response to NT4 in the uninjected (contralateral) hemispheres of NT4-treated brains, confirming that trauma does not account for NT4-induced heterotopias.

5.6
Heterotopias are not Caused by Rescue of MZ Neurons from Cell Death

In keeping with the classic neurotrophin hypothesis (Levi-Montalcini 1976), NT4 might produce excess MZ neurons by rescuing cells in the MZ from naturally occurring cell death (apoptosis). We used two different methods to detect the cut ends of DNA that occur in apoptotic cells to look for evidence of cell death in the embryonic murine neocortex (E14) and in cortical slices cultured in the presence or absence of NT4. We found that very rare apoptotic cells are present in the neocortex of E14 mice, and none is evident in the preplate or the MZ. After 24 or 48 h in slice culture, end-labeling demonstrates dying cells in the ventricular, subventricular, and intermediate zones, but very few in the cortical plate, and only a rare dying cell in the MZ. NT4 does not rescue these cells from apoptosis.

Finally, we applied a selective inhibitor of the interleukin-1β converting enzyme (ICE) family of proteases to block cell death in order to determine whether rescue from apoptosis could produce accumulations of neurons in the marginal zone similar to the collections produced by NT4. Application of Boc-aspartyl(Ome)-fluoromethylketone(BAF), an inhibitor of ICE-like proteases, to cortical slice cultures produces a marked reduction in end-labeled cells in neocortex in both control and NT4-treated slices indicating that BAF rescued these cells from apoptosis. However, rescue by BAF does not produce collections of neurons in the MZ and does not impair the formation of NT4-induced collections.

6
What is the Source of the Excess Neurons that Form NT4-Induced Heterotopias?

The excess neurons that accumulate in the MZ as a consequence of exogenous NT4 are similar in many ways to those that form the normal MZ population. NT4-induced accumulations consist of neurons that are generated at the same time as normal MZ neurons, and many are phenotypically similar to Cajal-Retzius neurons. Early-generated neurons continue to accumulate in the MZ after the cortical plate forms in response to NT4, and their pattern of accumulation in vivo mirrors that of normal development. On the basis of these observations, we have postulated that NT4 causes an exaggeration of the normal in-migration of MZ neurons and that NT4 may be involved in regulating that migration.

Two transient populations of pleiomorphic cells, the Cajal-Retzius (CR) neurons and the neurons of the subpial granule cell layer (SGL), are present in the MZ of the human fetus. The CR neurons are the first to arrive in the preplate and they continue to accumulate even after the cortical plate has formed. They are most numerous laterally, in the human temporal lobe, where they are arranged in rows just beneath the pia (Meyer and Gonzalez-Hernandez 1993; Meyer and Goffinet 1998). Neurons of the SGL arrive after the first CR cells, and after the early cortical plate has formed. The subpial layer of SGL neurons is quite similar to the dense subpial band of neurons produced in the rat after in vivo administration of NT4 (compare Fig. 3 in Gadisseux et al. 1992 with our Fig. 2E). As in the human SGL, the NT4-induced subpial neurons vary in their expression of maturational markers (Gadisseux et al. 1992).

Evidence derived from morphological observations indicates that SGL neurons originate in a proliferative subventricular zone in the frontal pole, just behind the olfactory bulb, then move to the subpial layer and migrate tangentially in the MZ across the entire cortex (Gadisseux et al. 1992; Meyer and Goffinet 1998). Similarly, in the embryonic rat a large cohort of early generated cells adjacent to the olfactory ventricle extends subpially along the rostrocaudal axis in the MZ of pyriform cortex and encroaches on the MZ of neocortex (Valverde and Santacana 1994; De Carlos et al. 1996). Recent morphological studies in the rat demonstrate neurons analogous to human subpial granule cells that emerge from the olfactory subventricular zone to

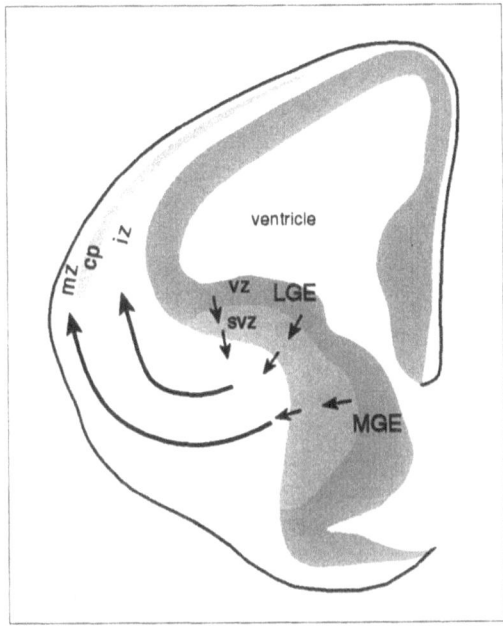

Fig. 3. Long-range tangential migration of cortical neurons. Diagrammatic representation of a coronal section of the early murine telencephalic vesicle. Postmitotic GABAergic neurons move from the forming striatum into the marginal zone (*mz*) and intermediate zone (*iz*) of neocortex (*large arrows*). Neurons move from the proliferative ventricular and subventricular zones (*vz* and *svz*) of the medial and lateral ganglionic eminences (*MGE* and *LGE*) into the forming striatum (*small arrows*); it is not yet clear whether all neurons destined for cortex traverse the developing striatum or move directly into cortex from the proliferative zones. In addition, neurons (not shown) destined for the marginal zone may also migrate tangentially from a subventricular zone near the olfactory bulb. (Adapted from Pearlman et al. 1998)

enter the neocortical MZ in a similar anterior and lateral distribution (Meyer et al. 1998). The strong homology between these observations on early MZ cells in normal mammals and the concentration of excess neurons in anterior and lateral regions of the MZ after NT4 application in vivo suggests that the olfactory subventricular zone may be a source of the excess neurons.

The olfactory subventricular zone cannot be the sole source of excess MZ neurons, however, since it is absent in our cortical slices, even though NT4 produces accumulations of neurons in the MZ that are more prominent laterally. Neurons originating in the olfactory subventricular zone that have already reached the ventrolateral MZ at the time of our cultures (E14 mouse) might be induced by NT4 to continue migrating tangentially into the MZ of neocortex. Alternatively, developing striatal regions, including the lateral and medial ganglionic eminences which express trkB (Fryer et al. 1996; J. E. Brunstrom and A. L. Pearlman, unpubl. observ.), are potential sources of these excess neurons.

Experimental evidence directly demonstrates that early-born neurons invade the neocortical MZ from the lateral ganglionic eminence (LGE), medial ganglionic eminence (MGE) and developing striatum (Fig. 3) (De Carlos et al. 1996; Anderson et al. 1997; Brunstrom et al. 1997; Lavdas et al. 1999). These regions also supply a tangentially migrating cohort of GABAergic neurons (DeDiego et al. 1994) to the neocortical intermediate zone (Anderson et al. 1997; Tamamaki et al. 1997; Lavdas et al. 1999) and subplate (Lavdas et al. 1999). The number of GABAergic neurons throughout the neocortex is dramatically reduced by separating the neocortex from subcortical structures (ganglionic eminences and striatum) during early embryonic development (Anderson et al. 1997), suggesting that the LGE and MGE are a major source of neocortical interneurons (Anderson et al. 1997; Lavdas et al. 1999). The migration of GABAergic neurons into the neocortex appears to be regulated by the transcription factors Dlx1, Dlx2 (Anderson et al. 1997), and Lhx6 (Lavdas et al. 1999).

Thus there is considerable evidence for a long tangential migration of neurons during neocortical development that is similar to the extensive tangential migrations of neurons during the normal development of both the olfactory bulb (Zigova et al. 1996) and the cerebellum (Miale and Sidman 1961). At present, the cellular or extracellular substrate for these migrations is poorly understood. In this context, it is intriguing that the large ECM-like protein reelin is expressed by neurons as they migrate from the rhombic lip to form the cerebellar anlage and by granule cells as they migrate in the cerebellar external granular layer. Reelin is also expressed in neurons of the developing striatum and embryonic cortical marginal zone (D'Arcangelo et al. 1995; Miyata et al. 1996; Schiffmann et al. 1997; Meyer et al. 1998), and has been identified on neurons that migrate into the MZ from the MGE (Lavdas et al. 1999).

In summary, our studies indicate that NT4 induces the tangential in-migration of excess neurons into the marginal zone of the embryonic neocortex. These excess neurons express both GABA and reelin (Brunstrom et al. 1997), and thus display a phenotype that is remarkably similar to subpial granule neurons of the MZ in rodents and humans (Meyer and Goffinet 1998; Meyer et al. 1998). We propose that the olfactory subventricular zone and developing striatal regions (including the LGE and MGE) are potential sources for the in-migration of excess neurons produced by NT4. These observations suggest that NT4 is important in regulating the number and fate of neurons in the MZ, and may thus have a role in the formation of certain human cortical dysplasias.

Acknowledgements. Work from the authors' laboratories was supported by grants from the National Eye Institute, the National Institute of Neurological Diseases and Stroke, and the McDonnell Center for Cellular and Molecular Neurobiology, Washington University.

References

Allendoerfer KL, Cabelli RJ, Escandon E, Kaplan DR, Nikolics K, Shatz CJ (1994) Regulation of neurotrophin receptors during the maturation of the mammalian visual system. J Neurosci 14:1795–1811

Allendoerfer KL, Shatz CJ (1994) The subplate, a transient cortical structure: its role in the development of connections between thalamus and cortex. Annu Rev Neurosci 17: 185–218

Anderson SA, Eisenstat DD, Shi L, Rubenstein JLR (1997) Interneuron migration from basal forebrain to neocortex: dependence on Dlx genes. Science 278:474–476

Anton ES, Marchionni MA, Lee KF, Rakic P (1997) Role of GGF/Neuregulin signaling in interactions between migrating neurons and radial glia in the developing cerebral cortex. Development 124:3501–3510

Antonopoulos J, Pappas IS, Parnavelas JG (1997) Activation of the GABAA receptor inhibits the proliferative effects of bFGF in cortical progenitor cells. Eur J Neurosci 9:291–298

Averbuch-Heller L, Pruginin M, Kahane N, Tsoulfas P, Parada L, Rosenthal A, Kalcheim C (1994) Neurotrophin 3 stimulates the differentiation of motorneurons from avian neural tube progenitor cells. Proc Natl Acad Sci 91:3247–3251

Baird A (1994) Fibroblast growth factors: activities and significance of non-neurotrophin neurotrophic growth factors. Curr Opin Neurobiol 4:78–86

Barbacid M (1994) The Trk family of neurotrophin receptors. J Neurobiol 25:1386–1403

Barde YA, Edgar D, Thoenen H (1982) Purification of a new neurotrophic factor from mammalian brain. EMBO J 1:549–553

Barres BA, Raff MC, Gaese F, Bartke I, Dechant G, Barde YA (1994) A crucial role for neurotrophin-3 in oligodendrocyte development. Nature 367:371–375

Bartlett PF, Brooker GJ, Faux CH, Dutton R, Murphy M, Turnley A, Kilpatrick TJ (1998) Regulation of neural stem cell differentiation in the forebrain. Immunol Cell Biol 76: 414–418

Bayer S, Altman J, Russo R, Dai X, Simmons J (1991) Cell migration in the rat embryonic neocortex. J Comp Neurol 307:499–516

Bayer SA, Altman J (1991) Neocortical Development. Raven Press, New York

Beck KD, Powell-Braxton L, Widmer HR, Valverde J, Hefti F (1995) Igf1 gene disruption results in reduced brain size, CNS hypomyelination, and loss of hippocampal granule and striatal parvalbumin-containing neurons. Neuron 14:717–730

Behar T, Ma W, Hudson L, Barker JL (1994) Analysis of the anatomical distribution of GAD67 mRNA encoding truncated glutamic acid decarboxylase proteins in the embryonic rat brain. Brain Res Dev Brain Res 77:77–87

Behar TN, Li YX, Tran HT, Ma W, Dunlap V, Scott C, Barker JL (1996) GABA stimulates chemotaxis and chemokinesis of embryonic cortical neurons via calcium-dependent mechanisms. J Neurosci 16:1808–1818

Behar TN, Schaffner AE, Tran HT, Barker JL (1995) GABA-induced motility of spinal neuroblasts develops along a ventrodorsal gradient and can be mimicked by agonists of GABA(a) and GABA(b) receptors. J Neurosci Res 42:97–108

Berkemeier LR, Winslow JW, Kaplan DR, Nikolics K, Goeddel D, Rosenthal A (1991) Neurotrophin 5: a novel neurotrophic factor that activates trkA and trkB. Neuron 7:857–866

Bittman K, Owens DF, Kriegstein AR, LoTurco JJ (1997) Cell coupling and uncoupling in the ventricular zone of developing neocortex. J Neurosci 17:7037–7044

Bohner AP, Akers RM, McConnell SK (1997) Induction of deep layer cortical neurons in vitro. Development 124:915–923

Bondy CA, Werner H, Roberts CT Jr, LeRoith D (1990) Cellular pattern of insulin-like growth factor-I (IGF-I) and type I IGF receptor gene expression in early organogenesis: comparison with IGF-II gene expression. Mol Endocrinol 4:1386–1398

Brickman YG, Ford MD, Small DH, Bartlett PF, Nurcombe V (1995) Heparan sulfates mediate the binding of basic fibroblast growth factor to a specific receptor on neural precursor cells. J Biol Chem 270:24941-24948

Brunstrom JE, Gray-Swain MR, Osborne PA, Pearlman AL (1997a) Neuronal heterotopias in the developing cerebral cortex produced by neurotrophin-4. Neuron 18:505-517

Brunstrom JE, Gray-Swain MR, Pearlman AL (1997b) GABA and Reelin are expressed by NT4-induced excess neurons in the marginal zone of developing neocortex. Soc Neurosci Abstr 23:80

Cabelli RJ, Allendoerfer KL, Radeke MJ, Welcher AA, Feinstein SC, Shatz CJ (1996) Changing patterns of expression and subcellular localization of TrkB in the developing visual system. J Neurosci 16:7965-7980

Cattaneo E, McKay R (1990) Proliferation and differentiation of neuronal stem cells regulated by nerve growth factor. Nature 347:762-765

Caviness VS Jr (1982) Neocortical histogenesis in normal and reeler mice: a developmental study based upon [3H] thymidine autoradiography. Dev Brain Res 4:293-302

Caviness VS, Takahashi T, Miyama S, Nowakowski RS, Delalle I (1996) Regulation of normal proliferation in the developing cerebrum: potential actions of trophic factors. Expt Neurol 137:357-366

Chao MV (1994) The p75 neurotrophin receptor. J Neurobiol 25:1373-1385

Chao MV, Hempstead BL (1995) p75 and Trk: a two-receptor system. Trends Neurosci 18:321-326

Coulier F, Pontarotti P, Roubin R, Hartung H, Goldfarb M, Birnbaum D (1997) Of worms and men: an evolutionary perspective on the fibroblast growth factor (FGF) and FGF receptor families. J Mol Evol 44:43-56

Culican SM, Baumrind NL, Yamamoto M, Pearlman AL (1990) Cortical radial glia: identification in tissue culture and evidence for their transformation to astrocytes. J Neurosci 10:684-692

D'Arcangelo G, Miao GG, Chen S-C, Soares HD, Morgan JI, Curran T (1995) A protein related to extracellular matrix proteins deleted in the mouse mutant reeler. Nature 374:719-723

D'Ercole AJ, Ye P, Calikoglu AS, Gutierrez-Ospina G (1996) The role of the insulin-like growth factors in the central nervous system. Mol Neurobiol 13:227-255

De Carlos JA, Lopez-Mascaraque L, Valverde F (1996) Dynamics of cell migration from the lateral ganglionic eminence in the rat. J Neurosci 16:6146-6156

DeDiego I, Smith-Fernandez A, Fairen A (1994) Cortical cells that migrate beyond area boundaries: characterization of an early neuronal population in the lower intermediate zone of prenatal rats. Eur J Neurosci 6:983-997

DeHamer MK, Guevara JL, Hannon K, Olwin BB, Calof AL (1994) Genesis of olfactory receptor neurons in vitro: regulation of progenitor cell divisions by fibroblast growth factors. Neuron 13:1083-1097

DiCicco-Bloom E, Black IB (1988) Insulin growth factors regulate the mitotic cycle in cultured rat sympathetic neuroblasts. Proc Nat Acad Sci USA 85:4066-4070

DiCicco-Bloom E, Friedman WJ, Black IB (1993) NT3 stimulates sympathetic neuroblast proliferation by promoting precursor survival. Neuron 11:1101-1111

DiCicco-Bloom E, Lu N, Pintar JE, Zhang J (1998) The PACAP ligand/receptor system regulates cerebral cortical neurogenesis. Ann N Y Acad Sci 865:274-289

Dono R, Texido G, Dussel R, Ehmke H, Zeller R (1998) Impaired cerebral cortex development and blood pressure regulation in FGF-2-deficient mice. EMBO J 17:4213-4225

Drago J, Murphy M, Carroll SM, Harvey RP, Bartlett PF (1991) Fibroblast growth factor-mediated proliferation of central nervous system precursors depends on endogenous production of insulin-like growth factor I. Proc Nat Acad Sci USA 88:2199-2203

Eckenstein FP (1994) Fibroblast growth factors in the nervous system. J Neurobiol 25:1467-1480

Eide FF, Vining ER, Eide BL, Zang KL, Wang XY, Reichardt LF (1996) Naturally occurring truncated trkb receptors have dominant inhibitory effects on brain-derived neurotrophic factor signaling. J Neurosci 16:3123-3129

Elshamy WM, Ernfors P (1996a) A local action of neurotrophin-3 prevents the death of proliferating sensory neuron precursor cells. Neuron 16:963-972

Elshamy WM, Ernfors P (1996b) Requirement of neurotrophin-3 for the survival of proliferating trigeminal ganglion progenitor cells. Development 122:2405-2414

Escandon E, Soppet D, Rosenthal A, Mendoza-Ramirez J, Szonyl E, Burton LE, Henderson CE, Parada LF, Nikolics K (1994) Regulation of neurotrophin receptor expression during embryonic and postnatal development. J Neurosci 14:2054-2068

Ferhat L, Represa A, Zouaouiaggoun D, Ferhat W, Benari Y, Khrestchatisky M (1997) Fgf-2 induces nerve growth factor expression in cultured rat hippocampal neurons. Eur J Neurosci 9:1282-1289

Florkiewicz RZ, Majack RA, Buechler RD, Florkiewicz E (1995) Quantitative export of FGF-2 occurs through an alternative, energy-dependent, non-ER/Golgi pathway. J Cell Physiol 162:388-399

Forsberg-Nilsson K, Behar TN, Afrakhte M, Barker JL, McKay RD (1998) Platelet-derived growth factor induces chemotaxis of neuroepithelial stem cells. J Neurosci Res 53:521-530

Frantz GD, McConnell SK (1996) Restriction of late cerebral cortical progenitors to an upper-layer fate. Neuron 17:55-61

Frodin M, Gammeltoft S (1994) Insulin-like growth factors act synergistically with basic fibroblast growth factor and nerve growth factor to promote chromaffin cell proliferation. Proc Natl Acad Sci USA 91:1771-1775

Fryer RH, Kaplan DR, Feinstein SC, Radeke MJ, Grayson DR, Kromer LF (1996) Developmental and mature expression of full-length and truncated TrkB receptors in the rat forebrain. J Comp Neurol 374:21-40

Gadisseux J-F, Goffinet AM, Lyon G, Evrard P (1992) The human transient subpial granular layer: an optical, immunohistochemical, and ultrastructural analysis. J Comp Neurol 324:94-114

Gaese F, Kolbeck R, Barde YA (1994) Sensory ganglia require neurotrophin-3 early in development. Development 120:1613-1619

Galli C, Meucci O, Scorziello A, Werge TM, Calissano P, Schettini G (1995) Apoptosis in cerebellar granule cells is blocked by high KCl, forskolin, and IGF-1 through distinct mechanisms of action: the involvement of intracellular calcium and RNA synthesis. J Neuroscience 15:1172-1179

Garner AS, Menegay HJ, Boeshore KL, Xie XY, Voci JM, Johnson JE, Large TH (1996) Expression of trkb receptor isoforms in the developing avian visual system. J Neurosci 16:1740-1752

Gasser UE, Hatten ME (1990) Central nervous system neurons migrate on astroglial fibers from heterotypic brain regions in vitro. Proc Natl Acad Sci USA 87:4543-4547

Ghosh A, Greenberg ME (1995) Distinct roles of bFGF and NT-3 in the regulation of cortical neurogenesis. Neuron 15:89-103

Giordano S, Sherman L, Lyman W, Morrison R (1992) Multiple molecular weight forms of basic fibroblast growth factor are developmentally regulated in the central nervous system. Dev Biol 152:293-303

Gotz R, Koster R, Winkler C, Raulf F, Lottspeich F, Schartl M, Thoenen H (1994) Neurotrophin-6 is a new member of the nerve growth factor family. Nature 372:266-269

Greene JM, Li YL, Yourey PA, Gruber J, Carter KC, Shell BK, Dillon PA, Florence C, Duan DR, Blunt A, Ornitz DM, Ruben SM, Alderson RF (1998) Identification and characterization of a novel member of the fibroblast growth factor family. Eur J Neurosci 10:1911-1925

Hallbrook F, Ibáñez CF, Persson H (1991) Evolutionary studies of the nerve growth factor family reveal a novel member abundantly expressed in Xenopus ovary. Neuron 6:845-858

Hanson MG Jr, Shen S, Wiemelt AP, McMorris FA, Barres BA (1998) Cyclic AMP elevation is sufficient to promote the survival of spinal motor neurons in vitro. J Neurosci 18: 7361–7371

Hatten ME (1985) Neuronal regulation of astroglial morphology and proliferation in vitro. J Cell Biol 100: 384–396

Hatten ME (1987) Neuronal inhibition of astroglial cell proliferation is membrane mediated. J Cell Biol 104: 1353–1360

Hatten ME (1990) Riding the glial monorail: a common mechanism for glial-guided neuronal migration in different regions of the developing mammalian brain. TINS 13: 179–184

Hatten ME, Heintz N (1998) Neurogenesis and migration. In: Zigmond M (eds) Fundamentals of Neuroscience. Academic Press, New York, pp 451–479

Hunter KE, Hatten ME (1995) Radial glial cell transformation to astrocytes is bidirectional: regulation by a diffusible factor in embryonic forebrain. Proc Natl Acad Sci USA 92: 2061–2065

Ip NY, Ibáñez CF, Nye SH, McClain J, Jones PF, Gies DR, Belluscio L, Le Beau MM, Espinosa R, Squinto SP, Persson H, Yancopoulos GD (1992) Mammalian neurotrophin 4: structure, chromosomal localization, tissue distribution, and receptor specificity. Proc Natl Acad Sci 89: 3060–3064

Johnson DE, Williams LT (1993) Structural and functional diversity in the FGF receptor multigene family. Adv Cancer Res 60: 1–41

Joseph SJ, Ford MD, Barth C, Portbury S, Bartlett PF, Nurcombe V, Greferath U (1996) A proteoglycan that activates fibroblast growth factors during early neuronal development is a perlecan variant. Development 122: 3443–3452

Kalchiem C, Carmeli C, Rosenthal A (1992) Neurotrophin 3 is a mitogen for cultured neural crest cells. Proc Natl Acad Sci 89: 1661–1665

Kilpatrick TJ, Bartlett PF (1993) Cloning and growth of multipotential neural precursors: requirements for proliferation and differentiation. Neuron 10: 255–265

Klein R, Conway D, Parada L, Barbacid M (1990a) The trkB tyrosine protein kinase gene codes for a second neurogenic receptor that lacks the catalytic kinase domain. Cell 61: 647–656

Klein R, Martin-Zanca D, Barbacid M, Parada L (1990b) Expression of the tyrosine kinase receptor gene trkB is confined to the murine embryonic and adult nervous system. Development 109: 845–850

Klein R, Nanduri V, Jing S, Lamballe F, Tapley P, Bryant S, Cordon-Cardo C, Jones K, Reichardt L, Barbacid M (1991) The trkB tyrosine protein kinase is a receptor for brain-derived neurotrophic factor and neurotrophin 3. Cell 66: 395–403

Klein R, Parada LF, Coulier F, Barbacid M (1989) TrkB, a novel tyrosine protein kinase receptor expressed during mouse neural development. EMBO J 8: 3701–3709

Klein R, Smeyne RJ, Wurst W, Long LK, Auerbach BA, Joyner AL, Barbacid M (1993) Targeted disruption of the trkB neurotrophin receptor gene results in nervous system lesions and neonatal death. Cell 75: 113–122

Knusel B, Rabin SJ, Hefti F, Kaplan DR (1994) Regulated neurotrophin receptor responsiveness during neuronal migration and early differentiation. J Neurosci 14: 1542–1554

Lamballe F, Smeyne R, Barbacid M (1994) Development of trkC, the neurotrophin-3 receptor, in the mammalian nervous system. J Neurosci 14: 14–28

Lauder JM (1993) Neurotransmitters as growth regulatory signals: role of receptors and second messengers. Trends Neurosci 16: 233–240

Lavdas AA, Grigoriou M, Pachnis V, Parnavalis JG (1999) The medial ganglionic eminence gives rise to a population of early neurons in the developing cerebral cortex. J Neurosci 19: 7881–7888

Lee MK, Tuttle JB, Rebhun LI, Cleveland DW, Frankfurter A (1990) The expression and posttranslational modification of a neuron-specific beta-tubulin isotype during chick embryogenesis. Cell Motil Cytoskel 17: 118–132

Lee SM, Danielian PS, Fritzsch B, McMahon AP (1997) Evidence that FGF8 signalling from the midbrain-hindbrain junction regulates growth and polarity in the developing midbrain. Development 124:959–969

Levi-Montalcini R (1976) The nerve growth factor: its role in growth, differentiation and function of the sympathetic adrenergic neuron. Prog Brain Res 45:235–258

Levitt P, Rakic P (1980) Immunoperoxidase localization of glial fibrillary acidic protein in radial glial cells and astrocytes of the developing rhesus monkey brain. J Comp Neur 193:815–840

Lillien L (1998) Neural progenitors and stem cells: mechanisms of progenitor heterogeneity. Curr Opin Neurobiol 8:37–44

Lindholm D, Carroll P, Tzimagiogis G, Thoenen H (1996) Autocrine-paracrine regulation of hippocampal neuron survival by IGF-1 and the neurotrophins BDNF, NT-3 and NT-4. Eur J Neurosci 8:1452–1460

Lindholm D, Castren E, Tsoulfas P, Kolbeck R, da Penha Berzaghi M, Leingartner A, Heisenberg C-P, Tesarollo L, Parada LF, Thoenen H (1993) Neurotrophin-3 induced by tri-iodothyronine in cerebellar granule cells promotes purkinje cell differentiation. J Cell Biol 122:443–450

Liu JP, Baker J, Perkins AS, Robertson EJ, Efstratiadis A (1993) Mice carrying null mutations of the genes encoding insulin-like growth factor I (Igf-1) and type 1 IGF receptor (Igf1r). Cell 75:59–72

Lo Turco JJ, Owens DF, Heath MJS, Davis MBE, Kriegstein AR (1995) GABA and glutamate depolarize cortical progenitor cells and inhibit DNA synthesis. Neuron 15:1287–1298

Lund PK, Moats-Staats BM, Hynes MA, Simmons JG, Jansen M, D'Ercole AJ, Van Wyk JJ (1986) Somatomedin-C/insulin-like growth factor-I and insulin-like growth factor-II mRNAs in rat fetal and adult tissues. J Biol Chem 261:14539–14544

Luskin MB, Shatz CJ (1985) Neurogenesis of the cat's primary visual cortex. J Comp Neurol 242:611–631

Maisonpierre PC, Belluscio L, Friedman B, Alderson RF, Wiegand SJ, Furth ME, Lindsay RM, Yancopoulos GD (1990a) NT-3, BDNF, and NGF in the developing rat nervous system: parallel as well as reciprocal patterns of expression. Neuron 5:501–509

Maisonpierre PC, Belluscio L, Squinto S, Ip NY, Furth ME, Lindsay RM, Yancopoulos GD (1990b) Neurotrophin-3: a neurotrophic factor related to NGF and BNDF. Science 247:1446–1451

Marin-Padilla M (1971) Early prenatal ontogenesis of the cerebral cortex (neocortex) of the cat (*Felis domestica*). A Golgi study. I. The primordial neocortical organization. Z Anat Entwicklungsgesch 134:117–145

Mason IJ (1994) The ins and outs of fibroblast growth factors. Cell 78:547–552

Mason IJ, Fuller-Pace F, Smith R, Dickson C (1994) FGF-7 (keratinocyte growth factor) expression during mouse development suggests roles in myogenesis, forebrain regionalisation and epithelial-mesenchymal interactions. Mech Dev 45:15–30

McConnell SK (1995) Constructing the cerebral cortex: neurogenesis and fate determination. Neuron 15:761–768

McConnell SK, Kaznowski CE (1991) Cell cycle dependence of laminar determination in developing neocortex. Science 254:282–285

McWhirter JR, Goulding M, Weiner JA, Chun J, Murre C (1997) A novel fibroblast growth factor gene expressed in the developing nervous system is a downstream target of the chimeric homeodomain oncoprotein E2A-Pbx1. Development 124:3221–3232

Meakin SO, Shooter EM (1992) The nerve growth factor family of receptors. Trends Neurosci 15:323–331

Meyer G, Goffinet AM (1998) Prenatal development of reelin-immunoreactive neurons in the human neocortex. J Comp Neurol 397:29–40

Meyer G, Gonzalez-Hernandez T (1993) Developmental changes in layer 1 of the human neocortex during prenatal life: a DiI-tracing and AchE and NADPH-d histochemistry study. J Comp Neurol 338:317–336

Meyer G, Soria JM, Martinez-Galan JR, Martin-Clemente B, Fairen A (1998) Different origins and developmental histories of transient neurons in the marginal zone of the fetal and neonatal rat cortex. J Comp Neurol 397:493–518

Meyer-Franke A, Kaplan MR, Pfrieger FW, Barres BA (1995) Characterization of the signaling interactions that promote the survival and growth of developing retinal ganglion cells in culture. Neuron 15:805–819

Miale IL, Sidman RL (1961) An autoradiographic analysis of histogenesis in the mouse cerebellum. Exp Neurol 4:277–296

Michaelson MD, Mehler MF, Xu H, Gross RE, Kessler JA (1996) Interleukin-7 is trophic for embryonic neurons and is expressed in developing brain. Developmental Biology 179:251–263

Middlemas DS, Lindberg RA, Hunter T (1991) trkB, a neural receptor protein-tyrosine kinase: evidence for a full-length and two truncated receptors. Mol Cell Biol 11:143–153

Minichiello L, Casagranda F, Tatche RS, Stucky CL, Postigo A, Lewin GR, Davies AM, Klein R (1998) Point mutation in trkB causes loss of NT4-dependent neurons without major effects on diverse BDNF responses. Neuron 21:335–345

Misson J-P, Edwards MA, Yamamoto M, Caviness VS Jr (1988) Identification of radial glial cells within the developing murine central nervous system: studies based upon a new immuno-histochemical marker. Dev Brain Res 44:95–108

Miyata T, Nakajima K, Aruga J, Takahashi S, Ikenaka K, Mikoshiba K, Ogawa M (1996) Distribution of a reeler gene-related antigen in the developing cerebellum: an immunohisto-chemical study with an allogeneic antibody CR-50 on normal and reeler mice. J Comp Neurol 372:215–228

Murphy M, Drago J, Bartlett PF (1990) Fibroblast growth factor stimulates the proliferation and differentiation of neural precursor cells in vitro. J Neurosci Res 25:463–475

Nadarajah B, Jones AM, Evans WH, Parnavelas JG (1997) Differential expression of connexins during neocortical development and neuronal circuit formation. J Neurosci 17:3096–3111

Nadarajah B, Makarenkova H, Becker DL, Evans WH, Parnavelas JG (1998) Basic FGF increases communication between cells of the developing neocortex. J Neurosci 18:7881–7890

Naruo K, Seko C, Kuroshima K, Matsutani E, Sasada R, Kondo T, Kurokawa T (1993) Novel secretory heparin-binding factors from human glioma cells (glia-activating factors) involved in glial cell growth. Purification and biological properties. J Biol Chem 268:2857–2864

Nielsen FC, Wang E, Gammeltoft S (1991) Receptor binding, endocytosis, and mitogenesis of insulin-like growth factors 1 and 2 in fetal rat brain neurons. J Neurochem 56:12–21

Nurcombe V, Ford MD, Wildschut JA, Bartlett PF (1993) Developmental regulation of neural response to FGF-1 and FGF-2 by heparan sulfate proteolgycan. Science 260:103–106

Ogawa M, Miyata T, Nakajima K, Yagyu K, Seike M, Ikenaka K, Yamamoto H, Mikoshiba K (1995) The reeler gene-associated antigen on Cajal-Retzius neurons is a crucial molecule for laminar organization of cortical neurons. Neuron 14:899–912

Ohbayashi N, Hoshikawa M, Kimura S, Yamasaki M, Fukui S, Itoh N (1998) Structure and expression of the mRNA encoding a novel fibroblast growth gactor, FGF-18. J Biol Chem 273:18161–18164

Ornitz DM, Xu J, Colvin JS, McEwen DG, MacArthur CA, Coulier F, Gao G, Goldfarb M (1996) Receptor specificity of the fibroblast growth factor family. J Biol Chem 271:15292–15297

Ornitz DM, Yayon A, Flanagan JG, Svahn CM, Levi E, Leder P (1992) Heparin is required for cell-free binding of basic fibroblast growth factor to a soluble receptor and for mitogenesis in whole cells. Mol Cell Biol 12:240–247

O'Rourke NA, Chenn A, McConnell SK (1997) Postmitotic neurons migrate tangentially in the cortical ventricular zone. Development 124:997–1005

O'Rourke NA, Sullivan DP, Kaznowski CE, Jacobs AA, McConnell SK (1995) Tangential migration of neurons in the developing cerebral cortex. Development 121:2165–2176

Orr-Urtreger A, Bedford MT, Burakova T, Arman E, Zimmer Y, Yayon A, Givol D, Lonai P (1993) Developmental localization of the splicing alternatives of fibroblast growth factor receptor-2 (FGFR2). Dev Biol 158:475–486

Ortega S, Ittmann M, Tsang SH, Ehrlich M, Basilico C (1998) Neuronal defects and delayed wound healing in mice lacking fibroblast growth factor 2. PNAS (USA) 95:5672–5677

Ozawa K, Urono T, Miyakawa K, Seo M, Imamura T (1996) Expression of the fibroblast growth factor family and their receptor family genes during mouse brain development. Mol Brain Res 41:279–288

Pabbathi VK, Brennan H, Muxworthy A, Gill L, Holmes FE, Vignes M, Haynes LW (1997) Catecholaminergic regulation of proliferation and survival in rat forebrain paraventricular germinal cells. Brain Res 760:22–33

Pearlman AL, Faust PL, Hatten ME, Brunstrom JE (1998) New directions for neuronal migration. Curr Opin Neurobiol 8:45–54

Pearlman AL, Sheppard AM (1996) Extracellular matrix in early cortical development. Prog Brain Res 108:117–134

Pearlman AL, Snider WD, Osborne PA, Brunstrom JE (1995) Neurotrophin-4 induces the production of c-fos in specific populations of developing cortical neurons that express trkB. Soc Neurosci Abst 21:546

Peters K, Ornitz D, Werner S, Williams L (1993) Unique expression pattern of the FGF receptor 3 gene during mouse organogenesis. Devel Biol 155:423–430

Peters K, Werner S, Chen G, Williams L (1992) Two FGF receptor genes are differentially expressed in epithelial and mesenchymal tissues during limb formation and organogenesis in the mouse. Development 114:233–243

Powell PP, Finklestein SP, Dionne CA, Jaye M, Klagsbrun M (1991) Temporal, differential and regional expression of mRNA for basic fibroblast growth factor in the developing and adult rat brain. Brain Res Mol Brain Res 11:71–77

Qian X, Davis AA, Goderie SK, Temple S (1997) FGF2 concentration regulates the generation of neurons and glia from multipotent cortical stem cells. Neuron 18:81–93

Qian X, Goderie SK, Shen Q, Stern JH, Temple S (1998) Intrinsic programs of patterned cell lineages in isolated vertebrate CNS ventricular zone cells. Development 125:3143–3152

Rajah R, Bhala A, Nunn SE, Peehl DM, Cohen P (1996) 7S nerve growth factor is an insulin-like growth factor-binding protein protease. Endocrinology 137:2676–2682

Rakic P (1972) Mode of cell migration to the superficial layers of fetal monkey neocortex. J Comp Neurol 145:61–84

Rakic P (1975) Timing of major ontogenetic events in the visual cortex of the rhesus monkey. In: Buchwald NA, Brazier M (eds) Brain Mechanisms in Mental Retardation. Academic Press, New York, pp 3–40

Rakic P, Cameron RS, Komuro H (1994) Recognition, adhesion, transmembrane signaling and cell motility in guided neuronal migration. Curr Opin Neurobiol 4:63–69

Ramon Y, Cajal S (1890) Sobre la existencia de celulas nerviosas especiales en la primera capa de las circunvoluciones cerebrales. Grac Med Catalana 3:737–739

Reid CB, Tavazoie SF, Walsh CA (1997) Clonal dispersion and evidence for asymmetric cell division in ferret cortex. Development 124:2441–2450

Ringstedt T, Linnarsson S, Wagner J, Lendahl U, Kokaia Z, Arenas E, Ernfors P, Ibáñez CF (1998) BDNF regulates reelin expression and Cajal-Retzius cell development in the cerebral cortex. Neuron 21:305–315

Rio C, Rieff HI, Qi PM, Corfas G (1997) Neuregulin and erbB receptors play a critical role in neuronal migration. Neuron 19:39–50

Rosen GD, Sherman GF, Richman JM, Stone LV, Galaburda AM (1992) Induction of molecular layer ectopias by puncture wounds in newborn rats and mice. Dev Brain Res 67:285–291

Ross ME (1996) Cell division and the nervous system: regulating the cycle from neural differentiation to death. Trends Neurosci 19:62–68

Rotwein P, Burgess SK, Milbrandt JD, Krause JE (1988) Differential expression of insulin-like growth factor genes in rat central nervous system. Proc Natl Acad Sci USA 85:265–269

Ryden M, Murray-Rust J, Glass D, Ilag L, Trupp M, Yancopoulos GD, McDonald NQ, Ibáñez CF (1995) Functional analysis of mutant neurotrophins deficient in low-affinity binding reveals a role for p75^{LNGFR} in NT-4 signalling. EMBO J 14:1979–1990

Schiffmann SN, Bernier B, Goffinet AM (1997) Reelin mRNA expression during mouse brain development. Eur J Neurosci 9:1055–1071

Schluesener HJ, Meyermann R (1994) Expression of BMP-6, a TGF-beta related morphogenetic cytokine, in rat radial glial cells. Glia 12:161–164

Segal RA, Greenberg ME (1996) Intracellular signaling pathways activated by neurotrophic receptors. Ann Rev Neurosci 19:463–489

Smallwood PM, Munoz-Sanjuan I, Tong P, Macke JP, Hendry SH, Gilbert DJ, Copeland NG, Jenkins NA, Nathans J (1996) Fibroblast growth factor (FGF) homologous factors: new members of the FGF family implicated in nervous system development. Proc Natl Acad Sci USA 93:9850–9857

Snider WD (1994) Functions of the neurotrophins during nervous system development: What the knockouts are teaching us. Cell 77:627–638

Stipp CS, Litwack ED, Lander AD (1994) Cerebroglycan: an integral membrane heparan sulfate proteoglycan that is unique to the developing nervous system and expressed specifically during neuronal differentiation. J Cell Biol 124:149–160

Strohmaier C, Carter BD, Urfer R, Barde YA, Dechant G (1996) A splice variant of the neurotrophin receptor trkb with increased specificity for brain-derived neurotrophic factor. EMBO J 15:3332–3337

Stylianopoulou F, Efstratiadis A, Herbert J, Pintar J (1988) Pattern of the insulin-like growth factor II gene expression during rat embryogenesis. Development 103:497–506

Takahashi T, Nowakowski RS, Caviness VS Jr (1992) BUdR as an S-phase marker for quantitative studies of cytokinetic behavior in the murine cerebral ventricular zone. J Neurocytol 21:185–197

Tamamaki N, Fujimori KE, Takauji R (1997) Origin and route of tangentially migrating neurons in the developing neocortical intermediate zone. J Neurosci 17:8313–8323

Tan SS, Kalloniatis M, Sturm K, Tam PP, Reese BE, Faulkner-Jones B (1998) Separate progenitors for radial and tangential cell dispersion during development of the cerebral neocortex. Neuron 21:295–304

Tanaka M, Sawada M, Yoshida S, Hanaoka F, Marunouchi T (1995) Insulin prevents apoptosis of external granular layer neurons in rat cerebellar slice cultures. Neurosci Lett 199:37–40

Timmusk T, Belluardo N, Metsis M, Persson H (1993) Widespread and developmentally regulated expression of neurotrophin-4 mRNA in rat brain and peripheral tissues. Eur J Neurosci 5:605–613

Tsoulfas P, Soppet D, Escandon E, Tessarollo L, Mendoza-Ramirez JL, Rosenthal A, Nikolics K, Parada LF (1993) The rat trkC locus encodes multiple neurogenic receptors that exhibit differential response to neurotrophin-3 in Pc12 cells. Neuron 10:975–990

Tsoulfas P, Stephens RM, Kaplan DR, Parada LF (1996) TrkC isoforms with inserts in the kinase domain show impaired signaling responses. J Biol Chem 271:5691–5697

Ulrich A, Schlessinger S (1990) Signal transduction by receptors with tyrosine kinase activity. Cell 61:203–212

Vaccarino FM, Schwartz ML, Hartigan D, Leckman JF (1995) Basic fibroblast growth factor increases the number of excitatory neurons containing glutamate in the cerebral cortex. Cereb Cortex 5:64–78

Valenzuela DM, Maisonpierre PC, Glass DJ, Rojar E, Nunez L, Kong Y, Gies DR, Stitt TN, Ip NY, Yancopoulos GD (1993) Alternative forms of rat TrkC with different functional capabilities. Neuron 10:963–974

Valverde F, Santacana M (1994) Development and early postnatal maturation of the primary olfactory cortex. Dev Brain Res 80:96–114

Verdi JM, Anderson DJ (1994) Neurotrophins regulate sequential changes in neurotrophin receptor expression by sympathetic neuroblasts. Neuron 13:1359–1372

Verdi JM, Groves AK, Farinas I, Jones K, Marchionni MA, Reichardt LF, Anderson DJ (1996) A reciprocal cell–cell interaction mediated by NT-3 and neuregulins controls the early survival and development of sympathetic neuroblasts. Neuron 16:515–527

Vicario-Abejon C, Johe KK, Hazel TG, Callazo D, McKay RDG (1995) Functions of basic fibroblast growth factor and neurotrophins in the differentiation of hippocampal neurons. Neuron 15:105–114

Voigt T (1989) Development of glial cells in the cerebral wall of ferrets: direct tracing of their transformation from radial glia into astrocytes. J Comp Neurol 289:74–88

Weise B, Janet T, Grothe C (1993) Localization of bFGF and FGF-receptor in the developing nervous system of the embryonic and newborn rat. J Neurosci Res 34:442–453

Werner H, Woloschak M, Adamo M, Shen-Orr Z, Roberts CT Jr, LeRoith D (1989) Developmental regulation of the rat insulin-like growth factor receptor gene. Proc Natl Acad Sci USA 86:7451–7455

Wilkinson GA, Farinas I, Backus C, Yoshida CK, Reichardt LF (1996) Neurotrophin-3 is a survival factor in vivo for early mouse trigeminal neurons. J Neurosci 16:7661–7669

Williams BP, Park JK, Alberta JA, Muhlebach SG, Hwang GY, Roberts TM, Stiles CD (1997) A PDGF-regulated immediate early gene response initiates neuronal differentiation in ventricular zone progenitor cells. Neuron 18:553–562

Yamaguchi TP, Conlon RA, Rossant J (1992) Expression of the fibroblast growth factor receptor FGFR-1/flg during gastrulation and segmentation in the mouse embryo. Devel Biol 152:75–88

Ye P, Xing YZ, Dai ZH, D'Ercole AJ (1996) In vivo actions of insulin-like growth factor-1 (Igf-1) on cerebellum development in transgenic mice: evidence that Igf-1 increases proliferation of granule cell progenitors. Brain Res Dev Brain Res 95:44–54

Zackenfels K, Oppenheim RW, Rohrer H (1995) Evidence for an important role of IGF-I and IGF-II for the early development of chick sympathetic neurons. Neuron 14:731–741

Zigova T, Betarbet R, Soteres BJ, Brock S, Bakay RAE, Luskin MB (1996) A comparison of the patterns of migration and the destinations of homotopically transplanted neonatal subventricular zone cells and heterotopically transplanted telencephalic ventricular zone cells. Dev Biol 173:459–474

Signalling from Tyrosine Kinases in the Developing Neurons and Glia of the Mammalian Brain

Elena Cattaneo[1] and Massimo Gulisano[2]

1
Introduction

The events that occur during the transition from proliferation to differentiation in the developing brain have attracted increasing attention, revealing the existence of multipotential central nervous system (CNS) stem cells and identifying the genes and factors that control their maturation and survival (Gage et al. 1995; Gage 1998; McKay 1997; Mehler 1997; Cattaneo and Pelicci 1998). Together with increasing our knowledge of the basic biology of brain development, this large body of data has also set the stage for new approaches to neurodegeneration, whereby genes encoding for pro-survival factors or in vitro expanded/differentiated CNS donor cells are vehiced to the diseased brain (Cattaneo and McKay 1991; Brustle and McKay 1996; Martinez-Serrano and Bjorklund 1997). Within a very brief period, this strong convergence of interests between basic and pre-clinical research has highlighted the way in which soluble growth factors, including the neurotrophins and cytokines, are molecules that have powerful effects on the ability of immature and mature brain cells to divide and/or to survive and differentiate (Eide et al. 1993; Mehler and Kessler 1994, 1997; Reichardt and Farinas 1997).

In recent years there has also been great progress in the identification and characterization of the intracellular signalling pathways that mediate the responses of CNS cells to these growth factors. Analysis of the mechanisms of signal transduction has begun to reveal the way in which signalling molecules act to specify the responsiveness of the cell to its environment. In certain cases, gene deletion analyses and pharmacological inhibition of signalling molecules have allowed description of the correlation between activation of a single pathway or molecule and particular biological responses. We have also learned how activity of the same signalling pathway can produce multiple

[1] Institute of Pharmacological Sciences, University of Milano, Via Balzaretti 9, 20133 Milano, Italy
[2] Laboratory of Molecular Biology, Department of Physiological Sciences, University of Catania, Viale Andrea Doria 6, 95125 Catania, Italy

Results and Problems in Cell Differentiation, Vol. 30
Goffinet and Rakic (Eds.): Mouse Brain Development
© Springer-Verlag Berlin Heidelberg 2000

(and often opposite) responses. The way in which a single factor can elicit such diverse responses has now also begun to be understood.

This chapter will review this growing field of research, with particular emphasis on the activity of three intracellular signalling systems that act during CNS development, namely the Ras-MAPK, the PI3-K and the JAK/ STAT. This review will reveal that, besides growth factors and growth factor receptors, signalling molecules act at critical crossroads to influence the transition from proliferation to differentiation in the brain.

2
Tyrosine Kinases During CNS Development

A central issue in the understanding of receptor tyrosine kinase signalling is whether different receptors activate different signal transduction pathways, and whether there are distinct pathways for proliferation and differentiation in the brain. Despite the complexity of such research, and the difficulties in assessing the role of specific pathways, recent reports have linked specific tyrosine phosphorylation events to cell differentiation in the brain.

The first evidence dates back to the early 1980s, when it was shown that protein tyrosine kinases are key elements in the process of differentiation of the nervous system. Pp60src, the protein product of the src gene, was the first protein tyrosine kinase found to be transiently expressed in developing neurons at the onset of terminal differentiation (Cartwright 1988). In particular, it was found that the activity of pp60src kinase increases in coincidence with the major period of neurogenesis in striatum and hippocampus, indicative of changes in protein tyrosine phosphorylation during maturation (Cartwright et al. 1988).

2.1
Growth Factors and Their Cell Surface Receptors

More recent investigations have focused on the action of tyrosine kinases that transduce from membrane embedded receptors. These cell surface receptors are divided into two major classes: (1) receptors presenting intrinsic tyrosine kinase (TK) activity; these are also known as receptor protein tyrosine kinases (RPTKs) (van der Geer et al. 1994; Segal and Greenberg 1996; Cattaneo and Pelicci 1998) and are exemplified by the epidermal growth factor receptor (EGFR), acidic and basic fibroblast growth factor receptor (FGFR), platelet derived growth factor (PDGFR) and Trk neurotrophin receptors; and (2) receptors like the cytokine receptors that lack intrinsic catalytic kinase domains and recruit cytoplasmic tyrosine kinases. Among these is the receptor for ciliary neurotrophic factor (CNTF) (Miyajima et al. 1992; Ihle 1995;

Ip and Yancopolous 1996; Taga and Kishimoto 1997; Cattaneo et al. 1999). Despite these differences, RPTKs and cytokine receptors share a common mechanism of receptor activation which involves receptor dimerization and autophosphorylation at specific tyrosine sites (Schlessinger and Ulrich 1992; Taniguchi 1995; Weiss and Schlessinger 1998).

In the developing mammalian brain both of these superfamilies of receptors (and their ligands) are highly expressed and play major roles (Reichardt and Farinas 1997; Mehler and Kessler 1997). The first evidence of their role during development came from studies done in the fruitfly, *Drosophila melanogaster*, where molecular analysis of mutations that affect cell differentiation have highlighted that several of these cell surface receptors act as early determinants of cell fate. Loss of function mutations in the sevenless gene of *Drosophila* (Harris et al. 1976) were indeed found to abolish the tyrosine kinase activity of a transmembrane receptor expressed in the developing omnatidia of the eye, resulting in abnormal differentiation of photoreceptors number 7 (reviewed in Basler and Hafen 1988; Rubin 1989). Instead of becoming photoreceptor number 7, these precursor cells differentiate into nonneuronal cells. Also, mutations in the *Drosophila* Ellipse gene, which encodes a homologue of the mammalian EGF receptor, result in the developmental failure of multiple cell types in the eye (Baker and Rubin 1989).

In mammals, classical biological approaches using cultured CNS cells have focused on analyses of the effects of the ligands for these growth factor receptors. It was found that growth factors and cytokines have a profound influence on the proliferation of CNS stem cells, and on the survival and differentiation of their postmitotic descendants.

These types of analyses immediately revealed that a single factor and its receptor are not endowed with a predetermined biological activity but, rather, that the final effect is dependent on the cellular context in which the receptor is expressed and on the degree of maturation of that given cell type. PDGF, for example, stimulates proliferation and prevents premature differentiation of glial progenitor cells of oligodendrocytes/type 2 astrocytes in rat optic nerve cultures (Barres et al. 1993). However, the same factor also stimulates neuronal differentiation by uncommitted neuroepithelial cells in vivo (Williams et al. 1997) and in vitro (Johe et al. 1996).

EGF is another factor that has been much studied. It was first found to stimulate neuritogenesis in primary neuronal cultures obtained from subcortical telencephalic regions (Morrison et al. 1987) and, more recently, to trigger the proliferation of immature stem/progenitor cells derived from the presumptive striatum (Reynolds and Weiss 1992; Reynolds et al. 1993). This evidence of multiple functions exerted by a single molecule in different cells or at different stages of development indicates that ligand-independent mechanisms function to diversify cell responsiveness. As we will discuss below, developmental changes in receptor number/subtypes or in transducing components may be involved in these multiple responses.

FGF molecules are potent mitogens for a wide variety of different cell types, both in tissue culture and in vivo, and are implicated in differentiation and migration of neuronal cells (Gensburger et al. 1987; Morrison et al. 1987; Cattaneo and McKay 1990; DeHamer et al. 1994; Gritti et al. 1996). FGFs are expressed in a strict temporal and spatial pattern during development and have important roles in brain patterning and limb formation. They bind in an overlapping pattern to four structurally related receptor tyrosine kinases (Galzie 1997). Different FGF family members activate the FGF receptor subtypes in an affinity dependent manner (Ornitz et al. 1996).

During embryonal development, FGF receptor signalling appears to be critical for the development of a wide range of organs and for patterning of the embryo. Indeed, targeted inactivation of the FGFR-1 gene leads to embryonal death prior to gastrulation and the embryos display severe growth retardation (Deng et al. 1994). Yamaguchi et al. (1994) generated FGFR-1-deficient embryos that remained capable of gastrulating and generating mesoderm, but mesodermal patterning was aberrant, leading to the conclusion that FGFR-1 is not needed for mesoderm formation per se. On the other hand, chimeric embryos with a low contribution of FGFR-1-deficient cells completed gastrulation and displayed various malformations of the limb buds and also partial duplication of the neural tube, indicating that FGFR-1 plays a role in neurulation (Celli et al. 1998).

FGF has also been shown to be critical for cell migration during brain development in vivo, through FGFR-1. When oligodendrocyte progenitors that expressed a dominant-negative version of FGFR-1 were transplanted into the brains of neonatal rats, the cells were unable to migrate and remained within the ventricles (Osterhout et al. 1997). FGFR-1-mediated migration appears to be regulated by other transduction pathways that are different from those in cells migrating, e.g., towards EGF or PDGF (Kundra et al. 1994; Hansen et al. 1996; LaVallee et al. 1998).

It is known that neurotrophins also act through RPTKs and, at least in vitro, can influence the survival of many classes of embryonic or neonatal peripheral or central neurons (Chao 1992; Reichardt and Farinas 1997). It should be kept in mind, however, that there is only partial in vivo confirmation of these effects. Thus, mice with the various neurotrophins (or their receptors) knocked out only exhibited deficits in sensory and sympathetic neurones, while motorneurones and other populations of central neurones were unaffected. These data do not indicate that these RPTKs have no role during brain development but, rather, suggest that CNS deficiencies in individual trophic factors can easily be compensated in vivo (Reichardt 1997). Mice mutant for more then one neurotrophin are therefore important for the analyses of their role during development (Minichiello and Klein 1996; Silos-Santiago et al. 1996).

The Eph (Ephrine) receptor tyrosine kinases are another class of RPTKs (the largest) which are emerging as the molecules that guide migration of cells and growth cones during embryonic development (Goodman and

Tessier-Lavigne 1997). Based on their concentration in embryonic regions containing ingrowing neuronal processes, the Eph receptors were indeed suspected, early on, to have a role in regulating aspects of axon growth. The most distinctive role of the Eph receptors appears to be their ability to mediate cell–cell repulsion through the binding of a ligand on an adjacent cell surface. The repulsive interactions are presumably mediated by transient receptor activation at the boundaries of complementary regions of high ligand or receptor expression. In contrast, overlapping expression patterns may regulate cell adhesion and cytoskeletal organization with possible consequences for the overall growth and fasciculation of neuronal processes.

A notable feature of Eph receptor signalling is that, upon receptor binding, responses may also be elicited in the ligand-expressing cells which are anchored to the membrane via either a phospholipid (PI) anchor or a transmembrane region (TM) and have a highly conserved cytoplasmic region, with multiple potential sites for tyrosine phosphorylation (Goodman and Tessier-Lavigne 1997). Analyses of *Nuk* function, a murine Eph receptor that binds TM ligands, revealed that both receptor and ligands are associated with a tyrosine kinase and that their interaction mediate bidirectional cell signalling (Holland et al. 1996). As a consequence, lack of *Nuk* in mice causes abnormal axonal connections (Orioli et al. 1996).

A better understanding of Eph receptor function requires the elucidation of their signalling properties. Only a few cytoplasmic signalling molecules that bind to the activated Eph receptors have been identified. Several of these molecules are known to transduce signals regulating cytoskeletal organization and neurite outgrowth and found to involve the PI3-K (Pandey et al. 1994) and other SH2 containing proteins (Holland et al. 1997).

Currently, it is unclear why there is a need for fourteen distinct Eph receptor genes, many of which appear to encode several variant forms with distinct functional properties, but it is tempting to speculate that such diversity is necessary to refine the spatial organization of embryonic structures.

Altogether, this large body of evidence on the expression and activities of various tyrosine kinases indicates that this class of proteins influences critical aspects of brain development.

3
Phospho-Tyrosines and Their SH2 Partners

The mechanisms by which activated growth factor receptors stimulate cytoplasmic and nuclear events are largely based upon phosphorylation cascades, which first occur at the cytoplasmic face of transmembrane receptors, and rapidly propagate onto specific signalling molecules inside the cell

(Pawson 1995). Tyrosine phosphorylation, in particular, comprises by far the largest proportion of all the phosphorylation events occurring in a given cell, the remainder being serine or threonine phosphorylation. This simple substitution of a phosphate group for a hydroxyl group on tyrosine residues results in modification of the activity, life span, or cellular localization of proteins. Given its importance, proper control and reversal of this kinase-catalyzed reaction is guaranteed by the activity of various phosphatases (see below).

Phosphotyrosil (P-Tyr) residues on proteins are, however, only one component of the scenario. These P-Tyr residues function as binding sites for intracellular signalling proteins containing domains of 100–180 aa known as the SH2 (from Src homology 2), or phosphotyrosil binding (PTB) domains (Pawson 1995).

The SH2 domain, originally identified in Src (homology 2 domain), was the first to be identified and served as a prototype to understand the role of these moduli in signal transduction (Koch et al. 1991). These domains are not restricted to particular types of signal transduction proteins, but are also present in protein kinases, phosphatases, transcription factors and Ras controlling proteins. They are also found in adaptor molecules with no enzymatic function, such as Shc and Grb2. When such SH2-containing proteins are recruited to the activated receptor, they can be phosphorylated for further regulation. Other domains, such as Src homology 3 (SH3, which recognizes proline-rich regions) and PH (Pleckstrin homology), localize PTKs and their substrates within specific compartments of the cell and provide specific contacts with other proteins in signalling cascades (Pawson 1995; Pawson and Schott 1997).

Cellular responses thus largely depend on the recognition between these two components. A genetic test of this concept came from transgenic mice where the P-Tyr residues on the Met receptor, which normally bind a large number of SH2 molecules, were converted to Phe residues. This resulted in the phenotype that was a null mutation, despite the fact that the activity of the kinase domain is unaltered (Maina 1996).

Subsequent work showed that different SH2s bind to distinct phosphotyrosine-containing regions of the RPTK. Thus, specificity of recognition occurs because SH2 domains recognize not only phosphotyrosine, but also the three residues immediately C-terminal to the P-Tyr. Therefore, although all SH2 domains bind to phosphotyrosyl proteins, they are not promiscuous in choosing partners and, currently, two groups of SH2 domains can be distinguished. One group of SH2-containing proteins (such as the one in proteins like Src and Grb2) prefer sequences with the pTyr-hydrophilic-hydrophilic-hydrophobic motif, while another group (SH2s found in PLC-gamma) select pTyr-hydrophobic-X-hydrophobic (where "X" stays for any aminocid) as a general motif (Pawson and Schott 1997).

Despite the existence of such strong specificity, it should be kept in mind that the recognition is not absolute, and that there may be more than one

SH2-containing protein within a cell that has high affinity for a particular ligand. Therefore, in vivo, the ability of an SH2 domain to engage a particular phosphoprotein may be critically dependent on the local concentration of proteins and on the modulating effect of other domains found on interacting proteins.

As an example of the importance of these domains in signal transduction and development, it has been shown that a mutation of one of the two SH3 domains found in SEM-5 blocks vulval development in *C. elegans* (Mayer et al. 1993).

4
Controlling the Activity of the Ras-MAPK Pathway

4.1
The Players

Several independent approaches, such as microinjection of neutralizing antibodies, Ras overexpression and genetic analyses in *Drosophila* and *C. elegans*, have implicated Ras in signal transduction downstream of RPTKs. Like other G proteins, Ras cycles between an active GTP state and an inactive GDP-bound state, thereby acting as a molecular switch (Campbell et al. 1998). The molecules activating Ras have been identified and found to include the SH2/PTB-containing adaptor protein like Shc (from Src homologue and collagen homologue) which binds to the activated RPTK and becomes itself phosphorylated at, at least, three sites (reviewed in Cattaneo and Pelicci 1998). One or more phosphotyrosines on Shc then function as docking sites for the Grb2 adaptor, which is bound constitutively to a proline-rich region of the protein named Son of Sevenless (Sos), the GDP/GTP exchanger for Ras. These early tyrosine phosphorylation events direct relocalization of Sos to the membrane with the formation of GTP-bound Ras (Campbell et al. 1998).

Recent studies indicate that the MAPK are targets that lie directly downstream of Ras (Fig. 1). GTP-bound Ras activates the MAPK-kinase-kinase (MAPKKK) Raf1 (Pritchard and McMahon 1997). This phosphorylates and activates the MAPK-kinase-kinase (also known as MEK), a dual specificity protein kinase which, in turn, phosphorylates and activates MAPK at its threonine and tyrosine activating sites (Marshall 1995). MAPK is a family of three enzymes, and at least one form (the extracellular regulated kinases, ERKs) can translocate into the nucleus where it can phosphorylate and activate various transcription factors (Segal and Greenberg 1996). The importance of Ras in RPTK signalling is revealed by studies performed using dominant negative Ras mutant, where MAPK activation and responses to a variety of growth factors were abolished (Marshall 1995).

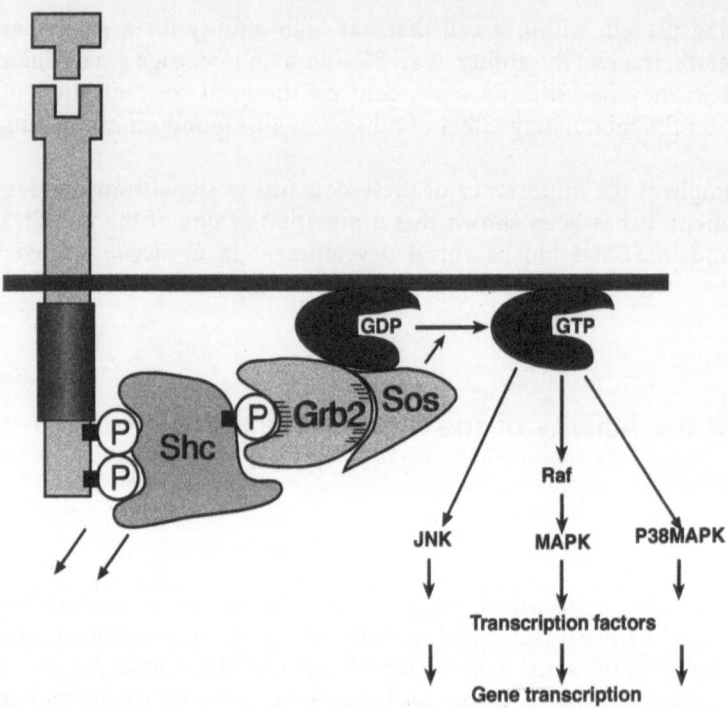

Fig. 1. The Shc-Ras-MAPK cascade activated by RPTKs is illustrated. The kinase domain of the activated RPTKs phosphorylates the receptor itself and components of specific pathways. Activated RPTKs transduce via adaptors like Shc and Grb2. The latter is constitutively associated with Sos, an exchanger of GDP/GTP, which activates Ras. Subsequent steps involve three kinase levels (a MAPKKK-like Raf1, then a MAPKK-like MEK and, finally, the MAPKs). MAPKs ultimately translocate to the nucleus where they phosphorylate transcription factors to generate immediate (*c-fos* gene expression) and delayed gene transcription responses. The other two MAPK members (JNK and p 38) activated by RPTKs (through other G proteins different from Ras) are shown. (Adapted from Cattaneo and Pelicci 1998). P = site of phosphorylation

Besides the ERKs, Ras also activates two other MAPK pathways, one involving the Jun N-terminal kinase, JNKs (also termed SAPKs), and the p38-MAPK. These kinases have been found to be associated with both cell apoptosis and survival. Thus, it has been shown that NGF withdrawal from cultured sympathetic neurons triggered cell death via activation of the JNK and phosphorylation of the c-Jun transcription factor. On the other hand, in the adult nervous system, c-Jun is also involved in neuroprotection and regeneration. It has thus been proposed that expression of the c-Jun transcription factor is an early and consistent component in the effective response of neurons to various forms of injury, and functions as a prerequisite for other transcriptional components to exert their specific decision about death, survival, or regeneration (Herdegen et al. 1997).

4.2
MAPK: Proliferation or Differentiation?

Extensively studied in non-neuronal cells and originally proposed to be the major mitogenic signalling cascade initiated by the EGFR family of kinases, activation of the Ras-MAPK pathway in neural cells has been linked to both proliferation and differentiation. Particularly, analyses of EGF signalling have been mostly carried out in tumoral PC12 cells derived from pheochromocytoma (Greene and Tischler 1976). In these cells, it was found that EGF promotes tyrosine phosphorylation of the receptor and of Shc adaptors, leading to a short-lived rise in RasGTP and transient MAPK stimulation and, thereby, to proliferative activity by the cells. In an apparently contrasting manner, the same signalling pathway was found to be activated by NGF, a factor that is known to induce neurite outgrowth in PC12 cells (Greene and Kaplan 1995). This controversy began to be solved with the discovery of quantitative differences in the way in which the Ras-MAPK pathway is activated by EGF or NGF, and it was demonstrated that the differentiative effect of NGF depends on a prolonged (hours) activation of the Ras-MAPK (Marshall 1995).

On the basis of these data, a model has been proposed that accounts for the different consequences of transient versus sustained activation of the ERKs (Marshall 1995). This model considers that sustained MAPK activation leads to their persistent nuclear accumulation, resulting in phosphorylation of transcription factors and changes in gene expression which causes differentiation. More in-depth analyses of these differences in response have led to the finding that the effects were strictly dependent on the number of receptors expressed on the cell membrane. Overexpression of the EGFR in PC12 cells overrode the limited ERK activation by EGF, and produced differentiative effects. Similarly, NGF elicited cell proliferation in PC12 subclones that only presented a limited number of NGFR. Thus, while high EGFR number leads to proliferation of PC12 cells, activation of fewer EGFR is associated with cell differentiation. The finding that different receptor levels may change cellular responsiveness has recently been extended to the developing brain in a study by Burrows et al. (1997). The authors introduced extra EGFRs into early cortical progenitor cells of the ventricular zone and found that they could modulate responsiveness of these cells to EGF.

These data indicate, therefore, that the ability of a cell to proliferate or differentiate in response to a given factor may be dictated by the number of receptors expressed on its membrane, this being translated into different degrees of intracellular tyrosine phosphorylation and, ultimately, of ERK activation. More recent work on NGF-triggered PC12 cells has demonstrated that the differentiative effects of NGF rely on two phases of ERK activation: the one that is Ras-dependent, and a second that is dependent on another small GTPase, named Rap1. This latter pathway is independent of Shc-Grb2 binding to the NGF receptor and recruits B-raf downstream, not Raf1 (York et al. 1998).

The data described up to this point seem to link NGF-induced prolonged ERKs activation with neuronal differentiation. However, a degree of redundancy in the mechanisms by which NGF activates the ERKs has been described at least for PC12 cells, which involves the phospholipase Cγ(PLCγ) (Greene and Kaplan 1995). It has been found that mutations within TrkA that abrogate its interaction with Shc have very little effect on the ability of NGF to reduce MAPK activation and neurite outgrowth. In contrast, mutant Trk, that do not specifically bind Shc and PLCγ specifically, failed to elicit MAPK and neurite outgrowth (Obermeier et al. 1994; Stephens et al. 1994).

Despite this very large and fascinating body of evidence, the existence of similar mechanisms in primary neurones, or in vivo, is far from established. Interesting correlative data, that still require investigation, come from cultured cortical neurones, where the degree of activation of MAPK in response to a single factor was found to vary as a function of maturation of the donor tissue (Ghosh and Greenberg 1995). This may be explained by qualitative or quantitative changes at the receptor level (Knusel et al. 1994; Burrows et al. 1997), but may also imply a dynamic regulation of the expression and activity of upstream signalling molecules that occurs as cells mature and which, in turn, influences the activity of downstream components (e.g. MAPK) of the cascade (Cattaneo and Pelicci 1998).

4.3
Changing Adaptors for the Ras-MAPK Pathway: The Shc(s)

Another potentially powerful mechanism for generating diversity in biological responses to growth factors appears to involve Shc adaptor proteins. Recent findings obtained in vivo and in vitro have revealed that expression of these adaptor proteins is finely regulated during brain maturation (Conti et al. 1997).

As with many other intracellular signalling proteins, Shc (now renamed ShcA) proved to be only one member of a family of genes, which is now known to include two new homologues, ShcB/Sli and ShcC/Rai/N-Shc (reviewed in Cattaneo and Pelicci 1998). These three Shc(s) share elevated homology in both the C-terminus SH2 domain and the N-terminus PTB domain, the most divergent sequence being in the proline- and glycin-rich CH1 (collagene homology 1) region.

It has been demonstrated that ShcA proteins and mRNAs (ShcA gene encodes for three proteins) are sharply downregulated in coincidence with neurogenesis in the brain (Conti et al. 1997). Specifically, in the embryonic mouse brain, ShcA was localised to the germinal epithelium where the immature CNS stem cells located in it still actively divide. In the areas of the embryonic or postnatal brain where neurons were already postmitotic, ShcA was very low or not detected. The adult brain was consistently devoid of ShcA, and it was not expressed in mature quiescent glia. The only exception to this pattern was the olfactory epithelium, which is the only tissue that still undergoes active cell renewal in the adult. In vivo immunoprecipitation studies on the embryonic telencephalic vesicles revealed that the ShcA present in the actively dividing cells was subject to phosphorylation and Grb2 binding, suggesting that it is capable of activating the Ras-MAPK pathway in these cells in vivo (Conti et al. 1997). These changes in the expression and activity of ShcA as a function of neuronal maturation were confirmed in vitro in differentiating neuronal cultures.

These data indicate that changes in the levels of ShcA adaptors may influence the activity of the Ras-MAPK pathway during development. Given the existence of two new Shc members, ShcB and ShcC, it has recently been suggested that one or both of them could replace ShcA in mature neurons (Cattaneo and Pelicci 1998). Although this has yet to be demonstrated, it seems possible that changes in Shc levels at different stages of development may affect the activity of downstream components of signalling pathways, generating proliferation or differentiation.

Finally, as an additional strategy to diversify responses from a single receptor, it was found that mutations at the Shc binding site in TrkB (the receptor for BDNF, brain-derived neurotrophic factor, and NT4, neurotrophin 4) cause loss of NT4-dependent neurons without major effects on diverse BDNF responses (Minichiello et al. 1998).

5
Controlling Cell-Survival Via PI3-K

Induction of differentiation is not the only effect of NGF in PC12 cells. Studies using wortmannin, the phosphatidylinositol 3-kinase (PI3-K) inhibitor, revealed that the activated TrkA receptor expressed in PC12 cells also stimulates cell survival through the PI3K pathway, which is independent of Ras. The mechanisms by which activated PI3-K influences neuronal cell survival have recently been elucidated (Fig. 2).

PI3-K is an SH2-containing enzyme associated with a variety of receptor and non-receptor PTKs (Auger et al. 1989a; Auger et al. 1989b). The enzyme is a heterodimer that phosphorylates the 3' position on a variety of inositol lipids and serines, on protein substrates (Auger et al. 1989). It has been shown that exposure of various cell types (including cerebellar neurones) to survival factors such as the neurotrophins and IGF1 induces activation of the PI3-K and of its crucial mediator, a serine/threonine protein kinase named Akt (alias PKB, protein kinase B; Burgering and Coffer 1995; Franke et al. 1995). Recently, three of the downstream targets of Akt action have

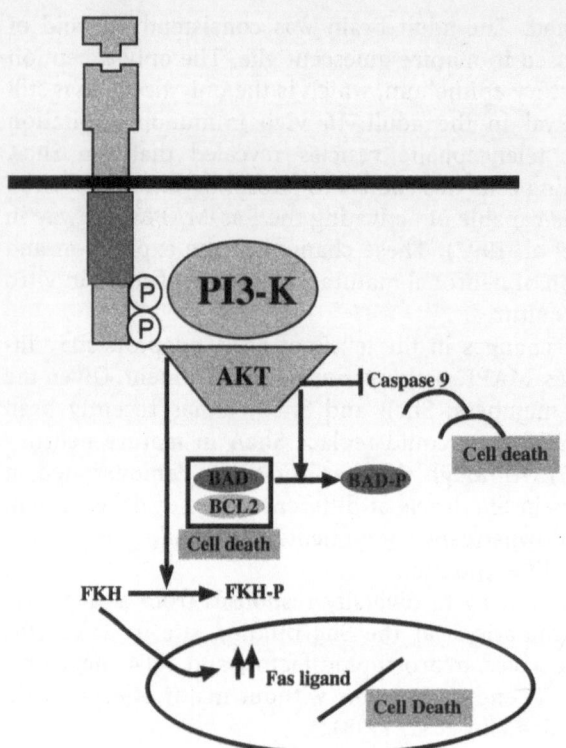

Fig. 2. The PI3-K signalling cascade. This pathway involves the dimeric enzyme PI3-K which activates Akt/PKB, the latter promoting cell survival via three mechanisms: (1) inhibiting caspase 9 activation; (2) phosphorylating pro-apoptotic BAD; (3) phosphorylating FKH transcription factor and, therefore, causing its retention in the cytoplasma. This prevents its nuclear translocation and the subsequent activation of cell death genes (like Fas ligand)

been identified. Two of them are components of the intrinsic cell death machinery. Thus, Akt directs phosphorylation of the Bcl2 family member BAD (Datta et al. 1997; del Peso et al. 1997) and of the protease caspase 9 (Cardone et al. 1998), thereby suppressing their pro-apoptotic function. This important activity of Akt has been further reinforced by the finding that a third mechanism of action includes the phosphorylation of transcription factors that control cell death genes. In the presence of survival factors, Akt phosphorylates FKHRL1, a member of the Forkhead family of transcription factors, leading to its association with, and sequestering by, the 14-3-3 cytoplasmic proteins (Brunet et al. 1999; Kops et al. 1999). Phosphorylated FKHRL1 thus remains apart from its nuclear transcriptional targets. In the absence of survival factors the PI3-K/Akt pathway is inactivated, FKHRL1 is unphosphorylated at its Akt sites and accumulates in the nucleus, where it can activate death genes like the Fas ligand. Although a large proportion of these data were obtained in cell lines (of non-neuronal origin), experiments performed on cultured cerebellar granule neurones have shown that survival depends both on activated Akt and on the presence of a wildtype-phosphorylatable version of FKHRL1 transcription factor (Brunet et al. 1999).

6
The JAK/STAT Pathway: A New Route to Proliferation and Differentiation in the Brain

Besides the intracellular activities of membrane-bound receptors containing intrinsic tyrosine kinases, there is another pathway that contributes signals from cytokine receptors at different stages of neuronal and glial maturation. Cytokine receptors distinguish themselves by the absence of the tyrosine kinase domain in their cytoplasmic tail (Miyajima et al. 1992). Despite that, they transduce intracellularly via phosphorylated substrates. This intriguing observation was explained with the discovery of a JAK (Just Another Kinase) family of cytoplasmic tyrosine kinases which are rapidly recruited to the activated receptors (Darnell et al. 1994; Taniguchi 1995). The JAKs, in turn, phosphorylate a class of transcription factors named STATs (from signal transducer and activator of transcription), which than dimerize and translocate into the nucleus where they activate the transcription of genes carrying STATs binding sites in their promoters. Four JAK members (named Jak1, Jak2, Jak3 and Tyk2) and seven mammalian members of the STAT family (namely Stat1, Stat2, Stat3, Stat4, Stat5a and Stat5b, and Stat6) are known to date, (reviewed in Cattaneo et al. 1999; Fig. 3).

Extensive information on the JAK/STAT pathway has been gained from studies on hematopoietic cells. It is now well established that the main mechanism of STAT activation in various cell types is through tyrosine phosphorylation at a single site in the carboxyl terminus of the protein. More recently, it has been found that the functions of the STATs may also be influenced by serine phosphorylation, which is necessary to achieve maximal STAT activation and DNA binding activity (Clark Lewis et al. 1991; Eilers et al. 1995; Wen et al. 1995; Zhang et al. 1995). Interestingly, the site of serine phosphorylation in Stat1 and Stat3 matches with the known MAP kinase recognition sequence (Clark Lewis et al. 1991), indicating that there may be cross-talk between the STATs and the MAPK pathway at this phosphorylation site (Wen et al. 1995).

Other studies have shown that despite their initial identification as cytokine signalling proteins, the JAKs and STATs also participate in signalling from activated RPTKs (Chin et al. 1996; Faris et al. 1996; Patel et al. 1996). Among them are receptors for molecules with well-known effects on developing and mature glia and neurons, like bFGF, PDGF and EGF. These data, together with the findings that the cytokine receptors (and RPTKs) and JAK/STAT are widely distributed in the developing and mature brain (see below; De Fraja et al. 1998), indicate that there may be a wide range of activities in the brain (some still unknown) in which these signalling proteins are involved. Cytokines may also signal through the Ras pathway (Sato et al. 1993; Bonni et al. 1997; Rajan and McKay 1998). Although the role of Ras in cytokine signalling is not fully understood, it may be assumed that the

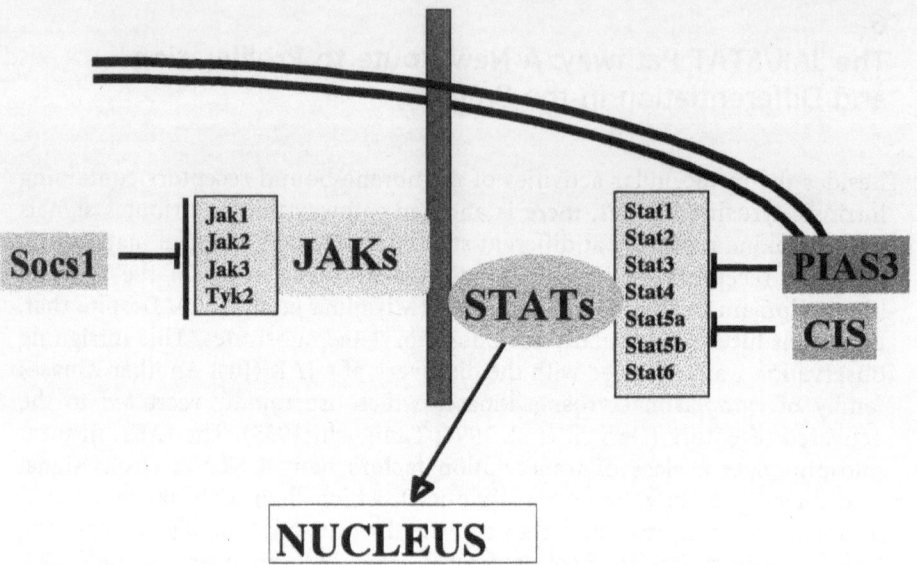

Fig. 3. The JAK/STAT signalling molecules and their inhibitors. The JAK/STAT transduces from activated cytokine receptors which do not possess an intrinsic tyrosine kinase domain and therefore recruit cytoplasmic JAKs. These phosphorylate the STAT transcription factors that translocate into the nucleus where they bind to specific DNA elements (DNA response elements) situated upstream of genes induced by cytokines. Two main subclasses of DNA response element are found in the promoters of genes activated downstream of the JAK-STAT pathway: interferon-stimulated response element (ISRE), a sequence located upstream of IFNα/β-inducible genes, and gamma activated sequence (GAS)-like sequences located upstream of IFNγ-inducible genes. These two classes of STAT binding elements are variations on the same basic motif. The GAS element was identified as the ligand of an IFNg-inducible Stat1-containing complex, known as gamma-activated factor (GAF). Since then, GAS-like elements which bind a wide range of STATs-containing complexes (not necessarily comprising Stat1) induced by other cytokines have been shown to consist of the same basic DNA motif, with subtle sequence differences in individual cases found in specific genes. On the other hand, ISREs appear to possess the ability to bind only Stat2-containing complexes, but vary in sequence composition between genes (Ihle 1995)

pleiotropic effects of these ligands require recruitment of more than one pathway.

Analyses of JAK/STAT presence and function in CNS cells have revealed that they occur at various stages of brain development. Some of these transcription factors and their upstream kinases are expressed with a certain degree of spatiotemporal specificity. For example, while Stat3 levels were remarkably constant at different embryonic and postnatal stages in various brain regions in vivo, levels of Stat6 were progressively decreased. Among the different brain regions analyzed, striatum was particularly rich in Stat5 and Jak2 during embryonic stages. Prompted by these observations, our group has investigated possible consequences of Stat5 activation in CNS cells. For this purpose, conditionally immortalized striatal-derived ST14A cells

(Cattaneo et al. 1996) were genetically engineered to express the IL3 receptor, which is known to transduce through Jak2 and Stat5 activation. Addition of IL3 to the culture medium elicited cell proliferation, in the absence of serum, via phosphorylated Jak2 and Stat5. Other work has shown that self-renewal of pluripotent embryonic stem cells also depends on activation of the STATs, in this case of Stat3 (Niwa et al. 1998). These data, therefore, define a role for some of the STATs during cell proliferation.

Stat3 (with Stat1 and the upstream Jak1) is also involved in transducing survival and differentiative signals from activated CNTFR (Rajan et al. 1996; reviewed in Cattaneo et al. 1999). The biological effects of this citokine are mediated by CNTFRα on responsive cells, which, once bound to CNTF, triggers sequential heterodimerization of the two signal transducing subunits of the CNTFR-LIF receptor β (LIFRβ) and gp130 (130 KDa glicoprotein). More recently, a role for Stat3 and Stat1 has been identified during gliogenesis, whereby differentiation of cerebral cortical precursor cells into GFAP expressing cells was found to require Stat3 and Stat1 signalling. Transfection of these primary cells with dominant-negative versions of these transcription factors (or of Jak1) abolished the expression of GFAP (Bonni et al. 1997). The promoter for GFAP is known to contain seven consensus sequences for activated STATs (Kahn et al. 1997). Deletion of these sequences prevented GFAP expression by the activated STATs (Bonni et al. 1997). These data indicate, therefore, that GFAP expression is under the control of phosphorylated-STATs and CNTF, and that the JAK/STAT signalling pathway critically controls cell fate during mammalian brain development.

More recently, cytokine-induced STAT activation was found to act in synergy with BMP2 (bone morphogenetic protein-2) signals (which involves the Smads transcription factors) to promote astrocytes differentiation from fetal neuronal progenitor cells. This synergy occurs at the level of the p300 transcriptional coactivator to which both Stat3 and Smad1 physically bind (Nakashima et al. 1999).

Beside the JAK/STAT, CNTF was also shown to influence the MAPK, although the role of MAPK activation by CNTF is less clear. One study showed that transfection of a dominant interfering form of MAPK Kinase (MEK) augmented CNTF induction of the GFAP promoter (Bonni et al. 1997). In another study, however, activation of both MAPKs and JAK/STAT pathways appeared to be positively coupled to astrocytic differentiation in vitro (Rajan and McKay 1998). It is possible that differences in the culture conditions and in CNTF doses used in the two studies may have accounted for the opposite results observed. Recent data do indeed indicate that different CNTF doses may elicit or inhibit GFAP expression, depending on quantitative differences in recruitment of members of these two pathways (Monville et al. 1998). The role of the JAK/STAT pathway has also been investigated in IL6-exposed PC12 cells, where IL6-induced phosphorylation of Stat3 was a negative regulator of MAPK-dependent neurite extension (Wu and Bradshaw 1996; Ihara et al. 1997).

Thus, it appears that the final biological response to a given factor is the result of a complex interplay between signalling pathways and interacting growth factors. Given the large number of JAKs and STATs members known, it is likely that new functions for these signal transducers in CNS cells will soon emerge.

The search for the additional functions of these kinases and of their downstream partners should, however, also consider the recently discovered inhibitors of JAKs and STATs (Fig. 3). These are molecules that block the interaction between activated intracellular tyrosine kinases and the STATs. One, identified with various names, is a JAK-binding protein that interacts with the JH1 domain of all of the four JAKs, thereby reducing their tyrosine kinase activity (Endo et al. 1997; Naka et al. 1997; Starr et al. 1997). Transcription of these inhibitory molecules was positively modulated by cytokines, suggesting that they may function in a negative feedback loop to regulate JAK/STAT signalling. Specific inhibitors of the STATs have also been identified (Yoshimura et al. 1995; Chung et al. 1997; Patel et al. 1998).

7
The Action of Phosphatases

It is now known that the net level of phosphate in a target substrate reflects the activity not only of the PTKs that catalyse phosphorylation, but also of the protein tyrosine phosphatases (PTPs) that are responsible for dephosphorylation of tyrosil residues. PTPs are a large and structurally diverse family of receptor-like and cytoplasmic enzymes which are found expressed in the embryonic, postnatal and adult brain as well as in cultured cells (Sahin and Hockfield 1993; Sahin et al. 1995; Hunter 1995; Shiozuka et al. 1995; Wang et al. 1995). They play a role in attenuation of signals generated by PTKs (Fischer et al. 1991). An initial demonstration of their role came with the finding that RPTK-mediated tyrosine phosphorylation can be greatly enhanced by treatment with the PTP inhibitor sodium orthovanadate (Hunter 1995). Among the various phosphatases now known, MKP-1, is a dual-specificity MAPK phosphatase which is itself induced by growth factor stimulation, presumably through the MAPK pathway. MKP-1 inactivates MAPKs by dephosphorylating both their phosphothreonine and their phosphotyrosine regulatory sites (Sun et al. 1993). Several additional MAPK phosphatases have been identified (Sun and Tonks 1994), some of which show specific expression in particular cellular districts. For example, SH-PTP2 has been localized to the brain where it is particularly associated with synaptic membranes (Suzuki et al. 1995).

Protein tyrosine phosphatases may also act positively to promote signalling events. Co-operation between PTPs and RPTKs is evident in the control of the Src family of cytoplasmic PTKs, which have an inhibitory phosphorylation site in their carboxyl termini (Sun 1994). Dephosphorylation of this by PTP promotes kinase activity and triggers the signalling cascade. Extensive analyses of PTP function in knock-out mice that lack specific PTP have shown that this leads to severe immunodeficiency and systemic autoimmune diseases, highlighting their roles in the control of signalling cascades.

8
Concluding Remarks

It is now established that growth factors act on their target cell via specific receptors which, in turn, recruit different signalling pathways and elicit different responses. Many of these growth factors, receptors, and their transducing molecules, have been identified. The molecular pathways leading to specific responses have been systematically dissected, both in vitro and in vivo. The panorama that emerges from these analyses is that the final biological response to a given factor is the result of a complex and highly dynamic interplay between different actors at different levels (Fig. 4). Growth factors and growth factor receptors are usually grouped into large families whose members share high sequence homology, which, on the one hand, is suggestive of potential redundancy and, on the other, permits different combinations between the factors and their receptors, also by virtue of their different binding affinities. There are at least three different intracellular signalling systems (namely, RAS/MAPK, PI3-kinase and JAK/STAT) downstream of the receptors that use tyrosine phosphorylation as leading motif. It appears that each receptor is capable of alternatively using either one or other of the signalling systems, depending on the period of the cell's life (i.e. the cell cycle) and its spatial position (e.g. zone of the embryo or as related to cell–cell interactions), thereby generating proliferation or differentiation.

The intriguing findings that a single growth factor can elicit such different responses as either mitogenesis or differentiation should now be considered in a different light. It is not only the message that the factor carries, but how that message is interpreted by the cell. During development, the dynamics of cell–cell interaction, morphogenesis and growth are too rapid to permit the exchange of an entire signalling cascade at each proliferative/differentiative state. As a result, the modulation of each transduction system at different stages of development becomes fundamentally important. Changing adaptor molecules or switching from one signalling pathway to another (already in use in the cell) permits the rapid responses that are required to successfully drive the complex machinery of development.

Conditional inactivation of single components of the system, and dominant negative experiments performed in mouse models, will provide new insights into how the cell decides to change the transduction system for a particular signal.

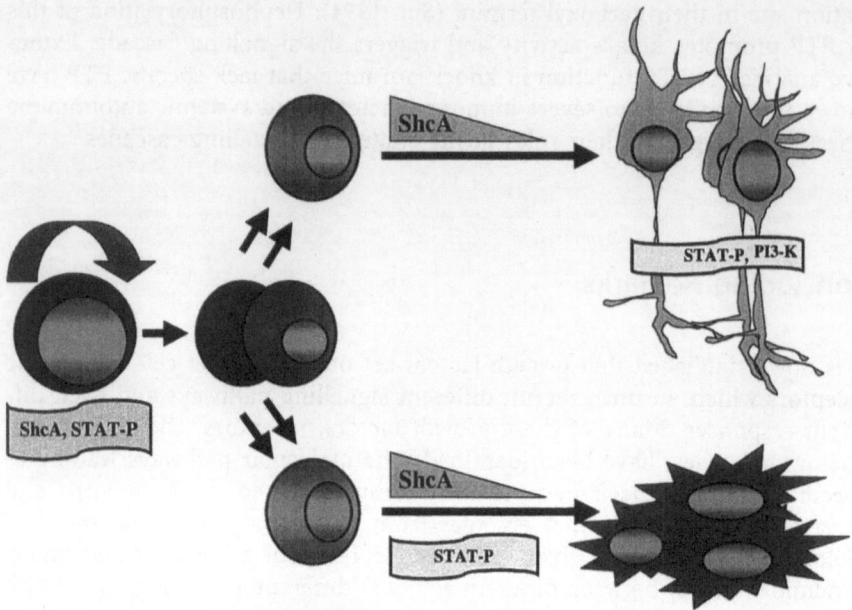

Fig. 4. Components of the signalling pathways implicated in proliferative and differentiative events of embryonic brain. Mature neurons and glia of the brain are all derived from a common precursor cell which actively divides (McKay 1997; Gage 1998). This proliferative ability is maintained through stimulation of pathways involving Shc-Ras-MAPK and JAK/STAT. In particular, high levels of ShcA are found in the dividing cells. JAK/STAT has also been shown to be activated in dividing cells. Changes in ShcA levels (and therefore presumably in activity of downstream components of the Ras-MAPK cascade) have been observed during differentiation, and may represent one of the prerequisites for proper cell differentiation in brain. JAK/STAT activation has also been found to play a role during glial differentiation and, together with activated PI3K, in maintenance of mature neuronal phenotypes

Acknowledgements. The work of the authors is supported by Associazione Italiana Ricerca Cancro, Telethon (Italy, #E840) and CNR (98.01050.CT04) to E.C. and EC Biotech (BIO4-CT98-0309), NATO Coll Res grant (NATO CRG 972252) and Telethon (Italy, #D.77) to M.G.

References

Auger KR, Carpenter CL, Cantley LC, Varticovski L (1989a) Phosphatidylinositol 3-kinase and its novel product, phosphatidylinositol 3-phosphate, are present in Saccharomyces cerevisiae. Biol Chem 264:20 181–20 184

Auger KR, Serunian LA, Soltoff SP, Libby P, Cantley LC (1989b) PDGF-dependent tyrosine phosphorylation stimulates production of novel polyphosphoinositides in intact cells. Cell 57:167–175

Baker NE, Rubin GM (1989) Effect on eye development of dominant mutation in *Drosophila* homologue of the EGF receptor. Nature 340:150–153

Barres BA, Schmid R, Sendtner M, Raff MC (1993) Multiple extracellular signals are required for long-term oligodendrocyte survival. Development 118:283-295

Basler K, Hafen E (1988) Sevenless and *Drosophila* eye development: a tyrosine kinase controls cell fate. Trends Genet 4:74-79

Bonni A, Sun Y, Nadal-Vicens M, Bhatt A, Frank DA, Rozovsky I, Stahl N, Yancopoulos GD, Greenberg ME (1997) Regulation of gliogenesis in the central nervous system by the Jak-Stat signaling pathway. Science 278:477-483

Brunet A, Bonni A, Zigmond MJ, Lin MZ, Juo P, Hu LS, Anderson MJ, Arden KC, Blenis J, Greenberg ME (1999) Akt promotes cell survival by phosphorylating and inhibiting a forkhead transcription factor. Cell 96:857-868

Brustle O, McKay RDG (1996) Neuronal progenitors as tools for cell replacement in the nervous system. Curr Opin Neurobiol 6:688-695

Burgering BM, Coffer PJ (1995) Protein kinase B (c-Akt) in phosphatidylinositol-3-OH kinase signal transduction. Nature 376:599-602

Burrows RC, Wancio D, Levitt P, Lillien L (1997) Response diversity and the timing of progenitor cell maturation are regulated by developmental changes in EGFR expression in the cortex. Neuron 19:251-267

Campbell SL, Khosravi-Far R, Rossman KL, Clark GJ, Der CJ (1998) Increasing complexity of Ras signaling. Oncogene 17:1395-1413

Cardone MH, Roy N, Stennicke HR, Salvesen GS, Franke TF, Stanbridge E, Frisch S, Reed JC (1998) Regulation of cell death protease caspase-9 by phosphorylation. Science 282:1318-1321

Cartwright CA, Simantov R, Cowan WM, Hunter T, Eckhart W (1988) pp60c-src expression in the developing rat brain. Proc Natl Acad Sci USA 85:3348-3352

Cattaneo E, McKay RDG (1990) Proliferation and differentiation of neuronal stem cells regulated by nerve growth factor. Nature 347:762-765

Cattaneo E, McKay RDG (1991) Identifying and manipulation of neuronal stem cells. Trend Neurosci 14:338-340

Cattaneo E, Pelicci PG (1998) Emerging roles for SH2/PTB-containing Shc adaptor proteins in the developing mammalian brain. Trends Neurosci 21:476-481

Cattaneo E, De-Fraja C, Conti L, Reinach B, Bolis L, Govoni S, Liboi E (1996) Activation of the Jak/Stat pathway leads to proliferation of ST14A central nervous system progenitor cells. J Biol Chem 271:23374-23379

Cattaneo E, Conti L, De-Fraja C (1999) Signalling through the Jak/Stat pathway in the developing brain. Trends Neurosci 22:365-369

Celli G, LaRochelle WJ, Mackem S, Sharp R, Merlino G (1998) Soluble dominant-negative receptor uncovers essential roles for fibroblast growth factors in multi-organ induction and patterning. EMBO J 17:1642-1655

Chao MV (1992) Neurotrophin receptors: a window into neuronal differentiation. Neuron 9:583-593

Chin YE, Kitagawa M, Su WC, You ZH, Iwamoto Y, Fu XY (1996) Cell growth arrest and induction of cyclin-dependent kinase inhibitor p21 WAF1/CIP1 mediated by STAT1. Science 272:719-722

Chung CD, Liao J, Liu B, Rao X, Jay P, Berta P, Shuai K (1997) Specific inhibition of Stat3 signal transduction by PIAS3. Science 278:1803-1805

Clark Lewis I, Sanghera JS, Pelech SL (1991) Definition of a consensus sequence for peptide substrate recognition by p44mpk, the meiosis-activated myelin basic protein kinase. J Biol Chem 266:15180-15184

Conti L, De Fraja C, Gulisano C, Migliaccio M, Govoni S, Cattaneo E (1997) Expression and activation of SH2/PTB containing Shc A adaptor protein reflects the pattern of neurogenesis in the mammalian brain. Proc Natl Acad Sci USA 94:8185-8190

Darnell JE Jr, Kerr IM, Stark GR (1994) Jak-STAT pathways and transcriptional activation in response to IFNs and other extracellular signaling proteins. Science 264:1415-1421

Datta SR, Dudek H, Tao X, Masters S, Fu H, Gotoh Y, Greenberg ME (1997) Akt phosphory-lation of BAD couples survival signals to the cell-intrinsic death machinery. Cell 91 : 231–241
De Fraja C, Conti L, Magrassi L, Govoni S, Cattaneo E (1998) Members of the Jak/Stat proteins are expressed and regulated during development in the mammalian forebrain. J Neurosci Res 54 : 320–330
DeHamer MK, Guevara JL, Hannon K, Olwin BB, Calof AL (1994) Genesis of olfactory receptor neurons in vitro: regulation of progenitor cell divisions by fibroblast growth factors. Neuron 13 : 1083–1097
del Peso L, Gonzalez-Garcia M, Page C, Herrera R, Nunez G (1997) Interleukin-3-induced phosphorylation of BAD through the protein kinase Akt. Science 278 : 687–689
Deng CX, Wynshaw-Boris A, Shen MM, Daugherty C, Ornitz DM, Leder P (1994) Murine FGFR-1 is required for early postimplantation growth and axial organization. Genes Dev 8 : 3045–3057
Eide FF, Lowenstein DH, Reichardt LF (1993) Neurotrophins and their receptors: current concepts and implications for neurologic disease. Exp Neurol 121 : 200–214
Eilers A, Georgellis D, Klose B, Schindler C, Ziemiecki A, Harpur AG, Wilks AF, Decker T (1995) Differentiation-regulated serine phosphorylation of STAT1 promotes GAF activation in macrophages. Mol Cell Biol 15 : 3579–3586
Endo TA, Masuhara M, Yokouchi M, Suzuki R, Sakamoto H, Mitsui K, Matsumoto A, Tanimura S, Ohtsubo M, Misawa H, Miyazaki T, Leonor N, Taniguchi T, Fujita T, Kanakura Y, Komiya S, Yoshimura A (1997) A new protein containing an SH2 domain that inhibits JAK kinases. Nature 387 : 921–924
Faris M, Ensoli B, Stahl N, Yancopoulos G, Nguyen A, Wang S, Nel AE (1996) Differential activation of the extracellular signal-regulated kinase, Jun kinase and Janus kinase-Stat pathways by oncostatin M and basic fibroblast growth factor in AIDS-derived Kaposi's sarcoma cells. AIDS 10 : 369–378
Fischer EH, Charbonneau H, Tonks NK (1991) Protein tyrosine phosphatases: a diverse family of intracellular and transmembrane enzymes. Science 253 : 401–406
Franke TF, Yang SI, Chan TO, Datta K, Kazlauskas A, Morrison DK, Kaplan DR, Tsichlis PN (1995) The protein kinase encoded by the Akt proto-oncogene is a target of the PDGF-activated phosphatidylinositol 3-kinase. Cell 81 : 727–736
Gage FH (1998) Stem cells of the central nervous system. Curr Opin Neurobiol 8 : 671–676
Gage FH, Ray J, Fisher LJ (1995) Isolation, characterization and use of stem cells from the CNS. Annu Rev Neurosci 18 : 158–192
Galzie Z, Kinsella AR, Smith JA (1997) Fibroblast growth factors and their receptors. Biochem Cell Biol 75 : 669–685
Gensburger C, Labourdette G, Sensenbrenner M (1987) Brain basic fibroblast growth factor stimulates the proliferation of rat neuronal precursor cells in vitro. FEBS Lett 217 : 1–5
Ghosh A, Greenberg ME (1995) Distinct roles for bFGF and NT-3 in the regulation of cortical neurogenesis. Neuron 15 : 89–103
Goodman CS, Tessier-Lavigne M (1997) Molecular mechanism of axon guidance and target recognition. In: Cowan WM, Jessel TM, Zipursky SL (eds) Molecular and cellular approaches to neural development. Oxford University Press, New York, pp 108–178
Greene LA, Kaplan DR (1995) Early events in neurotrophin signalling via Trk and p75 recep-tors. Curr Opin Neurobiol 5 : 579–587
Greene LA, Tischler AS (1976) Establishment of a noradrenergicclonal line of rat adrenal pheochromocytoma cells which respond to nerve growth factor. Proc Natl Acad Sci USA 73 : 2424–2428
Gritti A, Parati EA, Cova L, Frolichsthal P, Galli R, Wanke E, Faravelli L, Morasutti DJ, Roisen F, Nickel DD, Vescovi AL (1996) Multipotential stem cells from the adult mouse brain proliferate and self-renew in response to basic fibroblast growth factor. J Neurosci 16 : 1091–1100

Hansen K, Johnell M, Siegbahn A, Rorsman C, Engstrom U, Wernstedt C, Heldin CH, Ronn-strand L (1996) Mutation of a Src phosphorylation site in the PDGF beta-receptor leads to increased PDGF-stimulated chemotaxis but decreased mitogenesis. EMBO J 15:5299–5313

Harris WA, Starck WS, Walker JA (1976) Genetic dissection of the photoreceptor system in the compound eye of *Drosophila melanogaster*. J Physiol 256:415–439

Herdegen T, Skene P, Bahr M (1997) The c-Jun transcription factor-bipotential mediator of neuronal death, survival and regeneration. Trends Neurosci 20:227–231

Holland SJ, Gale NW, Mbamalu G, Yancopoulos JD, Henkemeyer M, Pawson T (1996) Bi-directional signalling through the Eph family Nuk and its transmembrane ligands. Nature 383:722–725

Holland SJ, Gale NW, Gish GD, Roth RA, Songyang Z, Cantley LC, Henkemeyer M, Yancopoulos GD, Pawson T (1997) Juxtamembrane tyrosine residues couple the Eph family receptor EphB2/Nuk to specific SH2 domain proteins in neuronal cells. EMBO J 16:3877–3888

Hopkins SJ, Rothwell NJ (1995) Cytokines and the nervous system I: expression and recogni-tion. Trends Neurosci 18:83–88

Hunter T (1995) Protein kinases and phosphatases: the yin and yang of protein phosphorylation and signaling. Cell 80:225–236

Ihara S, Nakajima K, Fukada T, Hibi M, Nagata S, Hirano T, Fukui Y (1997) Dual control of neurite outgrowth by STAT3 and MAPkinase in PC12 cells stimulated with interleukin-6. EMBO J 16:5345–5352

Ihle JN (1995) Cytokine receptor signalling. Nature 377:591–594

Ip NY, Yancopoulos GD (1996) The neurotrophins and CNTF: two families of collaborative neurotrophic factors. Annu Rev Neurosci 19:491–515

Johe KK, Hazel TG, Muller T, Dugich-Djordjevic MM, McKay RDG (1996) Single factors direct the differentiation of stem cells from the fetal and adult central nervous system. Genes Dev 10:3129–3140

Kahn MA, Huang CJ, Caruso A, Barresi V, Nazarian R, Condorelli DF, de Vellis (1997) Ciliary neurotrophic factor activates Jak/Stat signal transduction cascade and induces transcrip-tional expression of glial fibrillary acidic protein in glial cells. J Neurochem 68:1413–1423

Knusel B, Rabin SJ, Hefti F, Kaplan DR (1994) Regulated neurotrophin receptor responsiveness during neuronal migration and early differentiation. J Neurosci 14(3):1542–1554

Koch CA, Anderson D, Moran MF, Ellis C, Pawson T (1991) SH2 and SH3 domains: elements that control interactions of cytoplasmic signaling proteins. Science 252:668–674

Kops GJPL, de Ruiter ND, De Vries-Smits AMM, Powell DR, Bos JL, Boudewijn JL, Burgering BMTh (1999) Direct control of the forkhead transcription factor AFX by protein kinase B. Nature 398:630–634

Kundra V, Escobedo JA, Kazlauskas A, Kim HK, Rhee SG, Williams LT, Zetter BR (1994) Regulation of chemotaxis by the platelet-derived growth factor receptor-beta. Nature 367:474–476

LaVallee TM, Prudovsky IA, McMahon GA, Hu X, Maciag T (1998) Activation of the MAP kinase pathway by FGF-1 correlates with cell proliferation induction while activation of the Src pathway correlates with migration. J Cell Biol 141:1647–1658

Maina F, Casagranda F, Audero E, Simeone A, Comoglio PM, Klein R, Ponzetto C (1996) Uncoupling of Grb2 from the Met receptor in vivo reveals complex roles in muscle devel-opment. Cell 87:531–542

Marshall CJ (1995) Specificity of receptor tyrosine kinase signaling: transient versus sustained extracellular signal-regulated kinase activation. Cell 80:179–185

Martinez-Serrano A, Bjorklund A (1997) Immortalized neural progenitor cells for CNS gene transfer and repair. Trends Neurosci 20:530–538

Mayer BJ, Ren R, Clark KL, Baltimore D (1993) A putative modular domain present in diverse signaling proteins. Cell 73:629–630

McKay R (1997) Stem cells in the central nervous system. Science 276:6671–6674

Mehler MF, Kessler JA (1994) Growth factor regulation of neuronal development. Dev Neurosci 16:180–195

Mehler MF, Kessler JA (1997) Hematolymphopoietic and inflammatory cytokines in neural development. Trends Neurosci 20:357–365

Mehler MF, Mabie PC, Zhang D, Kessler JA (1997) Bone morphogenetic proteins in the nervous system. Trends Neurosci 20:309–317

Minichiello L, Klein R (1996) TrkB and TrkC neurotrophin receptors cooperate in promoting survival of hippocampal and cerebellar granule neurons. Genes Dev 10(22):2849–2858

Minichiello L, Casagranda F, Tatche RS, Stucky CL, Postigo A, Lewin GR, Davies AM, Klein R (1998) Point mutation in trkB causes loss of NT4-dependent neurons without major effects on diverse BDNF responses. Neuron 21:335–345

Miyajima A, Kitamura T, Harada N, Yokota T, Arai K (1992) Cytokine receptors and signal transduction. Annu Rev Immunol 10:295–331

Monville C, Conti L, De-Fraja C, Marty S, Peschanski M, Cattaneo E (1998) CNTF induces phosphorylation of at least two signalling pathways in long-term cultured astrocytes. Soc Neurosci Abstr 219.10

Morrison RS, Sharma A, de Vellis J, Bradshaw RA (1986) Basic fibroblast growth factor supports the survival of cerebral cortical neurons in primary culture. Proc Natl Acad Sci USA 83:7537–7541

Morrison RS, Kornblum HI, Leslie FM, Bradshaw RA (1987) Trophic stimulation of cultured neurons from neonatal rat brain by epidermal growth factor. Science 238:72–75

Naka T, Narazaki M, Hirata M, Matsumoto T, Minamoto S, Aono A, Nishimoto N, Kajita T, Taga T, Yoshizaki K, Akira S, Kishimoto T (1997) Structure and function of a new STAT-induced STAT inhibitor. Nature 387:924–929

Nakashima K, Yanagisawa M, Arakawa H, Kimura N, Hisatsune T, Kawabata M, Miyazono K, Taga T (1999) Synergistic signaling in fetal brain by STAT3-Smad1 complex bridged by p300. Science 248:479–482

Niwa H, Burdon T, Chambers I, Smith A (1998) Self-renewal of pluripotent embryonic stem cells is mediated via activation of STAT 3. Genes Dev 12:2048–2060

Obermeier A, Bradshaw RA, Seedorf K, Choidas A, Schlessinger J, Ullrich A (1994) Neuronal differentiation signals are controlled by nerve growth factor receptor/Trk binding sites for Shc and PLC gamma. Embo J 13:1585–1590

Orioli D, Henkemeyer M, Lemke G, Klein R, Pawson T (1996) Sek4 and Nuk receptors cooperate in guidance of commissural axons and in palate formation. EMBO J 15:6035–6049

Ornitz DM, Xu J, Colvin JS, McEwen DG, MacArthut CA, Coulier F, Gao G, Goldfarb M (1996) Receptor specificity of the fibroblast growth factor family. J Biol Chem 271:15292–15297

Osterhout DJ, Ebner S, Xu J, Ornitz DM, Zazanis GA, McKinnon RD (1997) Transplanted oligodendrocyte progenitor cells expressing a dominant-negative FGF receptor transgene fail to migrate in vivo. J Neurosci 17:9122–9132

Pandey A, Lazar DF, Saltiel AR, Dixit VM (1994) Activation of the Eck receptor protein tyrosine kinase stimulates phosphatidylinositol 3-kinase activity. J Biol Chem 269:30154–30157

Patel BK, Wang LM, Lee CC, Taylor WG, Pierce JH, LaRochelle WJ (1996) Stat6 and Jak1 are common elements in platelet-derived growth factor and interleukin-4 signal transduction pathways in NIH 3T3 fibroblasts. J Biol Chem 271:22175–22182

Patel BKR, Pierce JH, LaRochelle WJ (1998) Regulation of interleukin 4-mediated signaling by naturally occurring dominant negative and attenuated forms of human Stat6. Proc Natl Acad Sci 95:172–177

Pawson T (1995) Protein modules and signalling networks. Nature 373:573–580

Pawson T, Scott JD (1997) Signaling through scaffold, anchoring, and adaptor proteins. Science 278:2075–2080

Pritchard C, McMahon M (1997) Raf revealed in life-or-death decisions. Nat Genet 16:214–215

Rajan P, McKay RDG (1998) Multiple routes to astrocytic differentiation in the CNS. J Neurosci 18(10):3620–3629

Rajan P, Symes AJ, Fink JS (1996) STAT proteins are activated by ciliary neurotrophic factor in cells of central nervous system origin. J Neurosci Res 43:403–411

Reichardt LF, Farinas I (1997) Neurotrophic factors and their receptors: roles in neuronal development and function. In: Cowan WM, Jessel TM, Zipursky SL (eds) Molecular and cellular approaches to neural development. Oxford University Press, New York, pp 220–263

Reynolds BA, Weiss S (1992) Generation of neurons and astrocytes from isolated cells of the adult mammalian central nervous system. Science 255:1646–1649

Reynolds BA, Tetzlaff W, Weiss S (1993) A multipotent EGF-responsive striatal embryonic progenitor cell produces neurons and astrocytes. J Neurosci 12:4565–4574

Rubin GM (1989) Development of *Drosophila* retina: inductive events studied at single cell resolution. Cell 57:519–520

Sahin M, Hockfield S (1993) Protein tyrosine phosphatases expressed in the developing rat brain. J Neurosci 13:4968–4978

Sahin M, Slaugenhaupt SA, Gusella JF, Hockfield S (1995) Expression of PTPH1, a rat protein tyrosine phosphatase, is restricted to the derivatives of a specific diencephalic segment. Proc Nat Acad Sci USA 92:7859–7863

Sato N, Sakamaki K, Terada N, Arai K, Miyajima A (1993) Signal transduction by the high-affinity GM-CSF receptor: two distinct cytoplasmic regions of the common beta subunit responsible for different signaling. EMBO J 12:4181–4189

Schlessinger J, Ullrich A (1992) Growth factor signaling by receptor tyrosine kinases. Neuron 9:383–391

Segal RA, Greenberg ME (1996) Intracellular signaling pathways activated by neurotrophic factors. Annu Rev Neurosci 19:463–489

Segal RA, Takahashi H, McKay RDG (1992) Changes in neurotrophins responsiveness during the development of cerebellar granule neurons. Neuron 9:1041–1052

Shiozuka K, Watanabe Y, Ikeda T, Hashimoto S, Kawashima H (1995) Cloning and expression of PCPTP1 encoding protein tyrosine phosphatase. Gene 162:279–284

Silos-Santiago I, Fagan AM, Garber M, Fritzsch B, Barbacid M (1997) Severe sensory deficits but normal CNS development in newborn mice lacking TrkB and TrkC tyrosine protein kinase receptors. Eur J Neurosci 9:2045–2056

Starr R, Willson TA, Viney EM, Murray LJ, Rayner JR, Jenkins BJ, Gonda TJ, Alexander WS, Metcalf D, Nicola NA, Hilton DJ (1997) A family of cytokine-inducible inhibitors of signalling. Nature 387:917–921

Stephens RM, Loeb DM, Copeland TD, Pawson T, Greene LA, Kaplan DR (1994) Trk receptors use redundant signal transduction pathways involving Shc and PLC-gamma 1 to mediate NGF responses. Neuron 12:691–705

Sun H, Tonks NK (1994) The coordinated action of protein tyrosine phosphatases and kinases in cell signaling. Trends Biochem Sci 19:480–485

Sun H, Charles CH, Lau LF, Tonks NK (1993) MKP-1 (3CH134), an immediate early gene product, is a dual specificity phosphatase that dephosphorylates MAP kinase in vivo. Cell 75:487–493

Suzuki T, Matozaki T, Mizoguchi A, Kasuga M (1995) Localization and subcellular distribution of SH-PTP2, a protein-tyrosine phosphate with Src homology-2 domains. Biochem Biophys Res Commun 211:950–959

Taga T, Kishimoto T (1997) Gp130 and the interleukin-6 family of cytokines. Annu Rev Immunol 15:797–819

Taniguchi T (1995) Cytokine signaling through nonreceptor protein tyrosine kinases. Science 268:251–255

Ullrich A, Schlessinger J (1990) Signal transduction by receptors with tyrosine kinase activity. Cell 61:203–212

van der Geer P, Hunter T, Lindberg RA (1994) Receptor protein-tyrosine kinases and their signal transduction pathways. Annu Rev Cell Biol 10:251–337

Wang H, Yan H, Canoll PD, Silvennoinen O, Schlessinger J, Musacchio JM (1995) Expression of receptor protein tyrosine phosphatase-sigma (RPTP-sigma) in the nervous system of the developing and adult rat. J Neurosci Res 41:297–310

Weiss A, Schlessinger J (1998) Switching signals on or off by receptor dimerization. Cell 94:277–280

Wen Z, Zhong Z, Darnell JE Jr (1995) Maximal activation of transcription by Stat1 and Stat3 requires both tyrosine and serine phosphorylation. Cell 82:241–250

Williams BP, Park JK, Alberta JA, Muhlebach SG, Hwang GY, Roberts TM, Stiles CD (1997) A PDGF-regulated immediate early gene response initiates neuronal differentiation in ventricular zone progenitor cells. Neuron 18:553–562

Wu YY, Bradshaw RA (1996) Induction of neurite outgrowth by interleukin-6 is accompanied by activation of Stat3 signaling pathway in a variant PC12 cell (E2) line. J Biol l Chem 271:13 023–13 032

Yamaguchi TP, Harpal K, Henkemeyer M, Rossant J (1994) FGFR-1 is required for embryonic growth and mesodermal patterning during mouse gastrulation. Genes Dev 8:3032–3044

York RD, Yao H, Dillon T, Ellig CL, Eckert SP, McCleskey EW, Stork PJS (1998) Rap1 mediates sustained MAP kinase activation induced by nerve growth factor. Nature 392:622–626

Yoshimura A, Ohkubo T, Kiguchi T, Jenkins NA, Gilbert DJ, Copeland NG, Hara T, Miyajima A (1995) A novel cytokine-inducible gene CIS encodes an SH2-containing protein that binds to tyrosine-phosphorylated interleukin 3 and erythropoietin receptors. EMBO J 14:2816–2826

Zhang X, Blenis J, Li HC, Schindler C, Chen-Kiang S (1995) Requirement of serine phosphorylation for formation of STAT-promoter complexes. Science 267:1990–1994

The Role of the p35/cdk5 Kinase in Cortical Development

Young T. Kwon and Li-Huei Tsai[1]

1
Introduction

Extensive studies on the mouse mutant *reeler* have revealed many of the fundamental characteristics of neocortical development (Caviness and Rakic 1978; Caviness 1982; Caviness et al. 1988). The finding of another spontaneously occurring mouse mutant, *scrambler*, which exhibits nearly identical phenotypes with *reeler* suggests that the gene products mutated in the strains, mdab-1 and reelin, respectively, act in a common signaling pathway during cortical development (Sweet et al. 1996; Gonzalez et al. 1997; Howell et al. 1997; Sheldon et al. 1997; Ware et al. 1997; Rice et al. 1998). However, the vast complexity of events that must occur to set up the architecture of the cerebral cortex leads to the idea that multiple proteins are essential during cortical development. The p35/cdk5 kinase complex is one such molecular entity. Mouse knockouts of both p35 and cdk5 lead to disruptions of cortical lamination (Ohshima et al. 1996; Chae et al. 1997). Interestingly, the phenotype of the embryonic cerebral wall and adult neocortex in these mice is suggestive of but distinct from that of *reeler* or *scrambler*, implying that a different essential function during cortical development may be disrupted in mice lacking p35 or cdk5 (Gilmore et al. 1998; Kwon and Tsai 1998).

2
cdk5

The cdk5 kinase was originally cloned as a cDNA homologous to the cell cycle regulator cdc2 (Meyerson et al. 1992). Independently, two groups purified a kinase activity from bovine brain extracts which later was identified as cdk5 (Lew et al. 1992; Kobayashi et al. 1993). The cdk5 kinase, which is 60% homologous to cdk2, does not appear to play a critical role in the cell division

---

[1] Howard Hughes Medical Institute and the Department of Pathology, Harvard Medical School, Building D2-342, 200 Longwood Avenue, Boston, Massachusetts 02115, USA

Results and Problems in Cell Differentiation, Vol. 30
Goffinet and Rakic (Eds.): Mouse Brain Development
© Springer-Verlag Berlin Heidelberg 2000

cycle (Van den Heuvel and Harlow 1993). Rather, an examination of cdk5 kinase activity in a variety of adult mouse tissues revealed that cdk5 was most prominently active in forebrain (Tsai et al. 1993), which is composed predominantly of postmitotic neurons and glia.

3
p35 Family Members

The distinct tissue specificity of cdk5 kinase activity led several groups to search for and identify an activator subunit of cdk5. Biochemical purification of cdk5 kinase activity led to the identification of a 23 to 25-kDa novel activator of cdk5 (Ishiguro et al. 1994; Lew et al. 1994). This protein, however, proved to be a truncated form of a 35-kDa protein, called p35, cloned independently based on its association with cdk5 in cultured cortical neurons (Tsai et al. 1994). Interestingly, p35 has no significant homology to cyclins, which are the prototypical activating subunits of cyclin-dependent kinases, although studies of tertiary structure indicate that p35 is likely to assume a cyclin-like fold (Brown et al. 1995; Tang et al. 1997). The p35 protein lacks obvious domain structures but has several cyclin-dependent kinase consensus phosphorylation sites, and a short proline-rich region in the N-terminal third of the protein which marks the N terminus of the p25 cleavage product. The p25 cleavage product is necessary and sufficient for activating cdk5 (Lew et al. 1994), suggesting that the N-terminal region upstream of the proline-rich region may function in a different capacity. Indeed, a myristoylation signal, for membrane targeting, has been discovered in the N terminus of p35 (Patrick et al. 1999).

To date, one mammalian homologue of p35 has been discovered. Screening for p35 from a rat hippocampal library led to the identification of p39 (Tang et al. 1995). p39 is 57% identical to p35 overall at the amino acid level, but 65% identical in the region of the protein corresponding to p25. p39 has non-homologous insertions near the N and C termini, whose function remains unknown. Furthermore, a truncated form of p39 was shown to activate cdk5 in vitro, suggesting that cdk5 is the kinase partner for p39 (Tang et al. 1995).

Evolutionarily, p35 appears to be conserved across eukaryotic species. In *Xenopus*, there are two known p35 homologues, called Xp35.1 and Xp35.2, which share a high degree of identity to mammalian p35 (Philpott et al. 1997, 1999). Interestingly, it has been shown that Xp35.1/cdk5 functions to regulate muscle differentiation and patterning, while Xp35.2 may participate in neural development (Philpott et al. 1997, 1999). Additional homologues of p35 have also been found in *C. elegans* and the yeast *Saccharomyces cerevisiae* (Huang et al. 1996), whose function remains to be elucidated.

4
Expression Patterns

In situ hybridization analysis of E12 and E15 mouse brains indicates that p35 is absent in proliferating neuronal precursors and found in postmitotic neurons of the developing cortex (Delalle et al. 1997). In primary cortical neurons, the p35/cdk5 kinase is present in the soma, processes, and growth cones (Nikolic et al. 1996). Within the growth cone, p35/cdk5 is present beyond the microtubule core and is often found in lamella and filopodia, suggesting that the p35/cdk5 kinase may function in the dynamic regulation of the actin cytoskeleton (Nikolic et al. 1996). Furthermore, biochemical subcellular fractionation indicates that p35 is predominantly associated with membranes, whereas cdk5 is present in both membrane and cytosol fractions (Nikolic et al. 1998).

Expression studies of p35 mRNA revealed exquisitely strict expression in the developing central nervous system of mouse embryos as early as E10, while cdk5 is present in the central nervous system and peripheral nervous system structures such as the dorsal root ganglia (Tsai et al. 1993, 1994; Ino et al. 1994; Tomizawa et al. 1996; Delalle et al. 1997; Zheng et al. 1998). In postnatal and adult rat brain, p35 protein is relatively abundant in most areas of the forebrain including the neocortex and hippocampus (Tomizawa et al. 1996; Delalle et al. 1997; Zheng et al. 1998). Expression studies of p39 mRNA also indicate expression in the developing and adult CNS (Cai et al. 1997; Zheng et al. 1998), but reveal that p39 is expressed in the PNS as well (Zheng et al. 1998). Interestingly, the peak of p39 expression in the CNS occurs postnatally (S. Humbert and L.-H. Tsai, unpubl. data), suggesting that p39 may participate in functions independent from p35/cdk5 in migration during embryonic cortical development.

5
Function of the cdk5 Kinase in Neurite Outgrowth

In order to assess the function of the cdk5 kinase in cells, inhibition studies were performed in a variety of neuronal cell model systems. Inhibition of p35/cdk5 kinase activity in rat primary cortical neurons by transfection of dominant-negative versions of cdk5 or antisense p35 leads to the inhibition of neurite outgrowth in these neurons (Nikolic et al. 1996). This inhibition could be relieved by co-expression of wildtype p35 or cdk5, suggesting that the p35/cdk5 kinase is required for neurite outgrowth. This observation was corroborated in a rat hippocampal cell line, where inhibition of the p35/cdk5 or p39/cdk5 kinase also leads to inhibition of neurite outgrowth (Xiong et al. 1997). Finally, suppression of p35 or cdk5 expression by antisense oligonucleotide treatment in cerebellar macroneurons results in reduced axonal elongation, further indicating that the cdk5 kinase plays an important role during axonal extension (Pigino et al. 1997; Paglini et al. 1998).

6
p35 and cdk5 Knockout Mice

Insight into the physiological function of the p35/cdk5 kinase has been obtained from gene targeting experiments of both p35 and cdk5 in mice. We generated mice lacking p35, the CNS specific cdk5 activator (Chae et al. 1997). The p35 mutant mice generated from heterozygous intercrosses are born at the expected 1:2:1 ratio, and are fertile and viable through adulthood. Indeed, p35 mutant mice appear outwardly normal, without ataxia or apparent motor deficits, and are visually indistinguishable from their wildtype and heterozygous littermates. However, p35 mutant mice are susceptible to seizures, and possess a lower threshold for lethal seizure activity. In addition, p35 mutant mice exhibit certain behavioral abnormalities, such as reduced aggressiveness in males and nurturing deficits in females (Y. Kwon, T. Chae and L.-H. Tsai, unpubl. data). A complete histological examination of p35 mutant animals found no apparent abnormalities in somatic tissues. However, a variety of affected structures are present in the brains of these mice.

Most prominently, the laminar structure of the neocortex was disorganized, without the characteristically ordered, six-layered cellular architecture (Fig. 1). Indeed, one meaningful suggestion for the nature of the abnormal lamination was the presence of large pyramidal neurons, typically in layer V, present superficially in the cortex in p35 mutant mice. Further analyses using bromodeoxyuridine (BrdU)-labeling "birthdating" studies indicated that the overall cellular pattern in the cortex was inverted: earlier-born neurons occupied superficial layers of the cortex, while later-born neurons occupied deep cortical layers (Fig. 1). Furthermore, Golgi staining revealed the abnormal orientation of neurons in the cortex of p35 mutant mice. The radially oriented apical dendrites of wildtype pyramidal neurons were absent and instead, short neuronal processes extended from the soma without a distinct pattern.

In an effort to determine the nature of the cortical lamination abnormality, we examined the development of the embryonic cerebral wall (Kwon and Tsai 1998). Indeed, disruptions in the stratified cellular organization were prominently evident at E15. The intermediate zone, which normally appears relatively cell-sparse, was entirely occupied with cells. This increase in cell density in the intermediate zone was complemented by a decrease in the thickness of the cortical plate layer, suggesting that migrating neurons were unable to migrate normally during development. Moreover, a BrdU-labeling analysis of the migration profile of early-born cortical neurons revealed that the earliest-born cortical neurons were able to migrate normally to form the

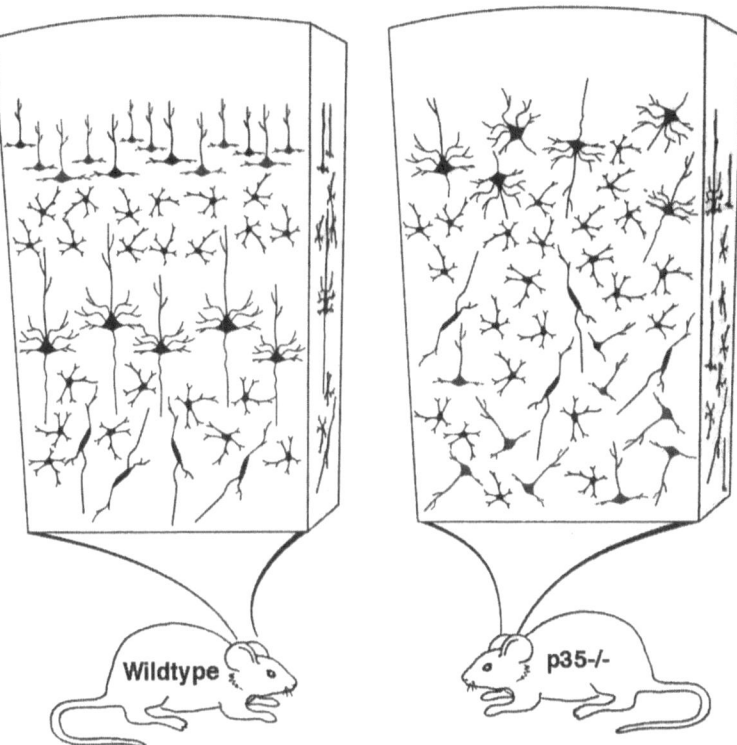

Fig. 1. Architecture of adult neocortex in wildtype and p35 mutant mice. In wildtype mice the neocortex is organized as a six-layered structure (*left*). In the cortex of p35 mutant mice (*right*), the normal lamination pattern is disrupted and lacks the organization of the wildtype cortex. In particular, large pyramidal neurons can be identified superficially in the neocortex (*top right*) of p35 mutant mice, whereas these neurons normally exist deeper in the wildtype cortex. In addition, orientation and shape of dendritic processes are altered in p35 mutant mice–short processes extend from the soma without a distinct pattern. Grey neurons (*top left, bottom right*) indicate BrdU-positive neurons after labeling E16/E17 proliferating neocortical precursors, and reveal the existence of an inversion of cortical layering in p35 mutant mice. (Adapted from an illustration by Amy Emmert in Strobel 1997)

cortical plate, while the next cohort was unable to migrate past their predecessors, and remained deeper in the cerebral wall (Fig. 2).

Although the cellular architecture of the cerebral wall was disrupted in p35 mutant mice, the marginal zone was clearly intact. The Cajal-Retzius neurons which populate the marginal zone were found to be present, as determined by calretinin and reelin immunoreactivity. Furthermore, analysis of E11 preplate precursors by BrdU-labeling analysis indicated that the preplate layer was split into the marginal zone and subplate by the first-born cohort of cortical neurons (Fig. 2). These studies suggested that the aspect of cortical development affected in p35 mutant mice may differ fundamentally from the defect in *reeler* mice.

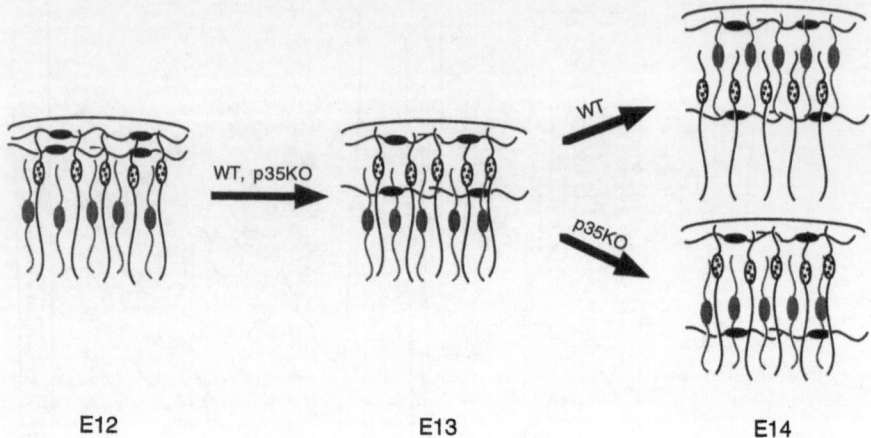

Fig. 2. Early events in neocortical development in wildtype and p35 mutant mice. Early during neocortical development (E12; *left*), the first wave of postmitotic neurons (*speckles*) approachs the preplate neurons (*black*). In both wildtype and p35 mutant mice, the earliest cohort of cortical neurons splits the preplate into the marginal zone and subplate (E13; *middle*). In normal development (E14; *top right*), the next cohort of cortical neurons (*grey*) migrates past their predecessors (*speckles*) to occupy a more superficial position in the cerebral wall. In contrast, the second cohort of cortical neurons (*grey*) is unable to migrate past their predecessors in p35 mutant mice (*bottom right*), leading to disruption of cortical lamination

Similar disruptions of embryonic cortical development were also observed in mice lacking cdk5 (Ohshima et al. 1996). The majority of cdk5-deficient mice die in utero after E16.5, with the remainder dying the day of birth. Histological analyses of E18.5 cdk5$^{-/-}$ mice exposed lesions in the brain and spinal cord, but not in other tissues, leaving the cause of death unknown. In the developing cortex and hippocampus, the normal stratification of neurons was absent, while the cerebellum lacked the foliation and layered structure found in normal mice. Other abnormalities were found in neurons of the brain stem and spinal cord, indicating a widespread influence of the cdk5 kinase in the nervous system.

Studies performed in cdk5 mutant mice to examine the migration defect of cortical neurons and the split of the preplate provided a complementary view of the defect in cortical lamination as in p35 mutant mice (Gilmore et al. 1998). The similarity of phenotypic characteristics in the developing cortex of p35 and cdk5 mutant mice suggests that p35 is the activator of cdk5 most responsible for the functions of the cdk5 kinase during cortical development. Together, the studies on p35 and cdk5 mutant animals indicate that the p35/cdk5 kinase plays a prominent role during the development of the cortex. On the other hand, the differences in viability and other phenotypes in p35 and cdk5 mutant mice suggest that other cdk5 activators are critically important in the development of the animal.

7
p35/cdk5 and *Reeler/Scrambler*

While mice lacking either p35 or cdk5 exhibit some similarities with *reeler/ scrambler* mice, the development of the embryonic cerebral wall in p35 and cdk5 mutant mice also shows significant differences from *reeler* and *scrambler*. In particular, the presence of the marginal zone and the split of the preplate by migrating cortical neurons represent fundamental differences from *reeler* and *scrambler* which imply that different aspects of cortical development may be affected when p35 or cdk5 is mutated. Indeed, our cumulative evidence suggests that the loss of the p35/cdk5 kinase affects the ability of cortical neurons to migrate past their predecessors, while reelin may participate in preventing migration of cortical neurons into the marginal zone.

Although we suggest that the p35/cdk5 kinase may influence migrating cortical neurons in a spatially distinct manner from reelin, it is possible that the phenotypic variations between p35 and cdk5 mutant mice and *reeler/ scrambler* mice are also defined by the temporal loss of function. For instance, reelin and mDab1 may be required to allow the initial wave of cortical neurons to split the preplate into the marginal zone and subplate, while the p35/cdk5 may act slightly later in development to allow the next cohort of cortical neurons to migrate past their predecessors. The initial defects in cortical architecture may undoubtedly influence further development, compounding the disrupted structure of the cortex.

Interestingly, non-cortical CNS structures in p35 and cdk5 mutant mice share both similarities to and differences from *reeler/scrambler*. The p35 mutant mice exhibit relatively mild cellular distortions in the hippocampus, olfactory bulb, and cerebellum, in contrast to severe disruptions present in those same structures in *reeler/scrambler*. Also, p35 mutant mice exhibit additional phenotypes not found in *reeler/scrambler*. Most strikingly, alterations of axonal fiber tracts were present in the cortex. In particular, the callosal axons which form the corpus callosum form a fiber bundle at the midline, but appear to defasciculate prematurely from the corpus callosum after crossing the midline and course obliquely through the cortex (Kwon et al. 1999). This callosal phenotype is not detected in *reeler* mice, suggesting that defasciculation of callosal axons is not an inherent manifestation of a disruption of cortical lamination. These results further suggest that defective axonal fasciculation and guidance may be primary responses to the loss of p35 in the cortex.

Intriguingly, cdk5 mutant mice exhibit a cerebellar disruption that resembles *reeler/scrambler*. The E18 cerebellum in cdk5$^{-/-}$ mice lacks foliation and exhibits a disrupted laminar structure, which is reminiscent of the *reeler* cerebellum. While early death of cdk5$^{-/-}$ mice precludes a conclusive determination of the similarity of cerebellar structures in the two mutant strains in the adult, the likeness suggests provocatively that cdk5 may operate together with reelin and mDab1 in cerebellar development. The relative intactness of the cerebellum in p35 mutant mice suggests that another activator of cdk5, perhaps p39, may be at work in this respect.

8
Substrates

Since the p35/cdk5 kinase is critically important for cortical lamination and other functions, identifying the substrates of the kinase has been an area of active investigation. Several putative substrates have been identified, including cytoskeletal elements such as tau, neurofilaments, and MAPs (reviewed in Lew and Wang 1995; Pigino et al. 1997), whose phosphorylation may influence the function of p35/cdk5 in neurite outgrowth. Other substrates such as Munc-18 and synapsin I allude to functions of the p35/cdk5 kinase in synaptic vesicle function (Matsubara et al. 1996; Shuang et al. 1998). Other known substrates include pak1 and the retinoblastoma protein (Lee et al. 1997; Nikolic et al. 1998). In addition, amino acid sequence analysis and degenerate peptide library approaches indicate that the p35/cdk5 kinase has a substrate specificity similar to cdc2 and cdk2 (Beaudette et al. 1993; Songyang et al. 1996).

9
Regulation

In addition to identifying substrates of p35/cdk5, another area of active investigation has been to understand the regulation of the p35/cdk5 kinase. It was shown initially that recombinant active kinase can be constituted in vitro by mixing bacterially produced p35 and cdk5 in the absence of post-translational modifications on p35 or cdk5 (Tsai et al. 1994). These and other experiments suggest that active kinase is regulated in part at the level of association of p35 and cdk5. Thus, molecular determinants which affect p35 production or degradation are likely to play major roles in the dynamic regulation of p35. Consistent with this idea, half-life studies indicate that the p35 protein has a half-life of approximately 20 min while cdk5 is a highly stable protein, suggesting that p35 is rate-limiting for active p35/cdk5 kinase (Patrick et al. 1998). Furthermore, the p35 protein is degraded by the ubiquitin-mediated degradation pathway, a process that is stimulated by phosphorylation, suggesting that p35 levels can be modulated dynamically in vivo by phosphorylation events (Patrick et al. 1998). Thus, clearance of the p35/cdk5 kinase by degradation of p35 is a possible mechanism of regulation of kinase activity during cortical development.

The localization of the p35/cdk5 kinase in growth cones and its role in regulating neurite outgrowth suggested that the p35/cdk5 kinase may be involved in the modulation of actin cytoskeletal dynamics. Indeed, p35 interacts with the small GTPase rac (Nikolic et al. 1998). This association requires rac to be activated, suggesting that the p35/cdk5 kinase is an effector of the rac protein. The presence of the p35/cdk5 kinase and activated rac leads to the hyperphosphorylation of another rac effector, pak1, which results in a reduction in pak1 kinase activity. Thus, the p35/cdk5 kinase may affect growth cone dynamics and neuronal motility by its association with rac and pak1.

The cdk5 kinase appears to be regulated in other ways as well. For instance, a two-hybrid screen with cdk5 led to the identification of a novel cdk5 interacting protein, cables (L. Zukerberg et al., submitted). Cables, which associates with both c-abl and cdk5, stimulates the tyrosine phosphorylation of cdk5. Tyrosine phosphorylated cdk5 exhibits greater kinase activity in p35/cdk5 complexes compared to unphosphorylated cdk5, implicating tyrosine kinases in the modulation of p35/cdk5 kinase activity.

10
Conclusion

The cortical lamination abnormalities found in gene knockouts of p35 and cdk5 have provided evidence that neuronal migration is disrupted by the loss of the p35/cdk5 kinase. Based on our cumulative studies, we propose that the p35/cdk5 kinase is necessary for the transmission or execution of signaling events which are critical for the migration of cortical neurons past their predecessors, and that multiple cellular pathways may influence p35/cdk5 in this process (Fig. 3).

However, many genes have been identified that clearly have important roles in cortical development. Lis1 and doublecortin were isolated as genes mutated in the human disorders type I lissencephaly and X-linked lissencephaly/double cortex syndrome, respectively (Reiner et al. 1993; des Portes et al. 1998; Gleeson et al. 1998). Lis1, which is a subunit of platelet-activating factor acetyl-hydrolase, associates with tubulins and regulates microtubule stability, suggesting that it may regulate cell motility during migration (Sapir et al. 1997). Doublecortin is a novel protein whose molecular function awaits to be defined. In addition, mutations in the Pak3 gene were identified in patients with non-syndromic X-linked mental retardation (MRX) syndrome (Allen et al. 1998). Recently, molecules implicated in the signaling pathway of the Rho family GTPases, Rho GDI and oligophrenin, were isolated by virtue of their involvement in X-linked mental retardation (Billuart et al. 1998; D'Adamo et al. 1998).

When taken together, the genetic and biochemical studies indicate that migration of cortical neurons is likely to be regulated by extracellular cues

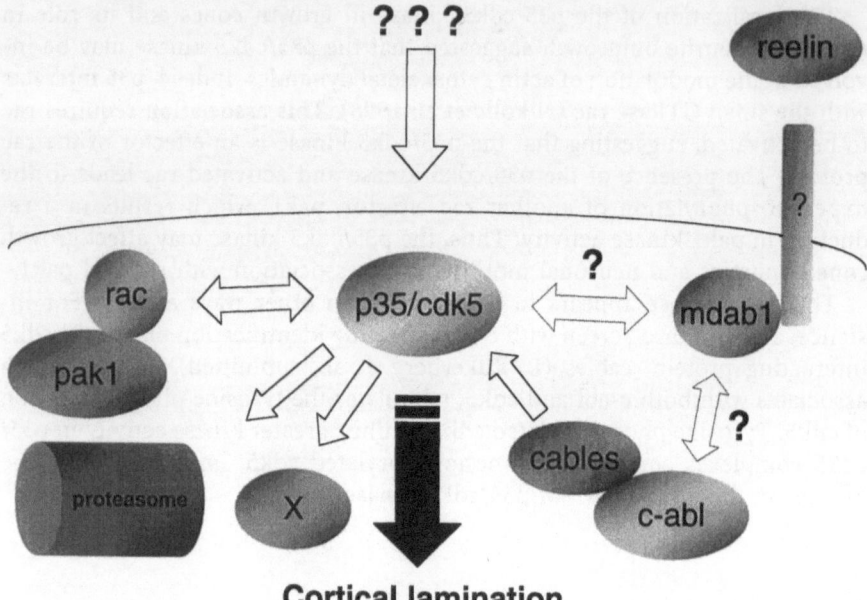

Cortical lamination

Fig. 3. Molecular model for the function of the p35/cdk5 kinase. Activity and function of the p35/cdk5 kinase during neocortical development is likely to be dynamically regulated in multiple ways. For instance, degradation mediated by the proteasome may be a mechanism for clearance of the p35/cdk5 kinase in migrating cortical neurons. In addition, association with rac and pak1 may influence function of the p35/cdk5 kinase in modulation of actin cytoskeletal dynamics in navigating growth cones. Furthermore, activity of the p35/cdk5 kinase may be regulated by tyrosine phosphorylation of cdk5, stimulated by cdk5 interacting protein cables. In all likelihood, other proteins (X) that regulate the p35/cdk5 kinase exist. Regulation of the kinase by its interactors may modulate the ability of the kinase to phosphorylate essential substrates during neocortical development, leading to the eventual disruption of neocortical structure. Reelin, which interacts with a yet unidentified receptor, signals through the phosphoprotein mDab1. This signaling pathway may interact with the p35/cdk5 kinase directly or indirectly

which signal through multiple intracellular signaling pathways, including those regulated by the rho family of GTPases and non-receptor tyrosine kinases c-abl and c-src, to coordinate and mobilize cytoskeletal proteins for the act of migration. The regulation of cdk5 kinase activity by c-abl tyrosine phosphorylation and the interaction of p35/cdk5 with rac and pak1 suggests that the p35/cdk5 kinase is likely to serve as an interface between the signaling pathways and cytoskeletal elements. The disrupted cortical lamination found in p35 and cdk5 mutant mice reveals a fundamental requirement for p35/cdk5 kinase activity during embryonic cortical development. Indeed, disruptions in the genes for molecules affecting cortical lamination identified thus far each represent rate-limiting steps in pathways that are likely to interact with each other to regulate neuronal migration. Future research should aim at defining the molecular action of each protein implicated in cortical assembly and the interplay between these proteins.

References

Allen KM, Gleeson JG, Bagrodia S, Partington MW, MacMillan JC, Cerione RA, Mulley JC, Walsh CA (1998) PAK3 mutation in nonsyndromic X-linked mental retardation. Nat Genet 20:25–30

Beaudette KN, Lew J, Wang JH (1993) Substrate specificity characterization of a cdc2-like protein kinase purified from bovine brain. J Biol Chem 268:20 825–20 830

Billuart P, Bienvenu T, Ronce N, des Portes V, Vinet MC, Zemni R, Crollius HR, Carrie A, Fauchereau F, Cherry M, Briault S, Hamel B, Fryns JP, Beldjord C, Kahn A, Moraine C, Chelly J (1998) Oligophrenin-1 encodes a rhoGAP protein involved in X-linked mental retardation. Nature 392:923–926

Brown NR, Noble ME, Endicott JA, Garman EF, Wakatsuki S, Mitchell E, Rasmussen B, Hunt T, Johnson LN (1995) The crystal structure of cyclin A. Structure 3:1235–1247

Cai XH, Tomizawa K, Tang D, Lu YF, Moriwaki A, Tokuda M, Nagahata S, Hatase O, Matsui H (1997) Changes in the expression of novel Cdk5 activator messenger RNA (p39nck5ai mRNA) during rat brain development. Neurosci Res 28:355–360

Caviness VSJ (1982) Neocortical histogenesis in normal and reeler mice: a developmental study based upon [3H]thymidine autoradiography. Dev Brain Res 4:293–302

Caviness VSJ, Rakic P (1978) Mechanisms of cortical development: a view from mutations in mice. Annu Rev Neurosci 1:297–326

Caviness VSJ, Crandall JE, Edwards MA (1988) The reeler malformation: implications for neocortical histogenesis. In: Peters A, Jones EG (eds) Cerebral cortex, vol 7. Plenum Press, New York, pp 59–89

Chae T, Kwon YT, Bronson R, Dikkes P, Li E, Tsai LH (1997) Mice lacking p35, a neuronal specific activator of Cdk5, display cortical lamination defects, seizures, and adult lethality. Neuron 18:29–42

D'Adamo P, Menegon A, Lo Nigro C, Grasso M, Gulisano M, Tamanini F, Bienvenu T, Gedeon AK, Oostra B, Wu SK, Tandon A, Valtorta F, Balch WE, Chelly J, Toniolo D (1998) Mutations in GDI1 are responsible for X-linked non-specific mental retardation. Nat Genet 19:134–139

Delalle I, Bhide PG, Caviness VS Jr, Tsai LH (1997) Temporal and spatial patterns of expression of p35, a regulatory subunit of cyclin-dependent kinase 5, in the nervous system of the mouse. J Neurocytol 26:283–296

des Portes V, Pinard JM, Billuart P, Vinet MC, Koulakoff A, Carrie A, Gelot A, Dupuis E, Motte J, Berwald-Netter Y, Catala M, Kahn A, Beldjord C, Chelly J (1998) A novel CNS gene required for neuronal migration and involved in X-linked subcortical laminar heterotopia and lissencephaly syndrome. Cell 92:51–61

Gilmore EC, Ohshima T, Goffinet AM, Kulkarni AB, Herrup K (1998) Cyclin-dependent kinase 5-deficient mice demonstrate novel developmental arrest in cerebral cortex. J Neurosci 18:6370–6377

Gleeson JG, Allen KM, Fox JW, Lamperti ED, Berkovic S, Scheffer I, Cooper EC, Dobyns WB, Minnerath SR, Ross ME, Walsh CA (1998) Doublecortin, a brain-specific gene mutated in human X-linked lissencephaly and double cortex syndrome, encodes a putative signaling protein. Cell 92:63–72

Gonzalez JL, Russo CJ, Goldowitz D, Sweet HO, Davisson MT, Walsh CA (1997) Birthdate and cell marker analysis of scrambler: a novel mutation affecting cortical development with a reeler-like phenotype. J Neurosci 17:9204–9211

Howell BW, Hawkes R, Soriano P, Cooper JA (1997) Neuronal position in the developing brain
    is regulated by mouse disabled-1. Nature 389:733-737
Huang QQ, Lee KY, Wang JH (1996) A novel yeast protein showing specific association with the
    cyclin-dependent kinase 5. FEBS Lett 378:48-50
Ino H, Ishizuka T, Chiba T, Tatibana M (1994) Expression of CDK5 (PSSALRE kinase), a neural
    cdc2-related protein kinase, in the mature and developing mouse central and peripheral
    nervous systems. Brain Res 661:196-206
Ishiguro K, Kobayashi S, Omori A, Takamatsu M, Yonekura S, Anzai K, Imahori K, Uchida T
    (1994) Identification of the 23 kDa subunit of tau protein kinase II as a putative activator of
    cdk5 in bovine brain. FEBS Lett 342:203-208
Kobayashi S, Ishiguro K, Omori A, Takamatsu M, Arioka M, Imahori K, Uchida T (1993) A
    cdc2-related kinase PSSALRE/cdk5 is homologous with the 30 kDa subunit of tau protein
    kinase II, a proline-directed protein kinase associated with microtubule. FEBS Lett 335:
    171-175
Kwon YT, Tsai LH (1998) A novel disruption of cortical development in p35$^{-/-}$ mice distinct
    from reeler. J Comp Neurol 395:510-522
Kwon YT, Tsai LH, Crandall JE (1999) Callosal axon guidance defects in p35$^{-/-}$ mice. J Comp
    Neurol 415:218-229
Lee KY, Helbing CC, Choi KS, Johnston RN, Wang JH (1997) Neuronal Cdc2-like kinase (Nclk)
    binds and phosphorylates the retinoblastoma protein. J Biol Chem 272:5622-5626
Lew J, Beaudette K, Litwin CM, Wang JH (1992) Purification and characterization of a novel
    proline-directed protein kinase from bovine brain. J Biol Chem 267:13 383-13 390
Lew J, Wang JH (1995) Neuronal cdc2-like kinase. Trends Biochem Sci 20:33-37
Lew J, Huang QQ, Qi Z, Winkfein RJ, Aebersold R, Hunt T, Wang JH (1994) A brain-specific
    activator of cyclin-dependent kinase 5. Nature 371:423-426
Matsubara M, Kusubata M, Ishiguro K, Uchida T, Titani K, Taniguchi H (1996) Site-specific
    phosphorylation of synapsin I by mitogen-activated protein kinase and Cdk5 and its effects
    on physiological functions. J Biol Chem 271:21 108-21 113
Meyerson M, Enders GH, Wu CL, Su LK, Gorka C, Nelson C, Harlow E, Tsai LH (1992) A family
    of human cdc2-related protein kinases. EMBO J 11:2909-2917
Nikolic M, Dudek H, Kwon YT, Ramos YF, Tsai LH (1996) The cdk5/p35 kinase is essential for
    neurite outgrowth during neuronal differentiation. Genes Dev 10:816-825
Nikolic M, Chou MM, Lu W, Mayer BJ, Tsai LH (1998) The p35/cdk5 kinase is a neuron-specific
    Rac effector that inhibits Pak1 activity. Nature 395:194-198
Ohshima T, Ward JM, Huh CG, Longenecker G, Veeranna, Pant HC, Brady RO, Martin LJ,
    Kulkarni AB (1996) Targeted disruption of the cyclin-dependent kinase 5 gene results in
    abnormal corticogenesis, neuronal pathology and perinatal death. Proc Natl Acad Sci USA
    93:11 173-11 178
Paglini G, Pigino G, Kunda P, Morfini G, Maccioni R, Quiroga S, Ferreira A, Caceres A (1998)
    Evidence for the participation of the neuron-specific CDK5 activator P35 during laminin-
    enhanced axonal growth. J Neurosci 18:9858-9869
Patrick GN, Zhou P, Kwon YT, Howley PM, Tsai LH (1998) p35, the neuronal-specific activator
    of cyclin-dependent kinase 5 (Cdk5) is degraded by the ubiquitin-proteasome pathway.
    J Biol Chem 273:24 057-24 064
Patrick GN, Zukerberg L, Nikolic M, Monte SDL, Dikkes P, Tsai LH (1999) Conversion of p35 to
    p25 deregulates cdk5 activity and promotes neurodegeneration. Nature 402:615-622
Philpott A, Porro EB, Kirschner MW, Tsai LH (1997) The role of cyclin-dependent kinase 5 and
    a novel regulatory subunit in regulating muscle differentiation and patterning. Genes Dev
    11:1409-1421
Philpott A, Tsai LH, Kirschner MW (1999) Neuronal differentiation and patterning in Xenopus:
    the role of cdk5 and a novel activator xp35.2. Dev Biol 20:119-132

Pigino G, Paglini G, Ulloa L, Avila J, Caceres A (1997) Analysis of the expression, distribution and function of cyclin dependent kinase 5 (cdk5) in developing cerebellar macroneurons. J Cell Sci 110:257–270

Reiner O, Carrozzo R, Shen Y, Wehnert M, Faustinella F, Dobyns WB, Caskey CT, Ledbetter DH (1993) Isolation of a Miller-Dieker lissencephaly gene containing G protein beta-subunit-like repeats. Nature 364:717–721

Rice DS, Sheldon M, D'Arcangelo G, Nakajima K, Goldowitz D, Curran T (1998) Disabled-1 acts downstream of reelin in a signaling pathway that controls laminar organization in the mammalian brain. Development 125:3719–3729

Sapir T, Elbaum M, Reiner O (1997) Reduction of microtubule catastrophe events by LIS1, platelet-activating factor acetylhydrolase subunit. EMBO J 16:6977–6984

Sheldon M, Rice DS, D'Arcangelo G, Yoneshima H, Nakajima K, Mikoshiba K, Howell BW, Cooper JA, Goldowitz D, Curran T (1997) Scrambler and yotari disrupt the disabled gene and produce a reeler-like phenotype in mice. Nature 389:730–733

Shuang R, Zhang L, Fletcher A, Groblewski GE, Pevsner J, Stuenkel EL (1998) Regulation of Munc-18/syntaxin 1A interaction by cyclin-dependent kinase 5 in nerve endings. J Biol Chem 273:4957–4966

Songyang Z, Lu KP, Kwon YT, Tsai LH, Filhol O, Cochet C, Brickey DA, Soderling TR, Bartleson C, Graves DJ, DeMaggio AJ, Hoekstra MF, Blenis J, Hunter T, Cantley LC (1996) A structural basis for substrate specificities of protein Ser/Thr kinases: primary sequence preference of casein kinases I and II, NIMA, phosphorylase kinase, calmodulin-dependent kinase II, CDK5, and Erk1. Mol Cell Biol 16:6486–6493

Strobel G (1997) Disordering the brain gives clues to brain disorders. Focus 1:1, 6

Sweet HO, Bronson RT, Johnson KR, Cook SA, Davisson MT (1996) Scrambler, a new neurological mutation of the mouse with abnormalities of neuronal migration. Mamm Genome 7:798–802

Tang D, Yeung J, Lee KY, Matsushita M, Matsui H, Tomizawa K, Hatase O, Wang JH (1995) An isoform of the neuronal cyclin-dependent kinase 5 (Cdk5) activator. J Biol Chem 270:26897–26903

Tang D, Chun ACS, Zhang M, Wang JH (1997) Cyclin-dependent kinase 5 (Cdk5) activation domain of neuronal Cdk5 activator. Evidence of the existence of cyclin fold in neuronal Cdk5a activator. J Biol Chem 272:12318–12327

Tomizawa K, Matsui H, Matsushita M, Lew J, Tokuda M, Itano T, Konishi R, Wang JH, Hatase O (1996) Localization and developmental changes in the neuron-specific cyclin-dependent kinase 5 activator (p35nck5a) in the rat brain. Neuroscience 74:519–529

Tsai LH, Takahashi T, Caviness VSJ, Harlow E (1993) Activity and expression pattern of cyclin-dependent kinase 5 in the embryonic mouse nervous system. Development 119:1029–1040

Tsai LH, Delalle I, Caviness VSJ, Chae T, Harlow E (1994) p35 is a neural-specific regulatory subunit of cyclin-dependent kinase 5. Nature 371:419–423

Van den Heuvel S, Harlow E (1993) Distinct roles for cyclin-dependent kinases in cell cycle control. Science 262:2050–2054

Ware ML, Fox JW, Gonzalez JL, Davis NM, Lambert de Rouvroit C, Russo CJ, Chua SC Jr, Goffinet AM, Walsh CA (1997) Aberrant splicing of a mouse disabled homolog, mdab1, in the scrambler mouse. Neuron 19:239–249

Xiong W, Pestell R, Rosner MR (1997) Role of cyclins in neuronal differentiation of immortalized hippocampal cells. Mol Cell Biol 17:6585–6597

Zheng M, Leung CL, Liem RK (1998) Region-specific expression of cyclin-dependent kinase 5 (cdk5) and its activators, p35 and p39, in the developing and adult rat central nervous system. J Neurobiol 35:141–159

The Reelin-Signaling Pathway and Mouse Cortical Development

Isabelle Bar, Catherine Lambert de Rouvroit, and André M. Goffinet[1]

1
Introduction

Reelin (Reln) is an extracellular matrix protein that plays a pivotal role in the patterning of the brain, as shown by the analysis of the Reln-null phenotype, the so-called reeler mouse. During the last 2 years, Reln has been shown to act on target cells together with partners that include Dab1, VLDLR and ApoER2, thus defining a Reln-signaling pathway. Although Reln and its partners act all over the central nervous system, it is at the level of the cerebral cortex, hippocampus and cerebellum that their action is best studied and the present discussion will focus on cortical development. A more detailed discussion of the reeler phenotype is provided in a recent review (Lambert de Rouvroit and Goffinet 1998a).

2
Overview of Early Cortical Development in Normal Mice

Before examining the role of Reln, Dab1, VLDLR and ApoER2 and the corresponding mutations, some features of early mammalian corticogenesis will be summarized; most of them are considered in detail in other chapters in this volume. The principal cell contingent of the cerebral cortex is the cortical plate (CP), which, in mice, appears on the 13.5–14th embryonic day (E13.5 or E14; day of insemination is E0). The telencephalic wall at that stage is composed of six concentric layers (Boulder Committee 1969; Caviness 1982). From the ventricle to the pial surface these are: (1) the ventricular zone, VZ; (2) the subventricular zone, sVZ; (3) the intermediate zone, IZ; (4) the sub(cortical)plate plate, sCP; (5) the cortical plate, CP; and (6) the marginal zone, MZ (Figs. 1, 2a).

---

[1] Neurobiology Unit, University of Namur School of Medicine, 61, rue de Bruxelles, B5000 Namur, Belgium

Results and Problems in Cell Differentiation, Vol. 30
Goffinet and Rakic (Eds.): Mouse Brain Development
© Springer-Verlag Berlin Heidelberg 2000

The ventricular zone is populated with the cell bodies of radial proliferating neuroepithelial precursors. The morphology of these cells varies according to their stage in the mitotic cycle, and the cell-cycle kinetics of the mouse cortical germinative zone is reviewed by Caviness et al. (this volume). Cells in interphase extend radially from the ventricle to the pial surface, where they come in contact with the basal lamina and contribute to the external limiting membrane, whereas cells in mitosis lose or retract their external process and are rounded near the ventricle, where they remain attached by the junctional complexes (Fig. 1A). For many years, it was assumed that the mitotic figures present in the subventricular zones gave rise to glial cells. More recently, it was shown that the sVZ gives birth not only to glial cells but also to a population of late-generated neurons that migrate towards the olfactory bulbs where they differentiate into interneurons (Luskin 1993). This late neuronal migration is directed almost tangentially within the IZ and seems to follow glial tunnels wherein neurons engage in chains. There is evidence that this migratory stream also gives rise to neurons that migrate tangentially over the molecular layer and could differentiate into Cajal-Retzius-like cells (Meyer et al. 1998, this vol.). The intermediate zone is traversed by immature cells involved in radial migration. At the early stage (E13–E14) migrating cells have a stellate shape (Shoukimas and Hinds 1978),

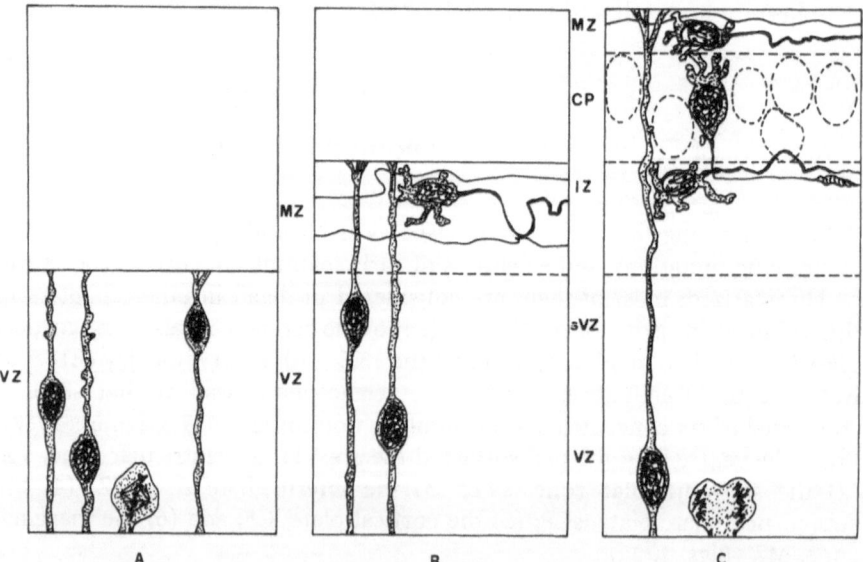

Fig. 1A–C. Schema of early events in cortical development. Until day 12 (E12) in the mouse, the cortical anlage is composed of a neuroepithelial, proliferating population (**A**). This is followed by a transient stage (**B**), at E12–E13, where the first postmitotic neurons settle at the periphery and form the preplate. The next stage (**C**), at E13–E14, is the appearance of the cortical plate (*CP*). *VZ* Ventricular zone; *MZ* marginal zone; *sVZ* subventricular zone; *IZ* intermediate zone

while at later stages their morphology is nearly bipolar. This variation in the shape of migrating neurons could be related to the length of the migration pathway and/or to the speed of radial migration (Nowakowski and Rakic 1981). Observations on Cdk5 and p35 mutant mice (Chae et al. 1997; Gilmore et al. 1998) provide support for this view. This work is summarized by Kwon and Tsai in this Volume, and will not be discussed further. Most migrating neurons in the IZ are apposed to radial extensions from the cells in the VZ. The characteristics of the radial extensions and their role as substrate for radial neuronal migration have been extensively documented in several species. In addition to radial migration, tangential migration in the early cortex of neurons generated in the ganglionic eminence has recently been convincingly demonstrated (Anderson et al. 1997; Tan et al. 1998). These neurons, which are GABA-positive and which could follow the route traced by early corticothalamic axons (Molnar 1998), probably give rise to inhibitory cortical interneurons.

2.1
The Preplate

In the mouse, a first population of postmitotic neurons appears at the level to the cortical anlage at day 12 (Fig. 1B). These early neurons are organized into a loose network called preplate. When the CP appears, preplate cells are divided into two contingents: some of them remain in the marginal zone and differentiate into neurons of layer I, including Reln-positive Cajal-Retzius cells but also Reln-negative pioneer cells (Meyer et al. 1998), while the others are displaced inward, below the CP, where they form the subplate and are thought to differentiate into neurons of layer VIb. Both preplate neurons and radial glial cells are extremely important in the early stages of cortical development. While radial fibers assist in radial migration, preplate neurons play an important role in hodological development. They extend the first axons to leave the cortex (McConnel et al. 1989; De Carlos and O'Leary 1992; Molnar and Blakemore 1995) and receive early thalamic afferents.

2.2
The Appearance of the Cortical Plate

In mice, the cortical plate (CP) appears at E13.5–E14 (Fig. 1C). With the exception of immigrant neurons from the ganglionic eminence (Anderson et al. 1997; Tan et al. 1998) and from the rostral forebrain which reach the cortex by tangential migration (Meyer et al. 1998), the CP grows mostly radially by adjunction of elements migrating from the ventricular zone. The CP gives rise to the major contingent of cortical cells, mostly of cortical laminae II–VIa.

The morphology of the early rodent CP has been extensively described (Shoukimas and Hinds 1978). Neurons of the cortical plate are polarized, radially oriented and parallel to each other. The dendritic bouquet emerges from the external cell pole, at the level of the future apical dendrite, and ramifies in the marginal zone. A few branches ramify also within the CP itself. As a rule, the axon originates from the inferior pole of the cell and extends radially in the subplate where it makes a sharp turn at a right angle before it runs horizontally towards the internal capsule where it is capped with prominent growth cones. Sidman and colleagues (Angevine and Sidman 1961; Miale and Sidman 1961) first demonstrated that, in the mouse cerebral cortex, inner cortical neurons (layer VI) are generated first, and that younger cells migrate beyond the previously established layers to settle at progressively more superficial levels. This is generally known as the 'inside–out' histogenetic pattern or gradient. With the exception of the early origin of the neurons of layer I, this principle is valid in all mammals, including man.

3
Cortical Phenotype in Reeler Mutant Mice

Reeler mutant mice, which are deficient in Reln, have been extensively studied for half a century (reviewed in Lambert de Rouvroit and Goffinet 1998a). The reeler mutation affects most of the brain, and particularly the cerebral cortex. The reeler cortical malformation is identical to that generated by mutations in the disabled-1 (Dab1) gene and to the alteration induced in mice that are doubly deficient in the lipoprotein receptors VLDLR and ApoER2. The morphological data on reeler cortical development summarized below are valid for all three mutations.

The cortical malformation is detected as soon as the CP develops, at E14. Although early neuroepithelial elements are minimally affected, studies with specific markers such as the RC2 antibody (Misson et al. 1991) reveal that they are not entirely normal in reeler mice (Hunter-Schaedle 1997). In coping with this observation, the presence of discontinuities in the external limiting membrane in reeler embryos was demonstrated long ago (Derer 1979). As Reln, Dab1, VLDLR or ApoER2 are not expressed in neuroepithelial cells, these anomalies of radial fibers may be secondary to the cortical malformation.

In mutant as in normal embryos, a first population of postmitotic neurons reaches the external level of the cortical anlage before the cortical plate appears. The mutant preplate does not appear morphologically different from its normal counterpart and its neurons are generated at the same time as in wild-type (Caviness 1982; Gonzales et al. 1997; Sheppard and Pearlman 1997). The pathognomonic feature of reeler mice is that the preplate is not divided into the marginal zone and subplate in mutant animals by the condensation of the CP. Instead, all preplate cells remain in a single superficial layer, called the superplate, and cortical plate neurons accumulate beneath them in a disorderly cortical plate (Caviness 1982; Fig. 2b). The neurons in the mutant cortical plate appear morphologically different from their normal counterparts. They are less elongated, less densely packed and less strictly radial. Instead of being perpendicular to the pial surface, their long axes assume a variable orientation. The apical dendrite is often oblique and can even run horizontally or be inverted. The axon emerges from the cell body at a variable angle, and not always from the inner pole of the cells; the initial segment is sometimes abnormally twisted. Instead of running radially through the CP and the subcortex until they reach the subplate, the mutant axons traverse the CP obliquely and collect at subcortical level before they resume their normal trajectory towards the internal capsule. The sharp angulation of the axon, so typical of normal CP

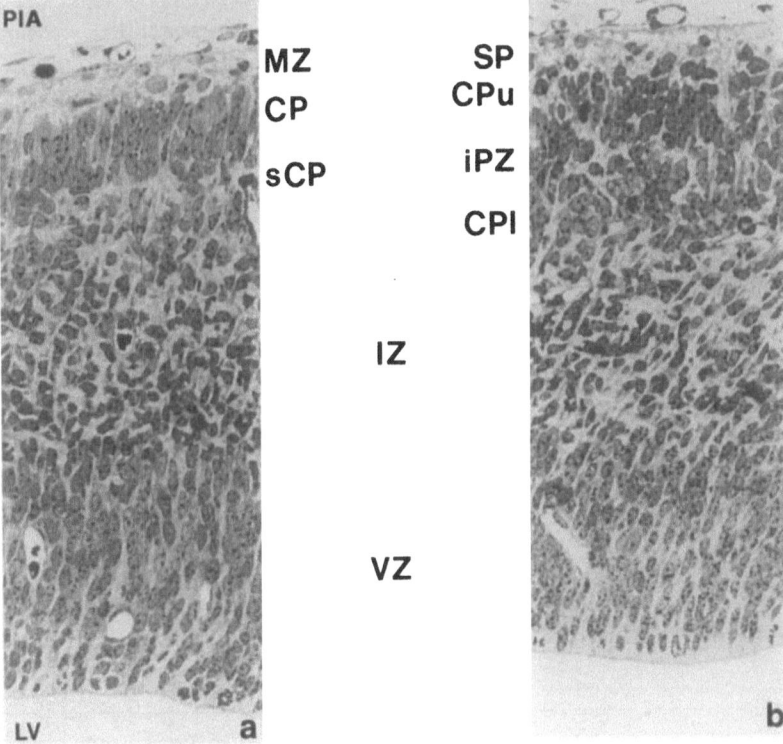

Fig. 2a,b. Early development (E14) of the cortical plate (*CP*) in normal (**a**) and reeler (**b**) mouse embryos. Neuronal precursors proliferate in ventricular zones (*VZ*). Postmitotic neurons leave the VZ and migrate across the intermediate zone (*IZ*) before they settle in the CP. In normal embryos, the CP divides the early population of the preplate into an external contingent in the marginal zone (*MZ*) and an inner contingent in the sub(cortical) plate (*sCP*). In embryos that are deficient in Reln, Dab1 or VLDLR/ApoER2, the plexus of the sCP forms externally and is called the superplate (*SP*). The CP is divided into an upper (*Cpu*) and a lower (*CPl*) part, separated by an intermediate plexus (*iPZ*). *PIA* Meningeal surface; *LV* lateral ventricle

cells, is often lacking in mutants. Fiber fascicles run through the cortical plate obliquely. The fiber plexus of the subplate is less clearly defined than in normal embryos and appears to be replaced by a superficial plexus in the "superplate". This observation, together with the finding that the intermediate plexiform zone contains bundles of thalamocortical axons (Pinto-Lord and Caviness 1979; Yuasa et al. 1994) suggests that the intermediate plexiform zone may represent partial subplate formation in mutant animals. Ultrastructurally, despite their abnormal orientation and their distorted shape, mutant neurons have a normal polarized morphology, with axonal and dendritic poles, and their differentiation is surprisingly normal.

In mutant embryos, the superplate replaces the marginal zone (MZ). Some cells have a horizontal orientation and features similar to those of resident neurons of the normal MZ, particularly Cajal-Retzius cells. Other neurons in the superplate are reelin-negative and different from Cajal-Retzius cells.

Several thymidine and BrdU studies (Caviness 1982; Gonzales et al. 1997; Trommsdorf et al. 1999) have convincingly demonstrated that CP neurons in Reln-, Dab1- or VLDLR/ApoER2-deficient animals are generated at the same time as their normal homologues, but that they do not migrate at the same level in the cortex. In mutants, the first generated neurons migrate to the superficial cortical level where they form a preplate as in normal embryos. However, when the cortical plate appears, the mutant preplate is not divided into two contingents as in the normal embryo. Instead, its two cell components, the marginal zone and subplate neurons, are both found externally in the superplate. In addition, CP neurons settle at progressively deeper levels in the cortex, which results in a gradient directed from outside to inside (Fig. 3). The intrinsic program of differentiation of the various cell classes is conserved: polymorphic neurons normally found at deep cortical level, are superficial in mutant cortex, whereas small pyramidal cells (layers II and III) are deeply located and large pyramids (layer V) are found in the superficial cortex. The small neurons of sensorial fields (e.g. the barrel field) are found at intermediate cortical level (Caviness et al. 1976). In mutant mice, cortical lamination is thus grossly inverted. The abnormal lamination of the mutant cortex is accompanied by a profuse anomaly of fiber pathways. Instead of running in the subplate and ascending radially through the cortex, fibers traverse the cortex obliquely, in fascicles, until they reach the vicinity of the pial surface (the

----

Fig. 3. Early cortical development. Condensation of the cortical plate at E14 divides the preplate (E12) into two cell contingents: pioneer neurons in the marginal zone and subplate cells. Reln-positive, Cajal-Retzius cells may derive from the preplate or reach the marginal zone by tangential migration. Early organization of the CP is defective in Reln-, Dab1- or (VLDLR + ApoER2)-mutant mice, but occurs normally in Cdk5- or p35-deficient mice. At a later stage, CP neurons migrate according to an inside–outside gradient. At E16, for example, older neurons are found at deep level, while younger cells settle more superficially. This gradient is defective in mice lacking Reln, Dab1, or (VLDLR + ApoER2), and also in Cdk5/p35 mutant animals (see Kwon and Tsai, this vol.)

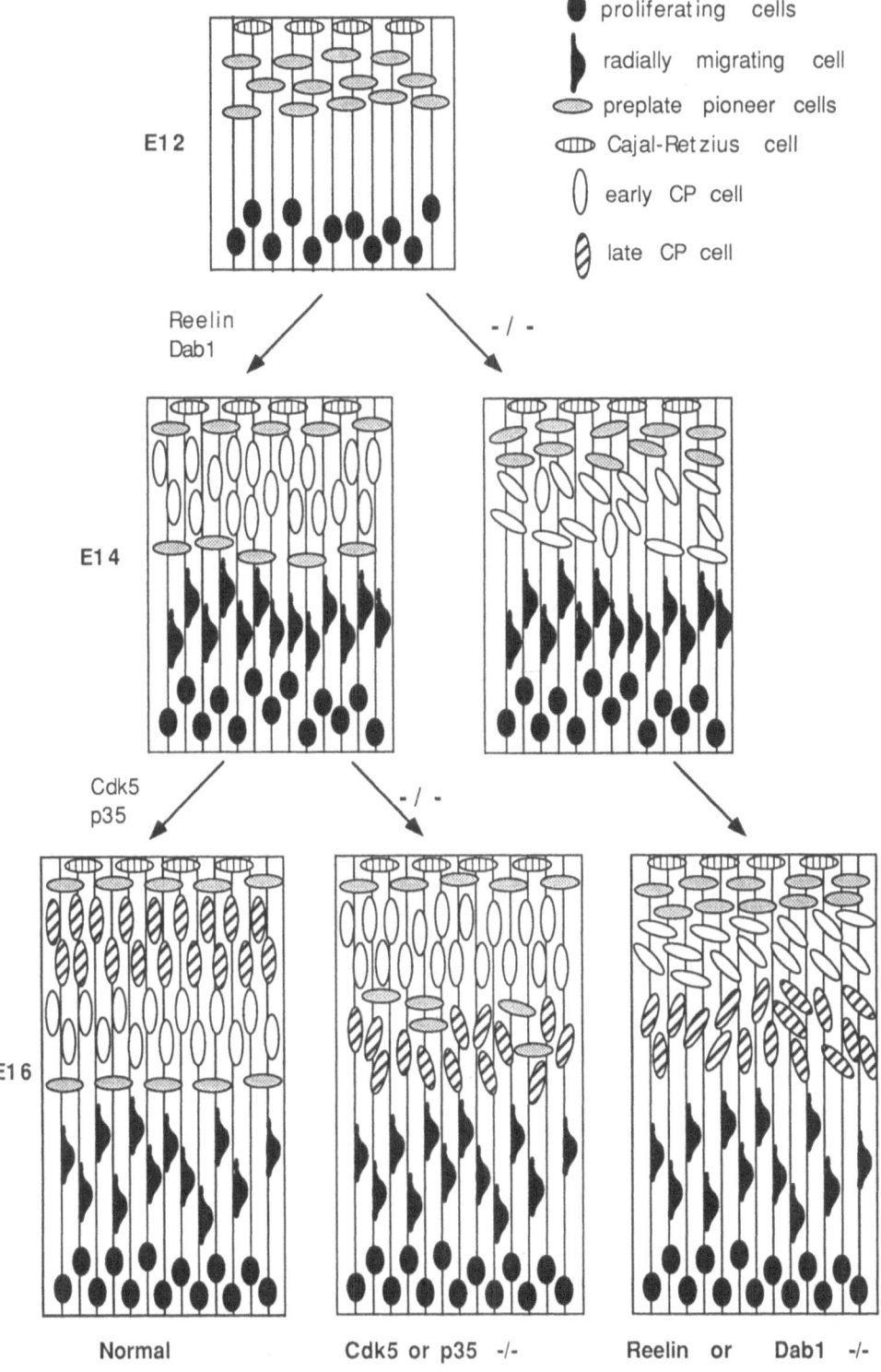

superplate), were they arch before turning towards their target (Caviness 1976). An interesting feature of the mutant cortex is that the specificity of connections is relatively preserved by the mutation (Molnar et al. 1998).

4
Reelin (Reln)

The sequence of the mouse Reln cDNA reveals several interesting features (D'Arcangelo et al. 1995; Royaux et al. 1997; Fig. 4). The length of the mRNA is 11689 nt (from the major transcriptional start site) and contains an open reading frame of 10383 nt. The stop codon is followed by approximately 1 kb of 3' untranslated sequence and a polyadenylation signal. Reln is composed of 3461 amino acids and has a predicted molecular mass of 388 kDa. The presence of a signal peptide and the absence of a transmembrane segment suggest that Reln is a secreted protein, which was confirmed by D'Arcangelo et al. (1997). Sequence comparisons revealed a 25% identity between the N terminus (about 250 residues) of Reln and that of F-spondin (Klar et al. 1992). After the first 500 amino acids, Reln consists of a series of eight consecutive repeat sequences. Each direct repeat, of 350–390 amino acids, contains two related subrepeats separated by a stretch of 30 amino acids with an EGF-like motif. The Reln sequence contains several potential N-glycosylation, amidation and myristylation sites distributed along the primary sequence, and a consensus (SGxG) sequence for O-xylosylation (glycanation) is contained within six of the eight EGF-like motifs. However, thus far, chondroitinase ABC digestion experiments failed to reveal the presence of a glycanated form (D'Arcangelo et al. 1997). The C terminus contains a basic stretch of 33 amino acids. The Reln repeats do not match any known sequence. The EGF-like domains are very similar to each other and related to those of the extracellular matrix proteins Tenascin C and X, Restrictin, and the integrin beta chain family.

4.1
The *Reln* Gene

The *Reln* gene (Fig. 4) is large, between 400 and 450 kb, principally due to the presence of large introns; the first intron is estimated to be 80–100 kb and

---

Fig. 4A,B. Genomic organization of reelin. A Physical map of the gene with localization of some of the genomic clones used for definition of exon/intron structure. Genomic sizes of repeat-encoding regions are *boxed* and to scale. B Structure of cDNA with numbering of the 65 exons, and protein with the eight repeats

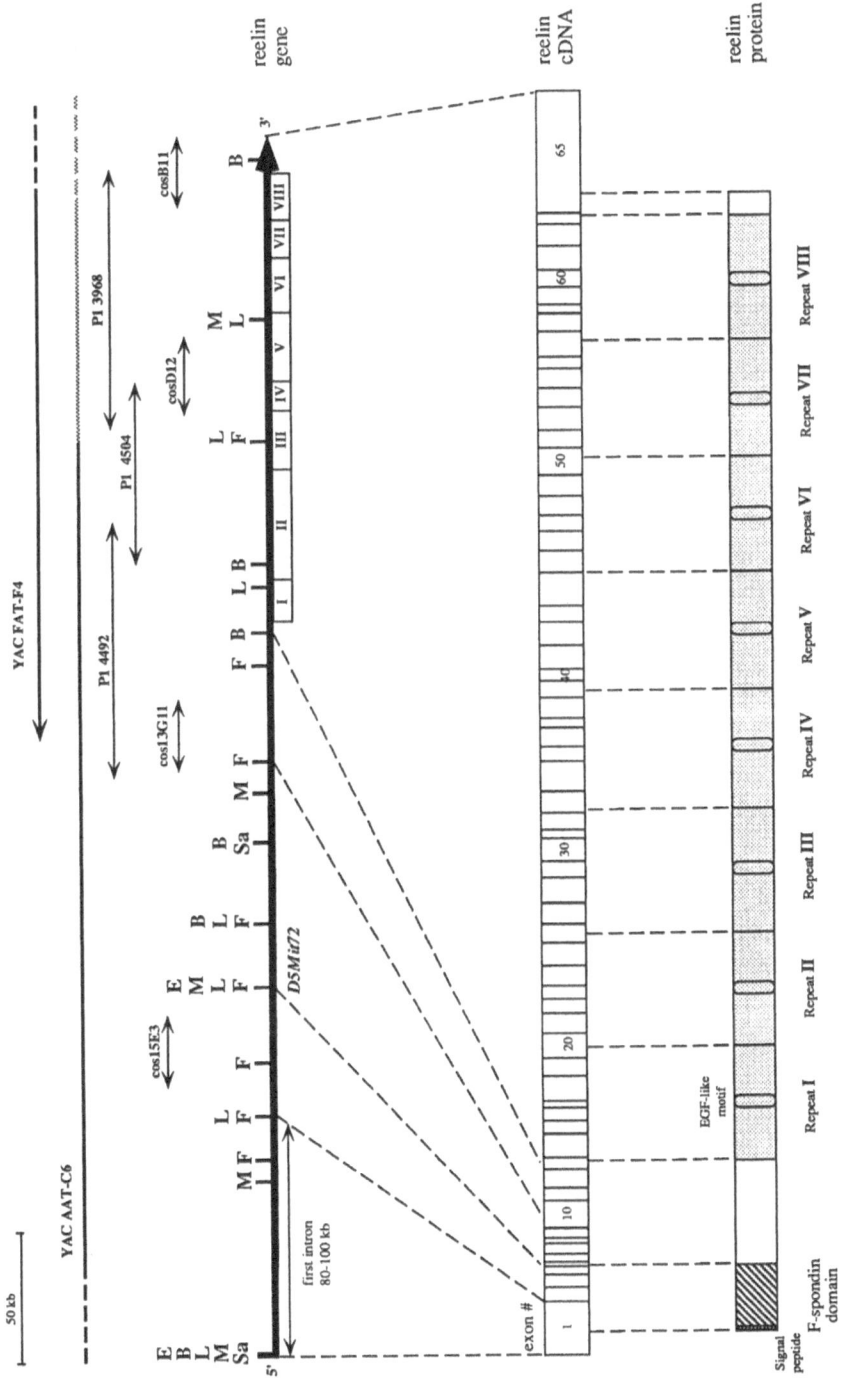

several are in the 10 to 20-kb range (Bar et al. 1995). Reln is composed of 65
exons, of which a total of 51 encode the eight Reln repeats. Exon 1 contains
the upstream untranslated region, the initiation codon, signal peptide and
part of the F-spondin domain. The rest of the F-spondin domain and the
unique segment are encoded by exons 1–12 and part of exon 13. The eight
successive repeats are each encoded by five to seven exons. The last two
exons are a six- nucleotide-long micro-exon and the terminal exon that encodes the arginine-rich 3′ end of the Reln protein. The eight Reln repeats are
thought to result from evolutionary duplications (Royaux et al. 1997).

About 10% of Reln RNA molecules from the embryonic brain lack the
hexanucleotide sequence AGTAAG, which was shown to correspond to a real
micro-exon (exon 64) flanked by two introns of 4 and 6 kb (Fig. 5). The
hexanucleotide sequence is absent from the traces of Reln mRNA detected in
liver and kidney tissue, showing that the microexon is constitutively expressed in neurons but skipped in other tissues. Although the functional
importance of this neuron-specific inclusion of the 6 nt micro-exon (coding
for the amino acids Ser-Val) is unclear, it is hard to believe that such a
complex splicing event would be irrelevant, the more so since it is present in
several species. A second alternative splicing event consists of the use of an
alternative polyadenylation site which affects between 10 and 25% of all Reln
messages (Lambert de Rouvroit et al. 1999b; Fig. 5). In genomic DNA, the
alternative terminal exon is contiguous with exon 63. The alternative
polyadenylation introduces two stop codons in the Reln reading frame immediately after the end of the preceding exon, which should result in the
production of a protein lacking the basic terminal sequence encoded by
the most frequently used terminal exon. It is reasonable to assume that the
removal of such a highly basic C-terminal segment affects the function of
the protein. Using in situ hybridization with a probe corresponding to the
alternative terminal exon, the expression of this alternative Reln form is
detected in Cajal-Retzius, the cells that express normal Reln. We have thus far
been unable to reveal the presence of the predicted truncated protein.

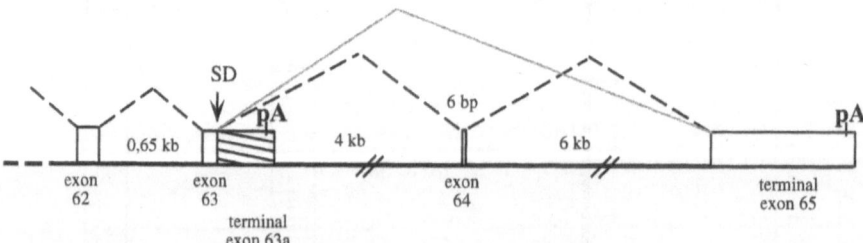

Fig. 5. Alternative splicing and alternative use of poly(A) sites found in the 3′ portion of the
reelin gene. Exons are represented by *open boxes* and introns by *lines* (not to scale). Alternative
splicing events are indicated by *dashed* or *gray lines*. *Hatched box* indicates the sequence
obtained by 3′ RACE [pA, poly(A) site: sequence AATAAA; SD, donor splice site]. (Adapted
from Royaux et al. 1997)

4.2
Reln mRNA Expression During Cortical Development

In their initial description of Reln, D'Arcangelo et al. (1995) noted heavy reelin mRNA expression in Cajal-Retzius neurons of the marginal zone (MZ) as well as in external granule cells of the cerebellum and a few other structures. Further studies of Reln mRNA expression during mouse cortical development (Schiffmann et al. 1997; Alcantara et al. 1998) confirmed that, at early developmental stages, E13–E14, the heaviest expression was found in Cajal-Retzius neurons located in the superficial tiers of the MZ. Heavily labeled neurons were also extremely abundant in the MZ of the hippocampal anlage. By contrast, no expression could be detected in the early cortical plate. At later stages, E17–E18, Reln expression remained the highest in Cajal-Retzius or similar cells, particularly in the hippocampus and dentate gyrus, while expression in the cortical plate remained negative. At birth, expression in the cortex became more widespread than at previous stages. Horizontal neurons in the MZ remained heavily labeled, but some signal was detected in the cortical ribbon. During the early postnatal period, Reln mRNA signal remained strong in Cajal-Retzius neurons in the neocortex the hippocampus and dentate gyrus. In contrast to previous stages, strongly labeled neurons were found scattered in the cortical ribbon. Reln-positive cortical cells were distributed throughout all cortical layers, although clearly more concentrated in the inner part of the cortex, in future laminae V and VI (Pesold et al. 1999). A strong expression was characteristically found in the depth of the entorhinal cortex (Alcantara et al. 1998). In the adult, some reelin expression was present in cortical interneurons (Pesold et al. 1999).

4.3
Reln Protein

Several antibodies are available to study the expression and cell biology of the reelin protein. Antibody CR-50 (Ogawa et al. 1995) is directed against the N terminal part of the protein and apparently able to block Reln function. Other monoclonal antibodies have been developed against both C terminal and N terminal epitopes (de Bergeyck et al. 1998), but no antibodies are available for internal epitopes. Among the N terminal antibodies, G10 is particularly useful because it allows studies of Reln in Western blots; its epitope maps between amino acids 189 and 245. Another interesting reagent is antibody 142, which labels Reln in widely divergent species (Meyer and Goffinet 1998; Meyer pers. Comm.).

Immunohistochemical studies of Reln expression using the CR50 or G10 antibodies yield similar results, and reveal Reln-immunoreactivity in cells that express the mRNA. The observation that the distribution of the Reln mRNA and protein are similar indicates that secreted Reln does not diffuse over large distances and is presumably not transported significantly by the axonal transport system, nor taken up by non-producing cell types. However, the staining observed with antireelin antibodies is due both to intracellular and extracellular protein. The pattern of intracellular staining appears granular, related to the secretion apparatus, and is found after tissue fixation and processing. The pattern of staining related to extracellular Reln immunoreactivity is more difficult to define, but appears more diffuse and pericellular. The hypothesis that Reln appears not to diffuse over large distances and is trapped in the local extracellular matrix should be presented with caution.

4.4
Studies of Reln Function

According to Ogawa et al. (1995) the reaggregation pattern formed by normal embryonic brain cells is different when they are incubated with and without the CR50 antibody. To a certain extent, incubation of normal cells with CR50 mimics the reeler phenotype. This result suggests that Reln produced by Cajal-Retzius cells is a local determinant that orchestrates some cell to cell interactions among cortical neurons in vitro. Consistent with this view, CR50 appears to modify the laminar development of the hippocampus after in vivo injection (Nakajima et al. 1997). Furthermore, in tissue culture, cerebellar Purkinje cells apparently stain with the CR50 antibody, even though they do not express any Reln mRNA, and the antibody CR50 disturbs the pattern formed by Purkinje cells in cerebellar aggregation cultures, suggesting that the antibody may block an interaction between Reln and a "receptor" on the surface of Purkinje cells (Miyata et al. 1997).

4.5
Reelin is Processed in vivo by a Metalloproteinase

When E15 embryonic brain extracts are analysed in Western blots (or with immunoprecipitation), several forms of Reln are detected. In addition to the expected band with a mass close to 400 kDa, corresponding to the size of Reln secreted by cells transfected with a full length cDNA (D'Arcangelo et al. 1997), at least one other form of 160–180 kDa is revealed with antibodies that recognize N-terminal epitopes, and not by antibodies against C-terminal sequences (Lambert de Rouvroit et al. 1999a; Fig. 6). A similar proteolytic processing of Reln is detected in explant cultures from E12–E13 mouse telencephalon, providing the opportunity to interfere with this processing in vitro. Among several proteinase inhibitors tested, only orthophenanthroline and dipicolinic acid, two metal chelators, were able to block Reln cleavage, showing that the activity involved is a metalloproteinase. Rather surprisingly, inhibitors of matrixins, ADAM-type metalloproteinases and neprilysins were

Fig. 6. Proteolytic processing of reelin occurs in normal embryos but not in reeler-Orleans mutant embryos in which reelin is not secreted. Using the G10 antibody that probes the N-terminal region of reelin, a prominent 160 to 180 kDa N-terminal fragment is detected in addition to the full-length protein. There is also a 300-kDa fragment, indicating presence of more than one cleavage site. (From Lambert de Rouvroit et al. 1999a)

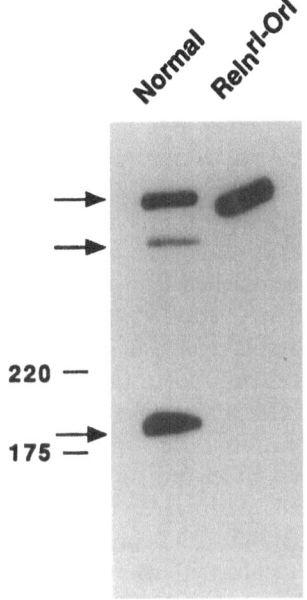

ineffective, suggesting that the enzyme belongs to another family. Further work is needed in order to characterize the enzyme involved in Reln processing and to assess the physiological importance of this phenomenon.

4.6
Reln and Axonal Growth

Heavy expression of Reln is found in horizontal neurons in the MZ of the hippocampus and experimental evidence was recently gathered suggesting that Reln assists in the guidance of some axons in the hippocampal formation (Del Rio et al. 1997; Frotscher1997; Ghosh 1997). When explants of the entorhinal cortex were cultivated with hippocampal slices, axons from the entorhinal cortex invaded the hippocampus and connected in the right layer, namely the outer molecular layer and the stratum lacunosum-moleculare. In the absence of CR cells, entorhinal–hippocampal connections did not form normally and the axons that did reach their target developed poorly. When normal preparations were treated with the antireelin CR50 antibody, a reduction in entorhino–hippocampal connections reminiscent of that seen in the absence of CR cells was also observed, and a similar defect was found in reeler mice. However, the alterations were more subtle than those seen after depletion of CR cells, suggesting that Reln is not the sole factor involved; in line with this, some in vitro actions of Cajal-Retzius cells on cerebellar development are not related to Reln production (Soriano et al. 1997).

5

Mouse *Disabled1* and the Scrambler/Yotari Mutations

Scrambler is a spontaneous mutation on chromosome 4, isolated in 1991. The yotari mutation appeared in a colony raised for a targeted null mutation of the inositol-1,4,5-trisphosphate receptor, and is allelic to scrambler. Phenotypes are identical to reeler (Goldowitz et al. 1997; Gonzales et al. 1997; Gallagher et al. 1998). While the scrambler/yotari gene was being approached using positional cloning (Sheldon et al. 1997; Ware et al. 1997), Howell et al. (1997a,b) showed that a mouse ortholog of the *Drosophila* disabled gene, named Disabled1 (Dab1), maps on mouse chromosome 4 and is expressed in the developing brain, and that the Dab1 knock-out has the reeler phenotype. This result expedited the identification of scrambler and yotari as Dab1 mutations (Sheldon et al. 1997; Ware et al. 1997). In scrambler, part of an IAP sequence is inserted at position 570 of the Dab1 cDNA. However, the IAP element is present both in normal and scrambler genomic DNA and the scrambler anomaly consists of the use of a cryptic acceptor site in an intron followed by splicing to an IAP cryptic donor site (Ware et al. 1997). The yotari mutation was studied only using RT-PCR, and it was demonstrated that yotari mRNA lacks nucleotides 570–927 from the normal sequence, corresponding to skipping of a few (probably four) exons (Sheldon et al. 1997). The precise genomic defects in scrambler yotari remain unknown.

Drosophila disabled was isolated in a screen for mutations that enhance the Abelson neurological phenotype in flies. The Dab1 gene encodes at least four transcripts encoding putative Dab1 proteins of 555, 217 and 271 residues (Fig. 7). The Dab217 message results from alternative polyadenylation; it is identical to Dab555 up to codon 199, then diverges and encodes a stop codon. The Dab271 message is similar to Dab555, except that it contains an additional exon of 270 nucleotides inserted between codons 241 and 242 of Dab555; this exon contains a termination codon. A fourth transcript, 555*, contains an additional exon inserted in frame at approximately position 980 (codon 239) of the 555 cDNA. The presence of other forms of mDab1 messages is not excluded, and in Northern blots three main Dab1 transcripts are found, with sizes of 5.5, 4.0 and 1.8 kb.

In addition to the original *Drosophila* disabled gene, Dab1 relatives include the mouse p96 gene, renamed Dab2, which is not mapped, and its human ortholog DOC2 (also abbreviated DAB2), which maps to human chromosome 5p13. A Dab ortholog, M110.5, is present in *C. elegans*. Blast similarity searches identify two other more distantly Dab-related mammalian proteins, Numb and Numb-like. The human DAB1 maps to chromosome 1p, in a region cosyntenic with mouse chromosome 4 (Lambert de Rouvroit et al. 1998a).

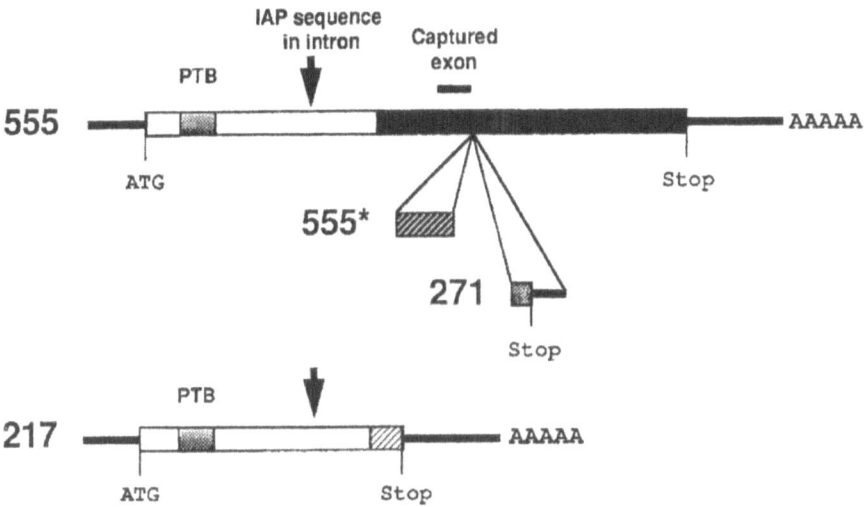

Fig. 7. Various forms of the mouse disabled-1 (Dab1) message. The N-terminal region of the protein contains a phosphotyrosine-binding cassette for protein–protein interactions (PTB). The main form of Dab1 mRNA contains a reading frame for a protein of 555 residues. Form 555* results from inclusion of an additional exon in frame within the 555 mRNA, while form 271 results from inclusion of an exon that contains a premature stop codon. Form 217 results from alternative polyadenylation. Also indicated are position of partial IAP insertion in scrambler mutants and exon captured by random sequencing during the positional cloning of scrambler. (Adapted from Howell et al. 1997)

The aminoterminal region of Dab1 contains a region distantly related to a domain known as phosphotyrosine binding (PTB) or protein interaction (PI), initially described in ShcA and IRS-1 proteins. Western blot analysis revealed that a p80 protein is the predominant Dab1 form in the brain and presumably corresponds to the Dab555 transcript. Dab1 p80 is tyrosine-phosphorylated during early brain development, at E10–E11; then the degree of tyrosine phosphorylation decreases progressively and is undetectable in the adult brain. The kinase(s) that participate(s) in Dab1 phosphorylation has not been identified. However, Dab1 was isolated using the yeast two-hybrid system with Src as a bait, and biochemical studies showed that Dab1-p80 in its tyrosine-phosphorylated state binds in vitro to the SH2 domain of Src, despite the fact that its sequence does not contain a typical SH2-binding site. Conversely, p80 is capable of binding other, yet unidentified phosphotyrosine protein via its PTB/PI domain. Taken together, these data suggested strongly that Dab1-p80 functions as a tyrosine kinase adaptor protein (Howell et al. 1997a). The significance of the association between Src and Dab1 remains unclear, however, as none of the knock-out mice for the intracellular tyrosine kinases Src, Abl and Fyn has a phenotype evocative of reeler.

Studies of Dab1 expression in the embryonic cortex and cerebellum at E13.5 were carried out by in situ hybridization and with anti-Dab1 antibodies. Both techniques reveal a strong expression in the neurons of the cortical plate and in cerebellar Purkinje cells (Sheldon et al. 1997; Gallagher et al. 1998; Rice et al. 1998). Interestingly, these two neuronal classes are the primary target of the reeler gene, in that they are the first elements to manifest the reeler phenotype (Goffinet 1984). These observations suggest strongly that Dab1 is a key element in the response of target neurons to the message delivered by Reln (Goffinet 1997; Fig. 8). Reln expression is upregulated in scrambler and yotari mice (Sheldon et al. 1997) and is found in

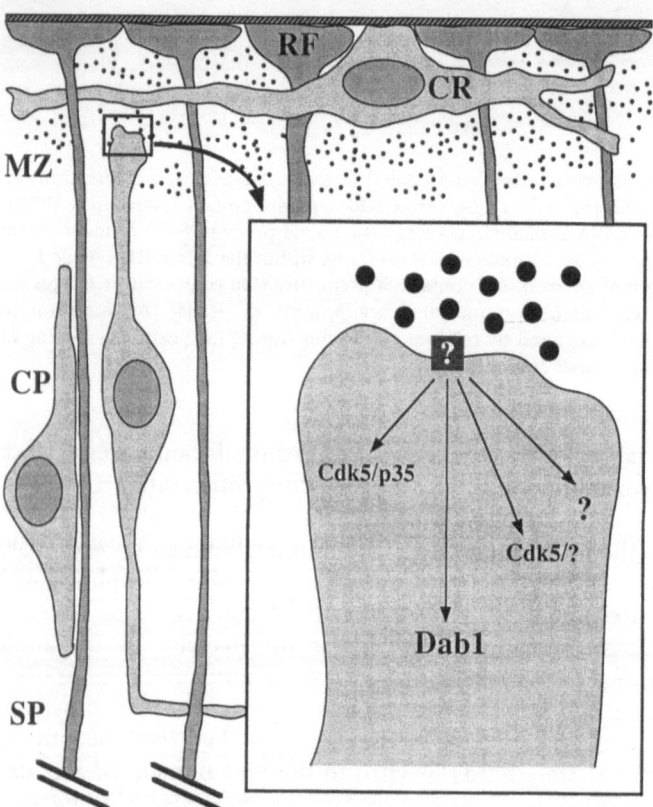

Fig. 8. Action of Reln on developing cortex. Immature neurons migrate along radial fibers (*RF*) before settling in the cortical plate (*CP*). Reelin (*dots*) is secreted by Cajal-Retzius cells (*CR*) in the marginal zone (*MZ*) and incorporated in the local extracellular matrix. End-migration cortical plate neurons respond to the presence of Reelin (which is sensed by their apical dendritic tip) by stopping migration, detaching from the RF and assuming a radial organization. *Inset* shows a neuronal tip. The mechanism by which it senses the presence of Reln is discussed further in Fig. 9. It requires Dab1, and possibly Cdk5 in complex with its p35 (and presumably other) activators, as discussed in Kwon and Tsai (this vol.). *PIA*: Meningeal surface. *SP*: subplate

Cajal-Retzius neurons, as in normal mice. Furthermore, addition of Reln to neuronal cultures increases tyrosine phosphorylation of Dab1 (Howell et al. 1999a).

Comparison of the expression of reelin and Dab1 indicates that the Reelin or Dab1 null phenotypes are more strongly expressed in regions where expression of reelin and Dab1 are topographically distinct, while the phenotype is more subtle in regions where expression of both proteins is concurrent. Of course, this simple hypothesis needs to be further studied, particularly with determination of the neuronal types that express Reln and Dab1 and by comparison of the promoters that drive Reln and Dab1 expression.

6
Very Low Density Lipoprotein Receptor and Apolipoprotein E Receptor Type 2

Two membrane lipoprotein receptors were recently shown to be necessary for the action of Reln (Trommsdorf et al. 1999). While very low density lipoprotein receptor (VLDLR) single mutants have some abnormalities of Purkinje cell layering, and apolipoprotein E receptor type 2 (ApoER2) mutants display subtle alterations of cortical development, the double mutants develop a drastic malformation identical to reeler or scrambler, suggesting that these two similar genes are largely redundant and that the subtle effects of single gene knock-out are mostly due to insufficient compensation by the paralog. VLDLR and ApoER2 are expressed in Dab1-positive neurons such as cortical plate and Purkinje cells, although ApoER2 expression is also present in the intermediate zone. VLDLR and ApoER2 belong to a five-member family that includes the low density lipoprotein receptor, of Nobel fame, responsible for the hepatic uptake of cholesterol-rich low density lipoprotein, megalin and lipoprotein receptor related protein (Stockinger et al. 1998; Van Leuven et al. 1998). Apparently, these three members do not share the effects of VLDLR and ApoER2 on brain development.

The fascinating finding that Reln, Dab1 and VLDLR/ApoER2 participate to a same genetic pathway fits in nicely with observations reported by using the yeast two-hybrid system with either Dab1 or the intracellular portion of the LDL receptor as baits. These studies showed that Dab1 interacts with the cytoplasmic tail of lipoprotein receptors and physically binds to the NPXY sequence implicated in the regulation of lipoprotein receptor-mediated endocytosis (Trommsdorf et al. 1998; Howell et al. 1999b). Dab1 also docks to the signalling protein Ship, to some phosphatidylinositols as well as to the amyloid precursor (APP) and the related proteins APLP1 and APLP2 (Howell et al. 1999b; Homayouni et al. 1999). APP is the precursor of the amyloid deposits in Alzheimer's dementia, and the ApoE4 allele is associated with an

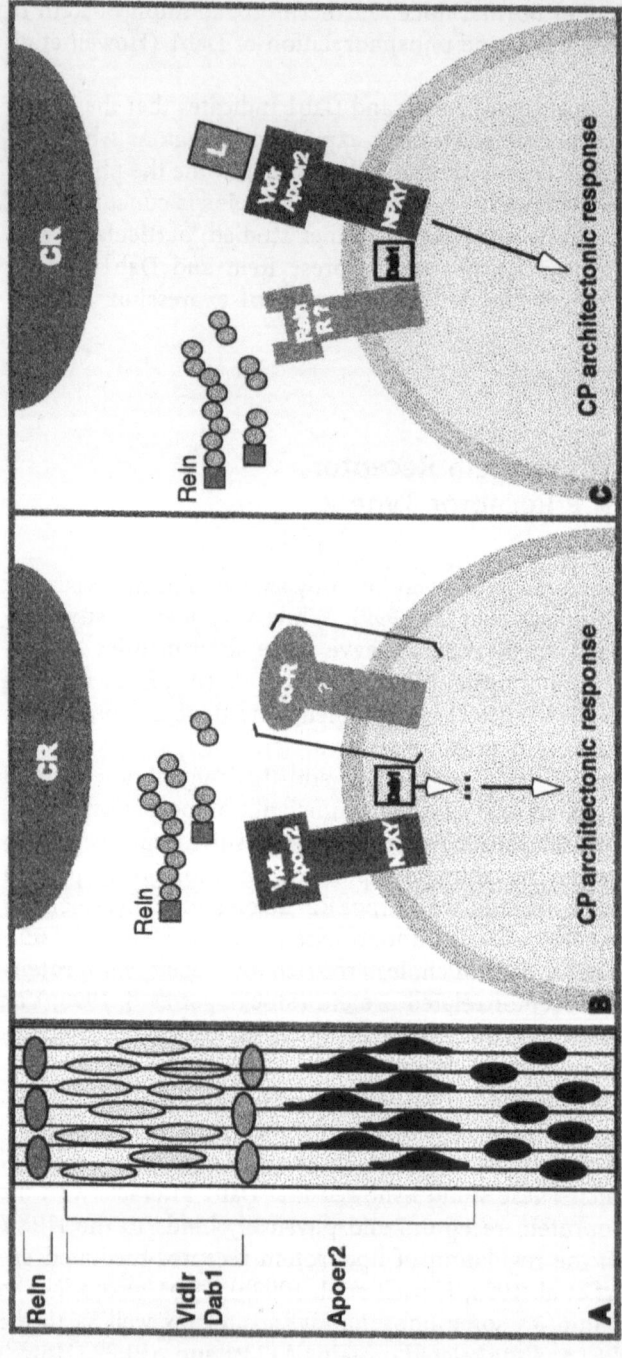

Fig. 9A–C. Lipoprotein receptors and the Reelin signaling pathway. A Neuron precursors (*black circles*) give rise to immature neurons that migrate (*black slugs*) in the cortical plate (*CP*), bordered externally by Reln-positive cells and internally by subplate. Development of the CP depends on Reln, as well as on Dab1, VLDLR and ApoER2 expressed in CP cells. **B** VLDLR and ApoER2 could react to presence of Reln or Reln fragments by docking Dab1 via their NPXY cytoplasmic sequence and translate the Reln message into cortical pattern. A coreceptor (*Co-R*) to Reln, for example a tyrosine kinase, could be present. *CR:* Cajal-Retzius cell. **C** Reln could act on an unidentified Reln receptor and, via Dab1, recruit VLDLR and ApoER2 that are then downstream of Dab1. VLDLR and ApoER2, together with an associated, putative extracellular ligand (*L*), would then instruct cortical architectonics. (Bar and Goffinet 1999)

increased predisposition to Alzheimer's disease. Although the significance of these observations remains unclear, they may not be merely coincidental and have pathophysiological implications.

Although ApoE, the best known ligand for the receptors VLDLR and ApoER2, is synthesised in the brain, probably in glial cells, it is unlikely to play a key role as ApoE-deficient mice develop normally. In addition to ApoE, lipoprotein receptors have a recognized ability to bind protean ligands such as thrombospondin-1, lipoprotein lipase, the serine protease urokinase-type plasminogen activator, and probably others. However, the ligands implicated in patterning the embryonic brain have not been clearly defined thus far.

Available data on Reln, Dab1 and lipoprotein receptors are compatible with at least two interpretations (Bar and Goffinet 1999; Cooper and Howell 1999; Fig. 9). A most parsimonious explanation is that VLDLR/ApoER2 serve as receptors to Reln or Reln fragments (Lambert de Rouvroit et al. 1999a) and, by docking Dab1, modulate associated tyrosine kinase(s) and thus govern the normal architectonic organization of neurons at the end of migration, presumably by acting on the neuronal cytoskeleton. A variant proposed by Trommsdorf et al. (1999) is that VLDLR/ApoER2 could bring Dab1 to a Reln co-receptor that has an associated tyrosine kinase activity. In support for this view, a physical association between VLDLR/ApoER2 and Reln or Reln fragments has been demonstrated (Hiesberger et al. 1999). A different view is that VLDLR/ApoER2 could act downstream of Reln and Dab1. Given the role of lipoprotein receptors in endocytosis, they could, for example, modify membrane trafficking, thereby shaping the normal phenotype of postmigratory neurons. In this scenario, the putative Reln receptor, its interaction with Dab1, and the lipoprotein receptor ligand(s) all remain to be characterized, and therefore this is an hypothesis that we consider rather unlikely.

Since the cloning of Reln (D'Arcangelo et al. 1995), a new Reln signaling pathway including Dab1, VLDLR and ApoER2 has been defined; the list is probably far from complete and several other partners remain to be defined. Future work will seek to identify the various components and to progressively unravel the interactions among them as well as with the other signaling pathways. Other studies will have to address the transcriptional regulation of Reln, Dab1, and lipoprotein receptor expression, which is poorly known. Answers to these questions should provide a better understanding of some basic mechanisms that govern brain development.

Acknowledgements. We thank C. Cortvrindt and C. Dernoncourt for their assistance. Work carried out in our laboratory was supported by grants ARC 183, FRSM 3.453395, and by the Fondation médicale Reine Elisabeth.

References

Alcantara S, Ruiz M, D'Arcangelo G, Ezan F, de Lecea L, Curran T, Sotelo C, Soriano E (1998) Regional and cellular patterns of reelin mRNA expression in the forebrain of the developing and adult mouse. J Neurosci 18:7779–7799

Anderson SA, Eisenstat DD, Shi L, Rubenstein JL (1997) Interneuron migration from basal forebrain to neocortex: dependence on Dlx genes. Science 278:474–476

Angevine JB, Sidman RL (1961) Autoradiographic study of cell migration during histogenesis of cerebral cortex in the mouse. Nature 192:766–768

Bar I, Goffinet AM (1999) Decoding the reelin signal. Nature 399:645–646

Bar I, Lambert de Rouvroit C, Krizman DB, Royaux I, Dernoncourt C, Ruelle D, Beckers MC, Goffinet AM (1995) A YAC contig containing the reeler locus with preliminary characterization of candidate gene fragments. Genomics 26:543–549

Boulder Committee (1969) Embryonic vertebrate central nervous system: revised terminology. Anat Rec 166:257–262

Caviness VJ Jr (1982) Neocortical histogenesis in normal and reeler mice: a developmental study based upon 3H-thymidine autoradiography. Dev Brain Res 4:297–326

Caviness VS Jr (1976) Patterns of cell and fiber distribution in the neocortex of the reeler mutant mouse. J Comp Neurol 170:435–448

Caviness VS, Frost DO, Hayes NL (1976) Barrels in somatosensory cortex of normal and reeler mutant mice. Neurosci Lett 3:7–14

Chae T, Kwon YT, Bronson R, Dikkes P, Li E, Tsai L-H (1997) Mice lacking p35, a neuronal specific activator of Cdk5, display cortical lamination defects, seizures and adult lethality. Neuron 18:29–42

Cooper JA, Howell BW (1999) Lipoprotein receptors: signaling functions in the brain? Cell 97:671–674

D'Arcangelo G, Miao GG, Chen SC, Soares HD, Morgan JI, Curran T (1995) A protein related to extracellular matrix proteins deleted in the mouse mutant reeler. Nature 374:719–723

D'Arcangelo G, Nakajima K, Miyata T, Ogawa M, Mikoshiba K, Curran T (1997) Reelin is a secreted glycoprotein recognized by the CR-50 monoclonal antibody. J Neurosci 17:23–31

de Bergeyck V, Naerhuyzen B, Goffinet AM, Lambert de Rouvroit C (1998) A panel of monoclonal antibodies against reelin, the extracellular matrix protein defective in reeler mutant mice. J Neurosci Meth 82:17–24

DeCarlos JA, O'Leary DDM (1992) Growth and targeting of subplate axons and establishment of major cortical pathways. J Neurosci 1194–1211

Del Rio JA, Heimrich B, Borrell V, Forster E, Drakew A, Alcantara S, Nakajima K, Miyata T, Ogawa M, Mikoshiba K, Derer P, Frotscher M, Soriano E (1997) A role for Cajal-Retzius cells and reelin in the development of hippocampal connections. Nature 385:70–74

Derer P (1979) Evidence for the occurrence of early modifications in the 'glia limitans' layer of the neocortex of the reeler mutant mouse. Neurosci Lett 13:195–202

Frotscher M (1997) Dual role of Cajal-Retzius cells and reelin in cortical development. Cell Tissue Res 290:315–322

Gallagher E, Howell BW, Soriano P, Cooper JA, Hawkes R (1998) Cerebellar abnormalities in the disabled (mdab1-1) mouse. J Comp Neurol 402:238–251

Ghosh A (1997) Axons follow Reelin routes. Nature 385:23–24

Gilmore E, Ohshima T, Gofinet AM, Kulkarni A, Herrup K (1998) Cyclin-dependent kinase 5-deficient mice demonstrate novel developmental arrest in cerebral cortex. J Neurosci 18:6370–6377

Goffinet AM (1984) Events governing organization of postmigratory neurons: studies on brain development in normal and reeler mice. Brain Res Rev 7:261–296

Goffinet AM (1997) Unscrambling a disabled brain. Nature 389:668–669

Goldowitz D, Cushing RC, Laywell E, D'Arcangelo G, Sheldon M, Sweet HO, Davisson M, Steindler D, Curran T (1997) Cerebellar disorganization characteristic of reeler in scrambler mutant mice despite presence of reelin. J Neurosci 17:8767–8776

Gonzales JL, Russo CJ, Goldowitz D, Sweet HO, Davisson MT, Walsh CA (1997) Birthdata and cell marker analysis of scrambler: a novel mutation affecting cortical development with a reeler-like phenotype. J Neurosci 17:9204–9211

Hiesberger T, Trommsdorff M, Howell BW, Goffinet A, Mumby MC, Cooper JA, Herz J (1999) Direct binding of reelin to VLDL receptor and ApoE receptor 2 induces tyrosine phosphorylation of disabled-1 and modulates tau phosphorylation. Neuron 24:481–489

Homayouni R, Rice DS, Sheldon M, Curran T (1999) Disabled-1 binds to the cytoplasmic domain of amyloid precursor-like protein 1. J Neurosci 19:7507–7515

Howell BW, Gertler FB, Cooper JA (1997a) Mouse disabled (mDab1): a Src binding protein implicated in neuronal development. EMBO J 16:121–132

Howell BW, Hawkes R, Soriano P, Cooper JA (1997b) Neuronal position in the developing brain is regulated by mouse disabled-1. Nature 389:733–737

Howell BW, Herrick TM, Cooper JA (1999a) Reelin-induced tyrosine phosphorylation of disabled 1 during neuronal positioning. Genes Dev 13:643–648

Howell BW, Lanier LM, Frank R, Gertler FB, Cooper JA (1999b) The Disabled 1 PTB domain binds to the internalization signals of transmembrane glycoproteins and to phospholipids. Mol Cell Biol 19:5179–5188

Hunter-Schaedle KE (1997) Radial glial cell development and transformation are disturbed in reeler forebrain. J Neurobiol 33:459–472

Klar A, Baldassare M, Jessell T (1992) F-spondin: a gene expressed at high levels in the floor plate encodes a secreted protein that promotes neural cell adhesion and neurite extension. Cell 69:95–110

Lambert de Rouvroit C, Goffinet AM (1998a) The reeler mouse as a model of brain development. Adv Anat Embryol Cell Biol 150:1–108

Lambert de Rouvroit C, Goffinet AM (1998b) Cloning of human DAB1 and mapping to chromosome 1p31-p32. Genomics 53:246–247

Lambert de Rouvroit C, de Bergeyck V, Cortvrindt C, Bar I, Eeckhout Y, Goffinet AM (1999a) Reelin, the extracellular matrix protein deficient in reeler mutant mice, is processed by a metalloproteinase. Exp Neurol 156:214–217

Lambert de Rouvroit C, Bernier B, Royaux I, de Bergeyck V, Goffinet AM (1999b) Evolutionarily conserved, alternative splicing of reelin during brain development. Exp Neurol 156:229–238

Luskin MB (1993) Restricted proliferation and migration of postnatally generated neurons derived from the forebrain subventricular zone. Neuron 11:173–189

McConnell SK, Gosh A, Shatz CJ (1989) Subplate neurons pioneer the first axon pathway from the cerebral cortex. Science 245:978–982

Meyer G, Goffinet AM (1998) Prenatal development of reelin-immunoreactive neurons in the human neocortex. J Comp Neurol 397:29–40

Meyer G, Soria JM, Martinez-Galan JR, Martin-Clemente B, Fairen A (1998) Different origins and developmental histories of transient neurons in the marginal zone of the fetal and neonatal rat cortex. J Comp Neurol 397:493–518

Miale I, Sidman RL (1961) An autoradiographic analysis of histogenesis in the mouse cerebellum. Exp Neurol 4:277–296

Misson JP, Austin CP, Takahashi T, Cepko CL, Caviness VS (1991) The alignment of migrating neural cells in relation to the murine neopallial radial glial fiber system. Cereb Cortex 1:221–229

Miyata T, Nakajima K, Mikoshiba K, Ogawa M (1997) Regulation of Purkinje cell alignment by reelin as revealed with CR-50 antibody. J Neurosci 17:3599–3609

Molnar Z (1998) Development of thalamocortical connections. Neuroscience intelligence unit series, RG Landes Bioscience Publishers, Austin, Texas

Molnar Z, Blakemore C (1995) How do thalamic axons find their way to the cortex? Trends Neurosci 18:389–397

Molnar Z, Adams R, Goffinet AM, Blakemore C (1998) The role of the first postmitotic cortical cells in the development of thalamocortical innervation in the reeler mouse. J Neurosci 18:5746–5765

Nakajima K, Mikoshiba K, Miyata T, Kudo C, Ogawa M (1997) Disruption of hippocampal development in vivo by CR-50 mAb against reelin. Proc Natl Acad Sci USA 94:8196–8201

Nowakowski R, Rakic P (1981) The site of origin and route and rate of migration of neurons to the hippocampal region of the rhesus monkey. J Comp Neurol 196:129–154

Ogawa M, Miyata T, Nakajima K, Yagyu K, Seike M, Ikenaka K, Yamamoto H, Mikoshiba K (1995) The reeler gene-associated antigen on Cajal-Retzius Neurons is a crucial molecule for laminar organization of cortical neurons. Neuron 14:899–912

Pesold C, Liu WS, Guidotti A, Costa E, Caruncho HJ (1999) Cortical bitufted, horizontal, and Martinotti cells preferentially express and secrete reelin into perineuronal nets, nonsynaptically modulating gene expression. Proc Natl Acad Sci USA 96:3217–3222

Pinto-Lord MC, Caviness VS Jr (1979) Determinants of cell shape and orientation: a comparative Golgi analysis of cell–axon interrelationships in the developing neocortex of normal and reeler mice. J Comp Neurol 187:49–70

Rice DS, Sheldon M, D'Arcangelo G, Nakajima K, Goldowitz D, Curran T (1998) Disabled-1 acts downstream of reelin in a signaling pathway that controls laminar organization in the mammalian brain. Development 125:3719–3729

Royaux I, Lambert de Rouvroit C, D'Arcangelo G, Demirov D, Goffinet AM (1997) Genomic organization of the mouse reelin gene. Genomics 46:240–250

Schiffmann SN, Bernier B, Goffinet AM (1997) Reelin mRNA expression during mouse brain development. Eur J Neurosci 9:1055–1071

Sheldon M, Rice DS, D'Arcangelo G, Yoneshima H, Nakajima K, Mikoshiba K, Howell B, Cooper JA, Goldowitz D, Curran T (1997) Scrambler and yotari disrupt the disabled gene and produce a reeler-like phenotype on mice. Nature 389:730–733

Sheppard AM, Pearlman AL (1997) Abnormal reorganization of preplate neurons and their associated extracellular matrix: an early manifestation of altered neocortical development in the reeler mutant mouse. J Comp Neurol 378:173–179

Shoukimas GM, Hinds JW (1978) The development of the cerebral cortex in the embryonic mouse: an electron microscopic serial section analysis. J Comp Neurol 179:795–830

Soriano E, Alvarado-Mallart RM, Dumesnil N, Del Rio JA, Sotelo C (1997) Cajal-Retzius cells regulate the radial glia phenotype in the adult and developing cerebellum and alter granule cell migration. Neuron 18:563–577

Stockinger W, Hengstschlager-Ottnad E, Novak S, Matus A, Huttinger M, Bauer J, Lassman H, Schneider WJ, Nimpf J (1998) The low density lipoprotein receptor gene family. J Biol Chem 273:32213–32221

Tan SS, Kalloniatis M, Sturm K, Tam PP, Reese BE, Faulkner-Jones B (1998) Separate progenitors for radial and tangential cell dispersion during development of the cerebral neocortex. Neuron 21:295–304

Trommsdorff M, Borg JP, Margolis B, Herz J (1998) Interaction of cytosolic adaptor adaptor proteins with neuronal apolipoprotein E receptors and the amyloid precursor protein. J Biol Chem 273:33556–33560

Trommsdorff M, Gotthardt M, Hiesberger T, Shelton J, Stockinger W, Nimpf J, Hammer RE, Richardson JA, Herz J (1999) Reeler/Disabled 1-like disruption of neuronal migration in knockout mice lacking the VLDL receptor and ApoE receptor 2. Cell 97:689–701

Van Leuven F, Thiry E, Stas L, Nelissen B (1998) Analysis of the human LRPAP1 gene coding for the lipoprotein receptor-associated protein: identification of 22 polymorphisms and one mutation. Genomics 52:145–151

Ware ML, Fox JW, Davis NM, Lambert de Rouvroit C, Russo C, Chua SC Jr, Goffinet AM, Walsh CA (1997) Aberrant splicing of a mouse disabled homolog, mDab1, in the scrambler mouse. Neuron 19:239–249

Yuasa S, Kitoh J, Kawamura K (1994) Interactions between growing thalamocortical afferent axons and the neocortical primordium in normal and reeler mutant mice. Anat Embryol 190:137–154

The Subpial Granular Layer in the Developing Cerebral Cortex of Rodents

Gundela Meyer[1], Rafael Castro[2], José Miguel Soria[3], and Alfonso Fairén[3]

1
Introduction

The subpial granular layer (SGL) is a transient layer of small, undifferentiated cells located immediately below the pial surface of the cerebral cortex. It was first described by Ranke (1910) in the marginal zone (MZ), the prospective cortical layer I, of human fetuses around midgestation. While Ranke believed the SGL to be composed mainly of glia, other authors such as Schaffer (1918) and Brun (1965) considered it a predominantly neuronal structure. The neuronal character of most of its constituent cells was later confirmed by Gadisseux et al. (1992) by using immunocytochemistry with neuron-specific antibodies.

In this chapter which deals with a possible SGL homologue in the rodent, we want to emphasize two features of the human SGL. First, its proposed origin from the periolfactory basal forebrain, from where it would spread through tangential subpial migration into the neocortical MZ (Brun 1965; Gadisseux et al. 1992). Second, its close relationship with the Cajal-Retzius cells (Fig. 1A), first observed by Schaffer (1918), and more recently stressed by Meyer and Gonzalez-Hernandez (1993) and Meyer and Goffinet (1998). The importance of the Cajal-Retzius cells in cortical development is partly due to the fact that they secrete the glycoprotein reelin, the product of the reeler gene, which is critically involved in the establishment of the normal lamination pattern of the cerebral cortex (D'Arcangelo et al. 1995, 1997; Ogawa et al. 1995; Schiffmann et al. 1997; Lambert de Rouvroit and Goffinet 1998). Meyer and Goffinet (1998) proposed that one of the possible functions of the SGL consists in providing additional reelin-producing cells as the cortex grows and the limited population of early-born preplate Cajal-Retzius

[1] Departamento de Anatomía, Facultad de Medicina, Universidad de La Laguna, 38071 La Laguna, Tenerife, Spain
[2] Departamento de Fisiología, Facultad de Medicina, Universidad de La Laguna, 38071 La Laguna, Tenerife, Spain
[3] Instituto de Neurociencias, CSIC, Universidad Miguel Hernandez, 03550 San Juan de Alicante, Spain

Results and Problems in Cell Differentiation, Vol. 30
Goffinet and Rakic (Eds.): Mouse Brain Development
© Springer-Verlag Berlin Heidelberg 2000

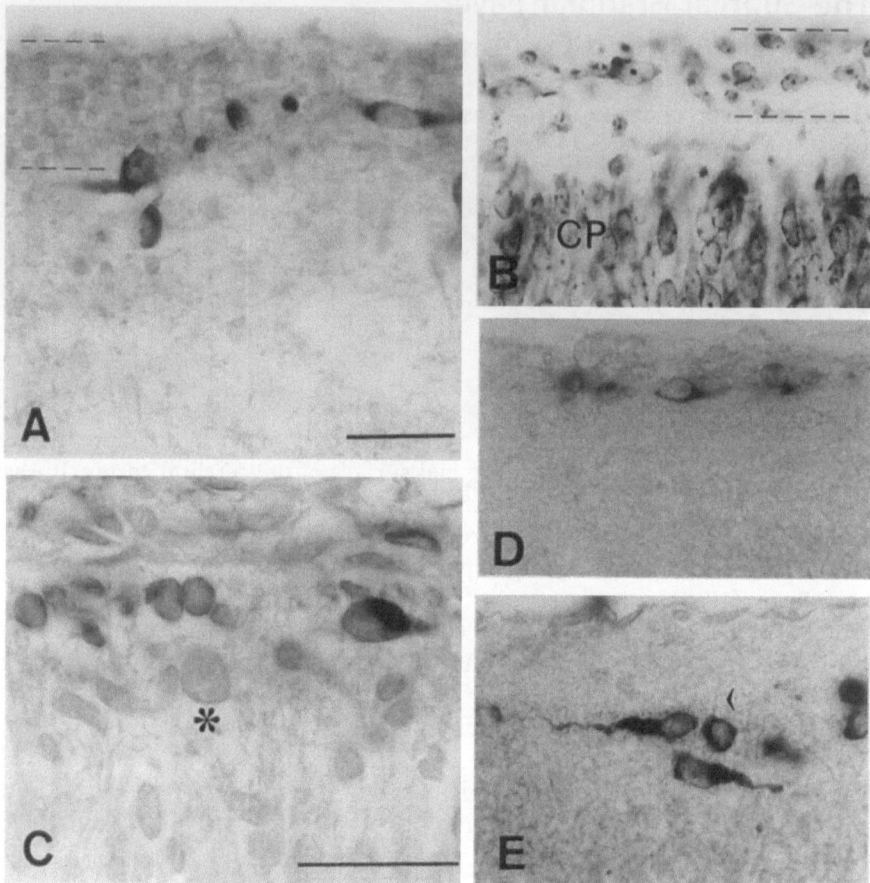

Fig. 1A–E. Subpial granular layer (SGL) and development of reelin-producing neurons of the marginal zone. **A** SGL of a human fetus at 20 gestational weeks. Reelin-immunoreactive granule cells are visible within SGL, while reelin-expressing Cajal-Retzius-like neurons lie within SGL as well as below it. **B** SGL stained with cresyl violet and indicated by *hatched line*, of a mouse at E18. SGL contains neurons of different sizes. *CP* Cortical plate. **C** Reelin-immunoreactive neurons of rat cortex at E16. Most reelin-expressing cells at this age have a granular morphology, but others have larger somata and are already differentiating into Cajal-Retzius cells. The large deep neuron (*asterisk*) corresponds to a reelin-negative pioneer neuron. **D** Reelin-expressing neurons at E18 in mouse cortex. At this age, Cajal-Retzius cells lie below the pia, and do not yet display the radially ascending branchlets that are characteristic of this cell type during early postnatal life. **E** Reelin-expressing Cajal-Retzius cells of a P5 rat. Most Cajal-Retzius cells have descended and lie now at about 20 μm from pial surface, to which they remain connected via ascending processes (*arrowhead*). *Bar* in **A**, for **B,D** and **E**: 20 μm; in **C**: 25 μm

cells (Marin-Padilla 1972, 1998) becomes dispersed over an expanding cortical surface. Whereas it appears plausible that the protracted migration period and dramatic surface expansion proper to the human brain would require a continuous supply of reelin-expressing cells, the small, fast developing rodent cortex was not suspected to have a subpial granular layer. In fact, ever since the first description by Ranke (1910), the SGL was regarded as a specific feature of the human cortex, with no nonprimate homologue.

While animal models are usually used to define homologies and to infer about the human brain, we chose the reverse approach (Meyer et al. 1998): we started with the human model, and tried to learn about the possible existence of the SGL in the rodent. We were specifically interested in the following questions: Does a homologue of the human SGL exist in the fetal rodent cortex? If so, where is its origin? How is it related to the reelin-producing cells of the marginal zone, the Cajal-Retzius cells? Are there other SGL cell populations unrelated to reelin-expression? And finally, what is the fate of the cells forming the rodent subpial granular layer?

2
Neuronal Populations of the Rodent Marginal Zone

Our study was carried out in prenatal and early postnatal rats, from embryonic day (E) 11 to postnatal day (P) 20, by using immunocytochemistry for the calcium-binding proteins calbindin and calretinin, for GABA and its synthesizing enzyme glutamic acid decarboxylase (GAD 67), neurofilaments and reelin (de Bergeyck et al. 1998). A birthdating analysis was carried out using BrdU injections at different early embryonic ages. The immunohistochemical data obtained in the rat were later confirmed in a parallel series of embryonic and postnatal mice (unpubl. data). Altogether, these studies did indeed provide evidence of the existence of a transient cell compartment below the pial surface, densely packed with small granular cells, which may well constitute the homologue of the human SGL (Fig. 1B).

2.1
Pioneer Neurons

Both the immunocytochemical results and the BrdU data revealed a variety of cells in the MZ which did not conform to the classical model of rodent corticogenesis, which is based on the Cajal-Retzius cell as the dominant cell type of the developing MZ. According to this concept, Cajal-Retzius cells are born very early at the preplate stage, and remain basically unchanged throughout corticogenesis (Marin-Padilla 1972, 1998). Instead, we distinguished two major cell populations with different birthdates, which thus appeared in the MZ at different moments of corticogenesis. We identified a population of pioneer neurons, which extended the first, pioneer axonal projection from the cortex, and a population of smaller granule cells which occupied the subpial compartment (Fig. 2). The pioneer neurons, born very early in development, at E11–E12, were characterized by their large size,

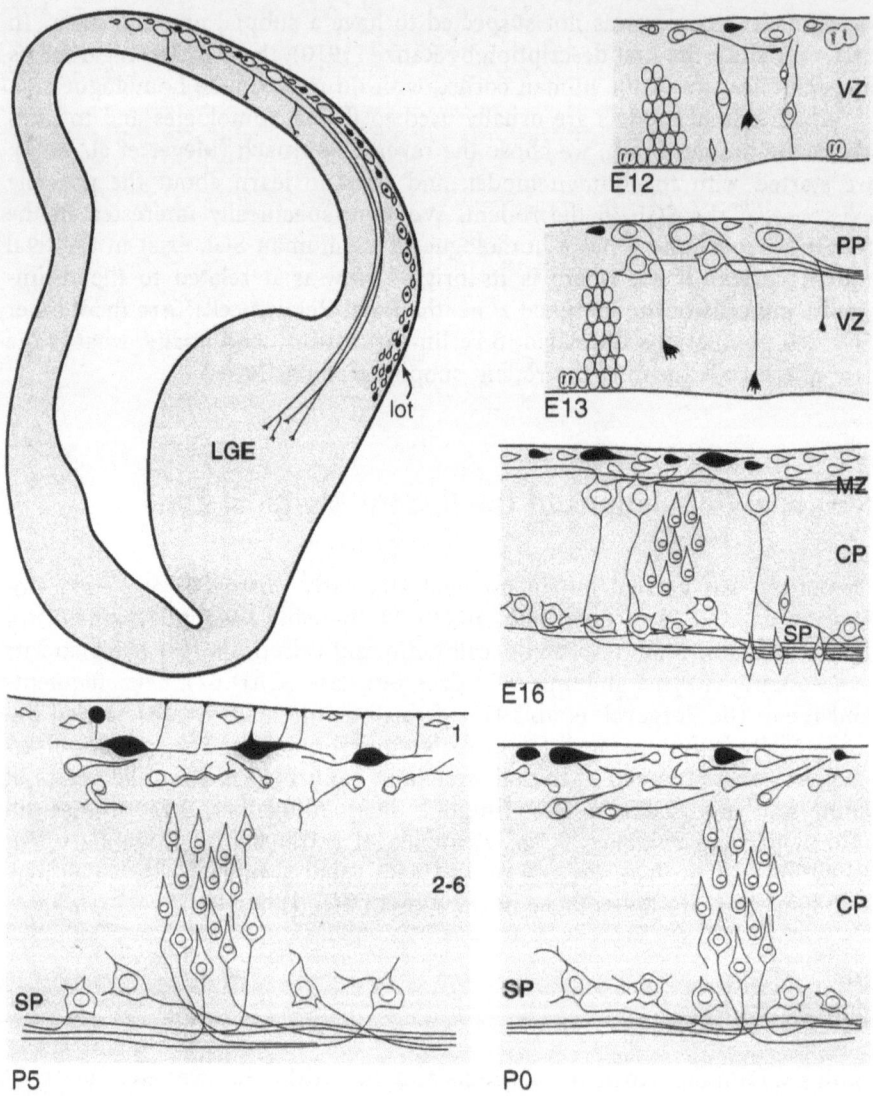

Fig. 2. The different cell classes in developing preplate (*PP*) and marginal zone (*MZ*), and their proposed axonal projection patterns. Large inset. A coronal section of rat brain that summarizes the axonal projections of pioneer neurons in PP at E12–14 into the ventricular zone (*VZ*) and nascent internal capsule. *LGE* Lateral ganglionic eminence; *lot* lateral olfactory tract. The drawings illustrate the changes in neuronal composition in PP and MZ from E12 to P5. *Black* neurons drawn represent reelin-expressing neurons which evolve from small granular neurons into Cajal-Retzius cells. Large neurons in the MZ with *white* somata represent pioneer neurons, a transient cell type that emits the first axonal projections of the developing cortex. Large subplate neurons are also indicated. Small *white* neurons in PP and MZ represent reelin-negative granule cells that form part of the subpial granular layer. *CP* Cortical plate; *IZ* intermediate zone; *SP* subplate

expression of the calcium-binding proteins calretinin or calbindin (Fig. 3A), and, most importantly, by their axonal projection pattern. Already at E12, they have sent axonal collaterals directly into the intermediate zone in the lateral part of the cortical vesicle and into the ventricular zone (Fig. 4A',D'); at E13 and E14, the axons formed part of the incipient internal capsule in the lateral ganglionic eminence. The pioneer cells did not express reelin or GABA; while initially they lay closely below the pia (Fig. 4A'), after the appearance of the cortical plate they became displaced into the deep MZ, and even towards the superficial cortical plate, until they disappeared shortly before birth. Pioneer cells are thus a transient cell type that does not survive into postnatal life. They share their axonal projections with the subplate neurons, which are born slightly later, at about E12–E14 (Raedler and Raedler 1978; Bayer and Altman 1991; De Carlos and O'Leary 1992; Valverde et al. 1995). A peculiar characteristic of the pioneer neurons of the MZ is that they form aggregates, with specific patterns of clustering within discrete cortical territories (Soria and Fairén 1999). These territories may constitute the earliest signs of regionalization thus far described in the developing cortex.

2.2
The Subpial Granular Layer

The second major cell population in the MZ, which we termed "subpial granule cells" on the basis of the cells' small soma size and superficial location, did not extend any recognizable axonal projection, but displayed a rather undifferentiated monopolar or bipolar morphology evocative of migrating neurons. The subpial granule cells were heterogeneous in their chronology of maturation as well as in their neurochemical properties (Fig. 3): partially overlapping subpopulations expressing calretinin, calbindin, reelin and GABA appeared in the MZ at different moments of prenatal life. The calbindin-positive granules appeared as early as E12 at the rostral pole of the telencephalic vesicle; they invaded the dorsal cortex from lateral to medial around E14. Calretinin-positive granules were visible in the rostrolateral cortex at E12, and similarly reached the dorsal cortical surface at E14. In turn, the GABA-expressing granules made their first appearance at E14 in the lateral preplate; they became visible in the dorsal cortex around E16, although they remained more numerous in the lateral cortex (Fig. 3B).

The fate of the different SGL members varied considerably: The calbindin-positive granules were a prominent SGL component from E14 to E20 (Fig. 3C). After birth, they descended deeper into the molecular layer, and seemed to differentiate into nonpyramidal neurons with a predominantly horizontal morphology. The calretinin-expressing granules were very abundant around birth and during the first postnatal days (Fig. 3D). From P2 to P5, they similarly descended from their subpial position, and changed into variable and rather irregular morphologies suggestive of incipient degeneration fol-

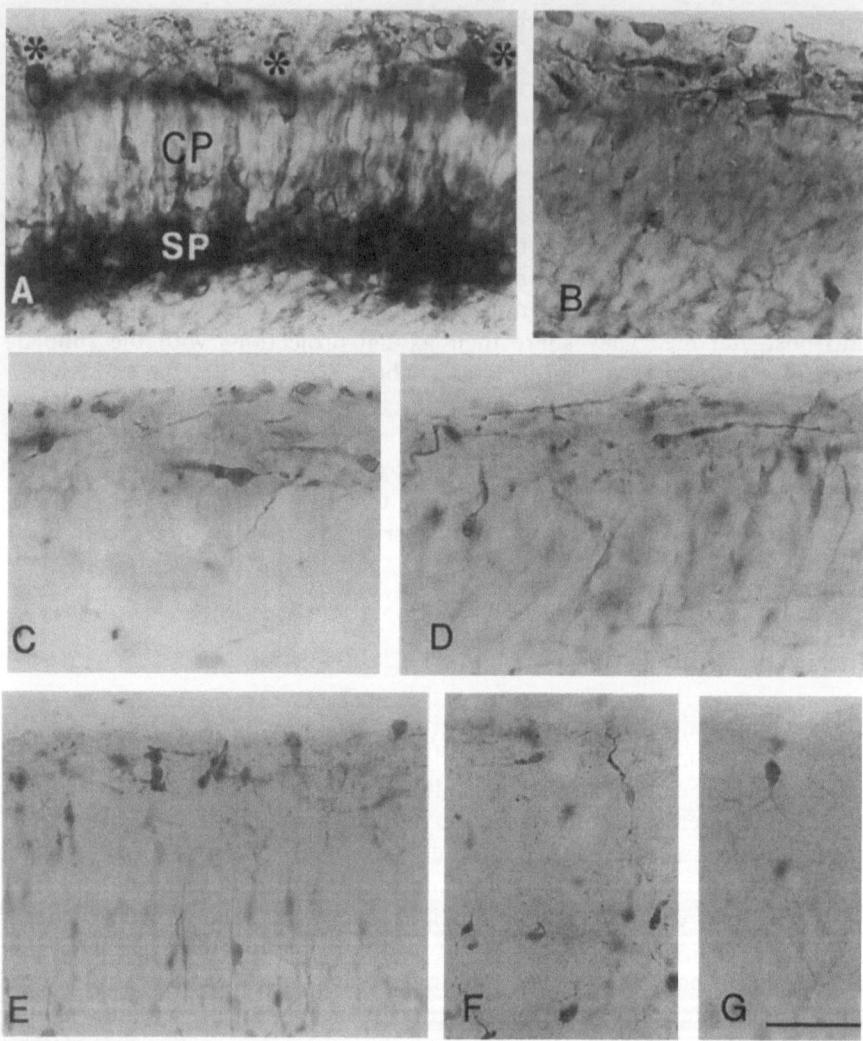

Fig. 3A–G. Pioneer neurons and subpial granule cells of rat cortex. **A** Calbindin-immunoreactive pioneer neurons at E16 (*asterisks*) reside in the marginal zone, from where they extend a pioneer axonal projection through cortical plate (*CP*) and subplate (*SP*) into the internal capsule. **B** GAD-immunoreactive granule cells in marginal zone at E16. **C** Calbindin-immunoreactive granule cells at E17. The most immature neurons occupy an immediate subpial position, while deeper neurons form a population of persisting horizontal interneurons of the molecular layer. These cells do not express reelin. **D** Calretinin-expressing granule cells at P0, most of which display a horizontal mono or bipolar morphology. **E** Calretinin-immunoreactive granule cells at P4 assume diverse morphologies. Part of these cells colocalize reelin and calretinin. **F** Dark shrunken somata and dendrites at P10 suggest degeneration of most calretinin-positive derivatives of the subpial granular layer. **G** A calretinin-positive subpial pyriform cell at P14. These cells, which colocalize GABA, reelin and calretinin, persist into adulthood, and are proposed to represent the latest derivative of the subpial granular layer to mature. *Bar* in **G**: 40 μm

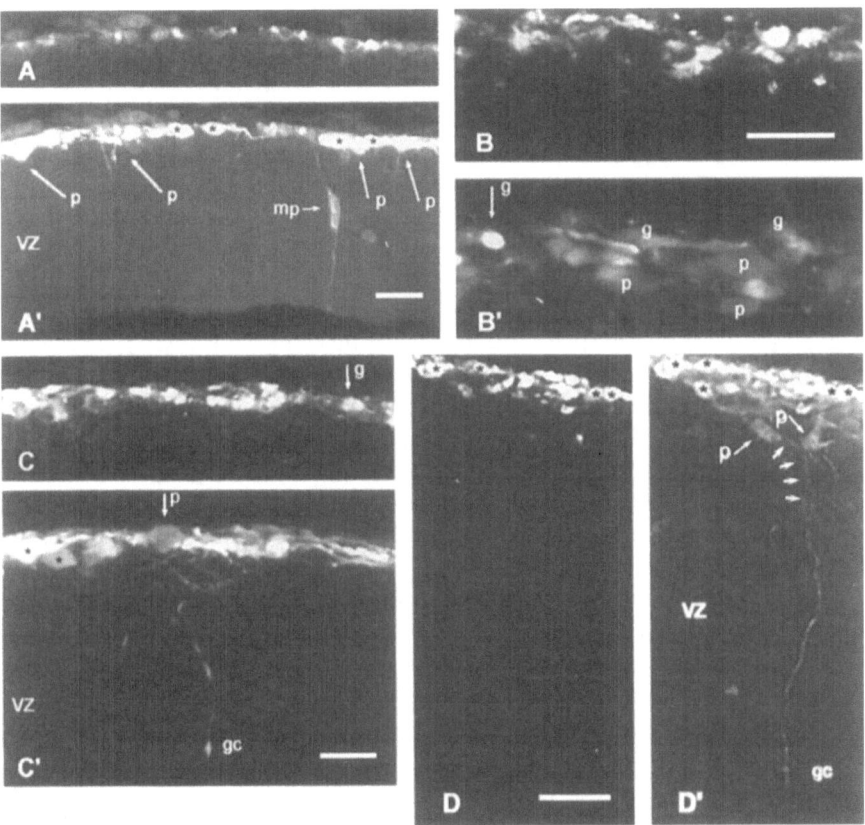

Fig. 4A–D. Pairs of confocal images (obtained in a Leica TCS-NT confocal laser scanning microscope) recorded through separate channels for fluorescence of Cy-2 and Cy-5, labeling reelin or one of the calcium-binding proteins calbindin and calretinin. All sections are from developing neocortex of rat, at E13. A–D Reelin-expressing SGL cells; A',C' and D' Calretinin-immunoreactive cells in SGL and VZ; B' Calbindin-immunoreactive cells in the MZ. A and A' *asterisks* in A' point to SGL neurons expressing both reelin and calretinin; *p* Pioneer neurons expressing calretinin; *mp* a migrating cell in VZ, with leading and trailing processes reaching the inner and outer borders of brain vesicle. These cells may develop into pioneer cells (see Meyer et al. 1998). B and B' Complete segregation between reelin- and calbindin-expressing cells. *p* calbindin-immunoreactive pioneer cells; *g* reelin-negative, calbindin-immunoreactive SGL cells. C,C',D and D' show SGL neurons expressing both reelin and calretinin (*asterisks*), and one example of an SGL cell expressing only reelin and not calretinin (*g* in C). Pioneer neurons (*p*) express calretinin. Axons of pioneer neurons enter VZ and their tips form growth cones (*gc*); axon shown in D' can be traced back to its cell of origin (*short arrows*). *Bars*: 25 μm

lowed by cell death (Fig. 3E,F). After P8, most calretinin-positive neurons had indeed disappeared from the molecular layer, and only a few cells with dark, shrunken dendrites remained. However, in the group of calretinin-positive neurons, we wish to mention a very conspicuous representative which began to differentiate, and to extend processes, after the first week of postnatal life;

we termed these cells "the subpial pyriform cells". They were located very superficially, immediately attached to the pial surface (Fig. 3G), so that we regarded them as the last derivative of the SGL to mature. These cells were remarkable also because they coexpressed calretinin, GABA and reelin, not only during development but also throughout adult life. The GABA-expressing SGL cells became undetectable toward the end of prenatal life. We were unable to determine whether they stopped GABA-expression or underwent cell death (see also Van Eden et al. 1989; Cobas et al. 1991; Del Rio et al. 1992).

The existence of a rodent SGL as a source of MZ neurons received experimental support from the findings of Brunstrom et al. (1997). These authors reported an excess of neurons invading the MZ after administration of neurotrophin 4, which formed heterotopic collections that distorted the underlying cortical architecture. The excess neurons shared an early birthdate, E13, with normal MZ neurons, and appeared in particularly high numbers in the rostral and lateral cortical areas. These results are consistent with an SGL population born in a generation area outside the neocortical neuroepithelium as proposed by our study.

So far we have not mentioned the reelin-expressing members of the SGL, because reelin is typically expressed by the granules that differentiate in the course of gestation into the cell type known as Cajal-Retzius cells. Cajal-Retzius cells dominate the MZ and molecular layer of the perinatal and early postnatal period (vide infra).

2.3
Reelin-Expressing Cajal-Retzius Cells of Rodent Cortex

The first reelin-expressing cells of the preplate were observed at E12 in the rat (in the mouse at E11), covering the entire cortical surface (Fig. 4A). At this early moment of corticogenesis, E12, E14, the reelin-expressing neurons displayed a granular appearance, which was in contrast to the large soma size of the pioneer cell populations (Fig. 1C). In the course of prenatal life, the reelin-expressing granule cells gradually differentiated into larger neurons, although they remained in an immediate subpial location (Fig. 1D). Only shortly after birth did they descend from their subpial position to a quite constant distance of about 20 μm from the pia, although preserving their pial connection through radially ascending branchlets (Fig. 1E). This postnatal morphology corresponds with the classical description of the rodent Cajal-Retzius cell. The fate of the rodent Cajal-Retzius cells is known from many studies (e.g. Derer and Derer 1990; Del Rio et al. 1995) which are confirmed by our own observations: most of them die after the first postnatal week, so that after P10 they are not present any more.

Colocalization experiments revealed that reelin was coexpressed with GABA in the prenatal cohorts of GABAergic subpial granule cells. In fact, almost all GABA-immunoreactive cells were also reelin-positive. In contrast, the calbindin-positive SGL population did not seem to express reelin (Fig. 4B, B'). On the other hand, calretinin was often colocalized with reelin in prenatal granule cells (Fig. 4C,C',D,D'); however, in the early postnatal week, when the reelin-expressing Cajal-Retzius cells had attained their maximum differentiation, calretinin and reelin were present in partly segregated cell populations. In particular, cells expressing only calretinin were located in the lower tier of layer I and showed morphological signs of degeneration. This seems to be different in the postnatal mouse, where calretinin and reelin were coexpressed by the Cajal-Retzius cells and no obvious signs of degeneration were seen during an equivalent period (J. R. Martínez-Galán, A. Fairén and M. Valdeolmillos, unpubl. observ.).

While most SGL derivatives, calretinin-positive granule cells and Cajal-Retzius cells, disappeared from the molecular layer after the first week of postnatal life, new reelin-positive neurons appeared in small numbers in the lower part of this layer, as well as in the cortical plate (see also Ikeda and Terashima 1997; Alcántara et al. 1998; Pesold et al. 1998). These deep reelin-expressing neurons are interneurons (Alcántara et al. 1998; Pesold et al. 1998).

We wish to point out that the reelin-competent human SGL appears after the emergence of the cortical plate, at 13/14 gestational weeks, but that small numbers of reelin-producing neurons are present in the preplate already at 5/6 weeks of gestation (Meyer et al. 1999). Similarly, in the rat the first reelin-expressing neurons were observed before the appearance of an SGL. Thus, the fact that reelin-expressing cells appear very early in the preplate, and have probably a local origin, does not contradict the finding that additional reelin-positive neurons are supplied by the SGL, and thus arrive in the MZ through tangential migration. The important point is that reelin-expression is not confined to a single cell type, defined as "the oldest neuron of the cortex" (Marin-Padilla 1998), but that reelin is provided through multiple mechanisms involving various cell populations as the cortex grows.

3
Possible Origin of the Rodent Subpial Granular Layer

So far, our results demonstrated the existence of several different cell populations in the rat MZ, some of which were indeed comparable to the SGL cells of the human cortex, while others, particularly the pioneer neurons, had not been reported previously. We were then interested in the origin of the subpial granule cells, especially whether they could be traced to a common origin from the retrobulbar ventricular compartment in the basal forebrain, such as proposed by Brun (1965) and Gadisseux et al. (1992) for the human SGL. We observed in the rat at E14 a very prominent fountainhead of cal-bindin- and calretinin-expressing neurons in the ventricular zone and marginal layer of an area immediately behind the olfactory bulb, the "retrobulbar generation compartment" (Fig. 5A). This fountainhead was continuous with a stream of granule cells which seemed to invade the neocortical MZ. We thus proposed that the rat SGL would spread from this retrobulbar compartment over the entire neocortex, similarly to what had been postulated for the human brain. In the mouse, the retrobulbar generation compartment was less conspicuous than in the rat, but we detected, nevertheless, at E13 a comparable stream of reelin- and calretinin-immunoreactive granule cells into the MZ (Fig. 5B–D), while the calbindin-positive component was much less prominent.

The rodent periolfactory basal forebrain thus seems to be a birthplace of granule cells destined for the subpial compartment of the cortical MZ (Fig. 6). However, more recent data suggest that it is probably not the sole place of origin of SGL neurons. Lavdas et al. (1999) mentioned the medial ganglionic eminence as a source of early-appearing neurons in the MZ; similarly, Tamamaki et al. (1997) showed a migration from the lateral ganglionic eminence into the cortical MZ. Cells in the MZ may thus derive from multiple sources, which also include an early local origin (Alcántara et al. 1998), perhaps involving local mitoses in the preplate (Valverde et al. 1995). It is also possible that the relative contribution of each source varies across species. While the retrobulbar generation area seems to be extremely developed in the human brain, it may be less significant for rodent cortical development. Our data suggest that there may be differences even between rat and mouse, with the contribution of the retrobulbar ventricle to the SGL being more substantial in the rat than in the mouse. Further research is certainly needed to dissect the different possible sources of SGL neurons, and to establish their relative relevance in different species.

4
Radial and Tangential Migration Pathways into the Cortex

For many years, radial migration along the radial glia substrate seemed to be the only route used by the cortical neurons on their way from their birthplace in the ventricular zone to their destination in their respective layer in the cortex (Rakic 1972). Only recently has it become evident that tangential migration is another, rather common way of translocation into the cortical plate (O'Rourke et al. 1992; De Diego et al. 1994; Tan et al. 1998). In addition, it is now known that a substantial proportion of cortical interneurons originate from outside the neocortical neuroepithelium, namely in the ganglionic eminences, from where they follow a nonradial migration route through the intermediate zone into the cortex (Anderson et al. 1997; Tamamaki et al. 1997). Thus, the proposed tangential pathway from the retrobulbar ventricular compartment into the neocortical MZ no longer appears exceptional.

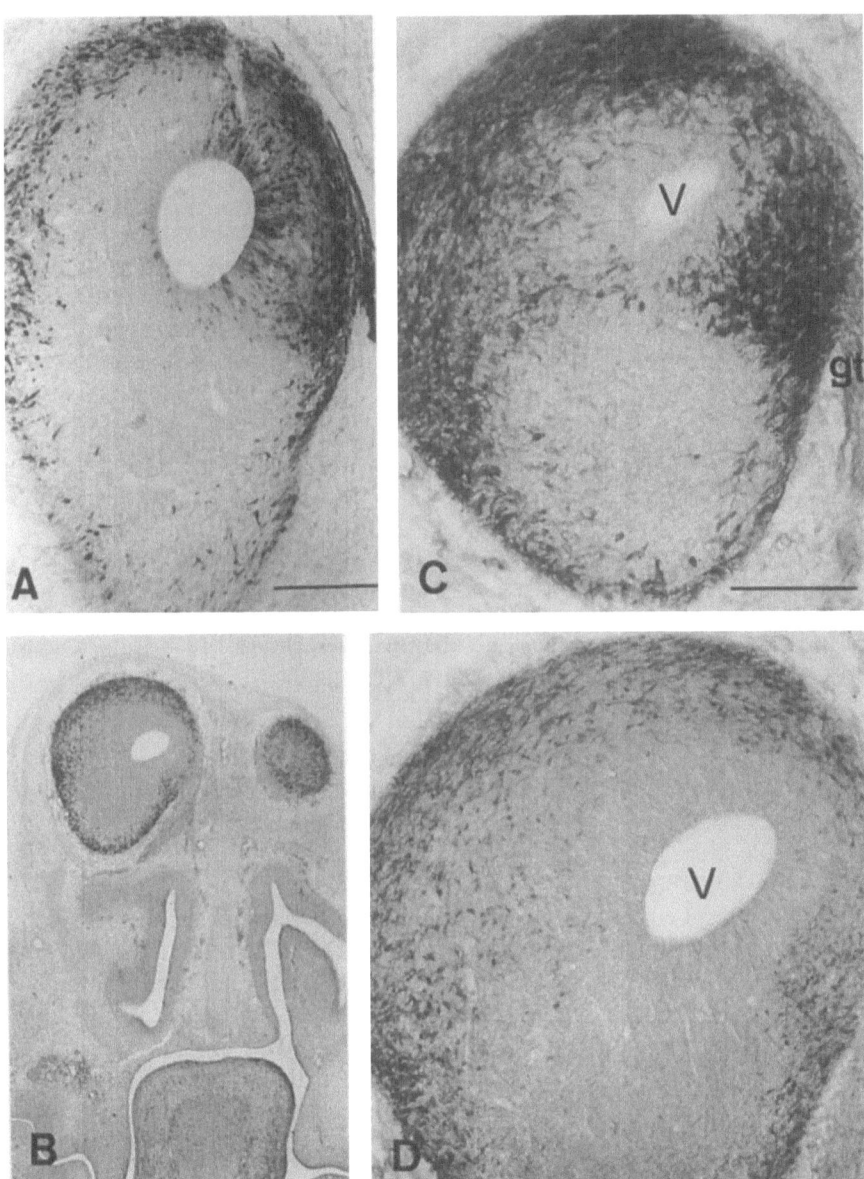

Fig. 5A–D. Proposed origin of the SGL from the retrobulbar area. A Rat retrobulbar area at E14, immunostained with calbindin. B–D: the mouse retrobulbar area at E13, immunostained with calretinin (B,C) and reelin (D). The low-magnification microphotograph in B shows the common plane of section. Medial sector of retrobulbar area contains large numbers of small cells that begin to express calbindin (in rat) and calretinin (in mouse) already in the ventricular zone, before entering the marginal zone which is particularly wide in this part of the brain. This medial sector (in A,C and D on right) is contacted by ganglion terminale (*gt* in C). *V* Rostral part of lateral ventricle. *Bar* in A: 140 µm; in C, for C,B and D: 120 µm

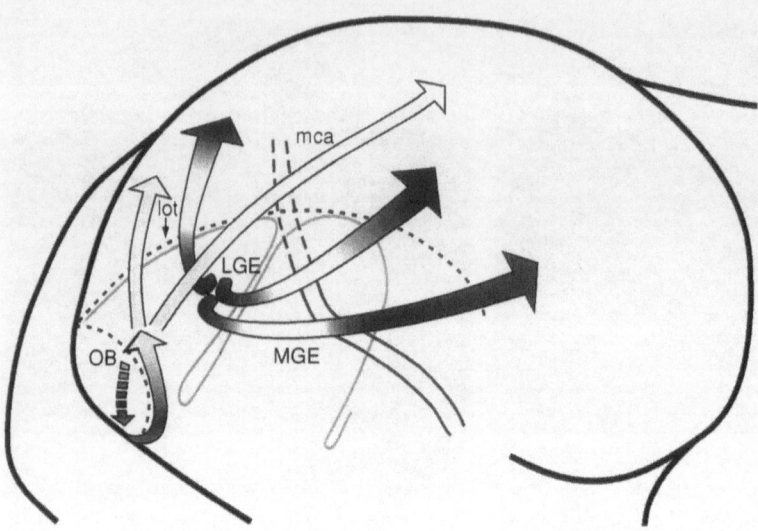

Fig. 6. Proposed migration pathway of SGL from periolfactory basal forebrain into neocortical marginal zone. This pathway may be paralleled by another migration stream originating in the lateral ganglionic eminence (LGE). The lateral olfactory tract (*lot*) and middle cerebral artery (*mca*) have been represented as topographical landmarks. *OB* olfactory bulb; *MGE* medial ganglionic eminence

Furthermore, as radial migration is arrested at the interface between cortical plate (CP) and marginal zone (Anton et al. 1996), tangential migration would provide an alternative and plausible mechanism for neurons to enter the MZ after the condensation of the CP.

On the other hand, the MZ may even constitute a migration route for neurons without a direct relationship with the development of the cortex, the emission of cortical pioneer projections, or the control of lamination via reelin. Recent results in the embryonic mouse (T. Hirata, pers. comm.) indicate that the proposed "guidepost cells" for mitral cell axons in the lateral olfactory tract, the so-called lot cells (Sato et al. 1998), migrate tangentially through the MZ from dorsal to ventral, to reach their final position around this fiber tract. The SGL may thus comprise more varieties of cell populations than previously believed, which may take unusual migration routes, and fulfill hitherto unsuspected functions.

5
Conclusions

We have described a possible rodent homologue of a transient neuronal layer, the SGL, hitherto considered specific to the human cortex. The SGL may fulfill important functions, which are necessary for building the complex architecture of the developing cortex. So far, we have been able to link certain cell populations within the SGL to the secretion of reelin, a gene product that is crucial for the establishment of the inside-out gradient of cortical lamination; many other relevant proteins may be expressed by cells in the SGL. The possibility that a homologous structure of the human SGL exists in the rodent opens a new way to explore experimentally this enigmatic cell layer, and to develop animal models for human pathologies related to migration disorders.

Acknowledgements. We thank Dr. André Goffinet for the gift of his anti-reelin-antibodies. G.M. was supported by grant DGICYT PB94-0582, and A.F. by grant DGICYT PB94-0219-CO2-01, Spain. R.C. also received support from the Unidad de Investigación, Residencia de la Candelaria, Santa Cruz de Tenerife.

References

Alcantara S, Ruiz M, D'Arcangelo G, Ezan F, De Lecea L, Curran T, Sotelo C, Soriano E (1998) Regional and cellular patterns of reelin mRNA expression in the forebrain of the developing and adult mouse. J Neurosci 18 : 7779–7799

Anderson SA, Eisenstat DD, Shi L, Rubenstein JLR (1997) Interneuron migration from basal forebrain to neocortex: Dependence on Dlx genes. Science 278 : 474–476

Anton ES, Cameron RS, Rakic P (1996) Role of neuron glial junctional domain proteins in the maintenance and termination of neuronal migration across the embryonic cerebral wall. J Neurosci 16 : 2283–2293

Bayer SA, Altman J (1991) Neocortical development. Raven Press, New York

Brun A (1965) The subpial granular layer of the foetal cerebral cortex in man. Its ontogeny and significance in congenital cortical malformations. Acta Pathol Microbiol Scand 179 (Suppl) : 1–98

Brunstrom JE, Gray-Swain MR, Osborne PA, Pearlman AL (1997) Neuronal heterotopias in the developing cerebral cortex produced by neurotrophin-4. Neuron 18 : 505–517

Cobas A, Fairén A, Alvarez-Bolado G, Sánchez MP (1991) Prenatal development of the intrinsic neurons of the rat neocortex: a comparative study of the distribution of GABA-immuno-reactive cells and the GABA$_A$ receptor. Neuroscience 40 : 375–397

D'Arcangelo G, Miao GG, Chen SC, Soares HD, Morgan JI, Curran T (1995) A protein related to extracellular matrix proteins deleted in the mouse mutant *reeler*. Nature 374 : 719–723

D'Arcangelo G, Nakajima K, Miyata T, Ogawa M, Mikoshiba K, Curran T (1997) Reelin is a secreted glycoprotein recognized by the CR-50 monoclonal antibody. J Neurosci 17 : 23–31

de Bergeyck V, Naerhuyzen B, Goffinet AM, Lambert de Rouvroit C (1998) A panel of mono-clonal antibodies against Reelin, the extracellular matrix protein defective in reeler mutant mice. J Neurosci Meth 82 : 17–24

De Carlos JA, O'Leary DDM (1992) Growth and targeting of subplate axons and establishment of major cortical pathways. J Neurosci 12 : 1194–1211

De Diego I, Smith-Fernández A, Fairén A (1994) Cortical cells that migrate beyond area boundaries: characterization of an early neuronal population in the lower intermediate zone of prenatal rats. Eur J Neurosci 6 : 983–997

Del Rio JA, Soriano E, Ferrer I (1992) Development of GABA-immunoreactivity in the neo-cortex of the mouse. J Comp Neurol 326 : 501–526

Del Rio JA, Martínez A, Fonseca M, Auladell C, Soriano E (1995) Glutamate-like immunore-activity and fate of Cajal-Retzius cells in the murine cortex as identified with calretinin antibody. Cerebr Cortex 5:13-21

Derer P, Derer M (1990) Cajal-Retzius cell ontogenesis and death in mouse brain visualized with horseradish peroxidase and electron microscopy. Neurosci 36:839-856

Gadisseux JF, Goffinet AM, Lyon G, Evrard P (1992) The human transient subpial granular layer: An optical, immunohistochemical, and ultrastructural analysis. J Comp Neurol 324:94-114

Ikeda Y, Terashima T (1997) Expression of reelin, the gene responsible for the reeler mutation, in embryonic development and adulthood in the mouse. Dev Dyn 210:157-172

Lambert de Rouvroit C, Goffinet AM (1998) The reeler mouse as a model of brain development. Adv Anat Embryol Cell Biol 150:1-108

Lavdas AA, Grigoriou M, Pachnis V, Parnavelas JG (1998) The medial ganglionic eminence gives rise to a population of early neurons in the developing cerebral cortex. J Neurosci 19:7881-7888

Marin-Padilla M (1972) Prenatal ontogenetic history of the principal neurons of the neocortex of the cat (Felis domestica). A Golgi study. II. Developmental differences and their signifi-cances. Z Anat Enwicklungesch 136:125-142

Marin-Padilla M (1998) Cajal-Retzius cells and the development of the neocortex. Trends Neurosci 21:64-71

Meyer G, Gonzalez-Hernandez T (1993) Developmental changes in layer I of the human neo-cortex during prenatal life: A DiI-tracing and AChE and NADPH-d histochemistry study. J Comp Neurol 338:317-336

Meyer G, Goffinet AM (1998) Prenatal development of Reelin-immunoreactive neurons in the human neocortex. J Comp Neurol 397:29-40

Meyer G, Goffinet AM, Fairén A (1999) What is a Cajal-Retzius cell? A reassessment of a classical cell type based on recent observations in the developing neocortical marginal zone. Cerebr Cortex 9:765-775

Meyer G, Soria JM, Martínez-Galan JR, Martín-Clemente B, Fairén A (1998) Different origins and developmental histories of transient neurons in the marginal zone of the fetal and neonatal rat cortex. J Comp Neurol 397:493-518

Ogawa M, Miyata T, Nakajima K, Yagyu K, Seike M, Ikenaka K, Yamamoto H, Mikoshiba K (1995) The reeler gene-associated antigen on Cajal-Retzius neurons is a crucial molecule for laminar organization of cortical neurons. Neuron 14:899-912

O'Rourke NA, Dailey ME, Smith SJ, McConnell SK (1992) Diverse migratory pathways in the developing cerebral cortex. Science 258:299-302

Pesold C, Impagnatiello F, Pisu MG, Uzunov DP, Costa E, Guidotti A, Caruncho HJ (1998) Reelin is preferentially expressed in neurons synthesizing γ-aminobutyric acid in cortex and hippocampus of adult rats. Proc Natl Acad Sci USA 95:3221-3226

Raedler E, Raedler A (1978) Autoradiographic study of early ontogenesis in rat neocortex. Anat Embryol 154:267-284

Rakic P (1972) Mode of cell migration to the superficial layers of fetal monkey neocortex. J Comp Neurol 145:61-84

Ranke G (1910) Beiträge zur Kenntnis der normalen und pathologischen Hirnrindenbildung. Zieglers Beitr 47:51-125

Sato Y, Hirata T, Ogawa M, Fujisawa H (1998) Requirement for early-generated neuron rec-ognized by monoclonal antibody Lot 1 in the formation of lateral olfactory tract. J Neurosci 18:7800-7810

Schaffer K (1918) Über normale und pathologische Hirnfurchung. Z Ges Neurol Psychiatr 38:1-77

Schiffmann SN, Bernier B, Goffinet AM (1997) Reelin mRNA expression during mouse brain development. Eur J Neurosci 9:1055-1071

Soria JM, Fairén A (1999) Specific tangential distribution of pioneer neurons in the rat marginal zone define early cortical territories. Cerebr Cortex (submitted)

Tamamaki N, Fujimori KE, Takauji (1997) Origin and route of tangentially migrating neurons in the developing neocortical intermediate zone. J Neurosci 17:8313–8323

Tan SS, Kalloniatis M, Sturm K, Tam PPL, Reese BE, Faulkner-Jones B (1998) Separate progenitors for radial and tangential cell dispersion during development of the cerebral neocortex. Neuron 21:295–304

Valverde F, De Carlos JA, López-Mascaraque L (1995) Time of origin and early fate of preplate cells in the cerebral cortex of the rat. Cereb Cortex 5:412–422

Van Eden CG, Mrzljak L, Voorn P, Uylings HBM (1989) Prenatal development of GABA-ergic neurons in the neocortex of the rat. J Comp Neurol 289:213–227

Development of Thalamocortical Projections in Normal and Mutant Mice

Zoltán Molnár[1,2] and Anthony J. Hannan[2]

1
Introduction

In vitro and in vivo studies have begun to reveal multiple mechanisms involved in the development of thalamocortical projections. Different cellular and molecular interactions dominate during the various stages of axon elongation, target selection, and establishment of layer- and area-specific connections. Recent work points to the crucial role of the early-developing thalamocortical projections and their interactions with the developing cortical circuitry in establishing the functional and structural organization of the cortex. In this chapter we give an overview of mutant mice, including knockout (KO) mice which have homozygous null gene mutations, which are providing interesting paradigms for distinct phases of thalamocortical development. We present selected mutants in two major groups. The first group involves mutations that directly influence some key mechanism of axon elongation and delivery (*reeler* mutant, L1 KO, TBR-1 KO, Pax-6 KO) or thalamocortical interactions (Adcy1 *Barrelless*, MAO-A KO, NMDA receptor mutant, PLC-β1 KO, GAP-43 KO). The second group (albino, anophthalmic mice, mice with extra vibrissae) includes mutations that do not affect thalamocortical outgrowth or the interactions between thalamic projections and cortical circuitry itself, but instead alter the sensory flow of information from the periphery and influence thalamocortical development. The number of available mutants is increasing rapidly and, combined with recent technology allowing gene expression to be experimentally manipulated in precise spatiotemporal patterns, will provide further understanding of the development and plasticity of thalamocortical connections.

The development of thalamocortical connectivity involves the sequential processes of axon elongation, fasciculation, transient side-branch formation, target selection, layer-specific cortical innervation and thalamic fibre inter-

---

[1] Institut de Biologie Cellulaire et de Morphologie, Rue du Bugnon 9, 1005 Lausanne, Switzerland and Department of Human Anatomy and Genetics, South Parks Road, Oxford, OX1 3QX, UK
[2] University Laboratory of Physiology, University of Oxford, Parks Road, Oxford OX1 3PT, UK

Results and Problems in Cell Differentiation, Vol. 30
Goffinet and Rakic (Eds.): Mouse Brain Development
© Springer-Verlag Berlin Heidelberg 2000

actions with cortical cells to form the mature area-specific cortical circuitry. The study of these developmental steps has involved approaches such as tracing developing connections, identifying and relating gene expression patterns and testing the causal relationships using in vitro organotypic or dissociated culture systems. With the development of functional genetics (see Tear 1999) the repertoire of techniques in developmental neurobiology has increased dramatically and current ideas can be readily tested in specific mouse mutants which have been derived from either spontaneous mutations or specific transgenic modifications. In this chapter we will attempt to integrate current ideas regarding the multiple mechanisms involved in thalamocortical development in rodents and present some examples of mutant and transgenic mice which provide excellent model systems for current and future research.

2
Neurogenesis and Formation of Mammalian Cortical Plate

The basic principles of cortical neurogenesis are very similar in all mammals. In the mouse, neurons are 'born' in the ventricular and subventricular zones between embryonic day 11 (E11) and E20. The first wave of postmitotic neurons migrate to the cortical surface to form the primordial plexiform zone (the preplate). The subsequently formed neurons migrate to the middle of the preplate, splitting it into the marginal zone at the top and the subplate at the bottom (nomenclature: Boulder Committee 1970; Marin-Padilla 1971; Luskin and Shatz 1985a,b; Valverde et al. 1995a,b). The cortical plate proper, sandwiched between the two components of the preplate (Luskin and Shatz 1985a), steadily thickens as subsequent waves of arriving neurons migrate through the existing cells of the plate and take up their position under the marginal zone in an inside-first, outside-last fashion (Angevine and Sidman 1961; Berry and Rogers 1965; Lund and Mustari 1977; Rakic 1977; Luskin and Shatz 1985a).

In the embryonic cortex of mammals the subplate forms a substantial layer (Kostovic and Rakic 1980; Luskin and Shatz 1985a). In the adult however, layer 1 and the white matter under the cortex contain few neurons (Cajal 1909, 1911). In the rat, thymidine birthdating at E12 labels only cells of the subplate and layer 1 (Lund and Mustari 1977; Valverde et al. 1989, 1995a, b; Bayer and Altman 1990, 1991). The first cells of the rat cortical plate itself are born on E14 and the last on E21. In mouse similar studies have demonstrated the early birth of the subplate and some of the marginal zone cells and their transient nature (Derer and Derer 1990). Thus cortical cell generation finishes before birth and the subventricular zone disappears by postnatal day 3 (P3), however, cell migration into the upper layers continues until the end of the first postnatal week (Berry and Rogers 1965; Lund and Mustari 1977; Miller

1988). The clonal relationship and areal dispersion of cortical neurons is beginning to be elucidated using retrovirus-mediated gene transfer experiments (Walsh and Cepko 1993; Mione et al. 1994; Götz et al. 1995) and transgenic techniques (McConnell 1995; Rakic 1995; Soriano et al. 1995; Tan et al. 1995; Levitt et al. 1997).

In the tangential plane of the developing cerebral cortex, there is another distinct neurogenetic gradient, with the anteroventral areas preceding the caudodorsal in maturity (Berry and Rogers 1965; Lund and Mustari 1977). Corticogenesis begins first in the ventromedial region adjacent to the paleocortex and archicortex, and there is a simple orderly wave of maturation, from rostral and ventral to caudal and medial.

3
Introduction to Development of Thalamic Nuclei

Two phases of thalamic growth have been distinguished based on the examination of Nissl-stained specimens: (1) an early phase of cellular differentiation during which the epithalamus, dorsal thalamus and ventral thalamus differentiate from one another, followed by (2) a later phase during which the individual nuclei differentiate within the three larger subdivisions.

In the first phase, the rat thalamus and hypothalamus can be distinguished after the appearance of a ventricular groove (anterior diencephalic wall) at E12 (Coggeshall 1964). At E13, a wedge-shaped enlargement appears in the diencephalic wall with further furrows defining the developing dorsal and ventral thalamus. These furrows become less apparent as the thalamus further increases in size. In the second phase, at E16–E17 in the rat, differentiation of nuclei commences in the epithalamus and ventral thalamus but not in the dorsal thalamus (Coggeshall 1964; Jones 1985). Within the thalamus nuclear differentiation starts posteriorly and laterally, and proceeds anteriorly and medially (Rose 1942a,b; Coggeshall 1964; Jones 1985). In view of the growing evidence that there are transient cell populations in the developing thalamus itself (Mitrofanis 1992), it is important to re-examine the development of the thalamus in relation to the birthdates of its cells in embryonic and early postnatal material rather than drawing conclusions from the mature brain.

Autoradiographic studies show that cellular proliferation precedes differentiation of the thalamic nuclei by several days. In the rat, virtually all thalamic neurons are born between E13 and E19 (McAllister and Das 1977; Altman and Bayer 1979), coinciding with the generation of preplate cells of the occipital cortex. The postmitotic neurons migrating outwards from the ventricular and subventricular zones come to rest in regions that are determined by their time of birth, and not by nuclear boundaries. The lateral geniculate nucleus (LGN) is generated over 2 days between E12 and E14 (Brückner et al. 1976; Lund and Mustari 1977). Jones (1985) describes three gradients of neurogenesis in the developing thalamus: (1) *posterior to anterior*, in that the cells of the posterior nuclei are born before those of anterior nuclei (referring to the relative position of nuclei in the adult thalamus); (2) *lateral to medial*, with lateral nuclei born before the medial ones (e.g. neurons in the reticular nucleus are born before those of the mediodorsal nucleus); (3) *ventral to dorsal* (Jones 1985). It has not yet been established whether these gradients exist within individual nuclei as well as across the thalamus as a whole (Altman and Bayer 1979; Hickey 1980; Shatz 1981, 1983; Hickey and Hitchcock 1984). It is also important to know how these gradients are distributed across the *embryonic* thalamus, since there are considerable changes in the orientation of the thalamus as a whole during development in some species (Rakic 1977; Jones 1985).

4
Overview of Thalamocortical Projections in the Adult Mouse

The thalamus is a relay nucleus through which most of the sensory information reaches the cerebral cortex (Jones 1985). In adult mouse, every dorsal thalamic nucleus projects to the cerebral cortex and all cortical areas receive projections from the dorsal thalamus (Caviness and Frost 1980; Jones 1983, 1985; Caviness 1988). The literature distinguishes two types of projections, specific and nonspecific (Jones 1985). The specific projections terminate within the borders of one or a few cortical fields, whilst the nonspecific or diffuse projections project over a wide area of the cortex. The same dorsal thalamic nucleus can establish both types of projections. A single thalamic nucleus can project to more than one cortical area and each cortical field can receive several thalamic inputs (Jones 1985). The topography of projections from a single nucleus can also change abruptly at the borders of neighbouring cortical areas (such as the reversal of the maps between primary and secondary visual areas). The complex adult organization suggests that various mechanisms are involved during development.

5
Timing and Early Pattern of Thalamic Axon Outgrowth

Until recently, the only information about the development of thalamocortical connectivity came from studies using degeneration techniques (e.g. Lund and Mustari 1977) or transneuronal transport of label from the eye (Rakic 1976). In recent years, however, the use of fluorescent carbocyanine dyes (Honig and Hume 1986) has revolutionised the tracing of embryonic pathways, since these dyes can be employed in fixed tissue (Godement et al.

1987). Although some aspects of the development of thalamocortical projections were addressed more than two decades ago (e.g. Lund and Mustari
1977; Wise and Jones 1978), the greater versatility and sensitivity of the new
techniques promised a greater depth of understanding.

In both rat and mouse, thalamocortical development follows the same
basic pattern, although in mouse development is more advanced at equivalent gestational stages (Molnár and Blakemore 1995a; Fig. 1). In mouse,
implantation of small carbocyanine dye (DiI) crystals into the dorsolateral
part of the thalamus reveals thalamic fibres reaching the internal capsule in
an organised fashion at E13.5. The thalamic fibres run through the primitive
internal capsule and under the developing corpus striatum, but do not reach
the intermediate zone of the most ventral cortical areas until 15.5 days of
gestation. In addition to staining thalamocortical fibres, DiI implantation in
the thalamus labelled numerous cells within the thalamic reticular nucleus
and a few within the internal capsule (Molnár et al. 1998a,b), in the so-called
perireticular nucleus (Mitrofanis 1992, 1994).

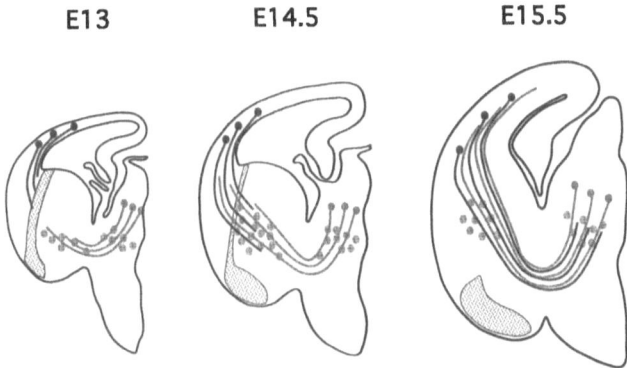

Fig. 1. Early development of reciprocal thalamocortical and corticofugal projections in embryonic mouse brains. Each schematic drawing represents an imaginary paracoronal section
through the right hemisphere, revealing the entire thalamocortical pathway. E13 Synchronous
outgrowth of the early thalamic and corticofugal fibres. *Grey dots* Cells that possess thalamic
projections in ventral thalamus and internal capsule. *Black dots* Thin stripe of transient cells
with *Pax 6* positive cells (Stoykova et al. 1997; Fernandez et al. 1998) extending between the
ventricular zone near the striato-cortical junction. The growing corticofugal axons pause at this
stripe of cells without sustantially penetrating them at this early stage. The thalamic axons
advance amongst cells of the internal capsule which have already developed dorsal thalamic
projections. E14.5 First encounter of thalamocortical and earliest corticofugal (preplate) projections outside the primitive internal capsule: the 'handshake'. Fibres from corresponding
regions of cortex and thalamus become closely associated with one another. Stripe of *Pax*-
6-positive cells becomes much thinner and an island appears around the future location of
amygdala. E15.5 Thalamic fibres establish a topographic map while 'waiting' in the subplate.
Corticospinal projections from layer 5 begin to deviate from corticothalamic projections and
form the cerebral peduncle (not shown). Corticofugal projections reach the ventral thalamus,
but do not yet enter the dorsal thalamus. The thin stripe of intermediate compartment has
disappeared, but *Pax-6* activity continues to be present at amygdala

Fibres from the mouse dorsolateral thalamus reach the intermediate zone beneath the ventral cortex at around E15, and the occipital cortex itself at E15.5, a day earlier than in rat. As they leave the diencephalon, the thalamocortical fibres form distinct fascicles, which open up in a fan-shaped pattern. Viewed in coronal sections the ordering of the fibres is such that the inferior lateral fascicles are destined for the ventral part of the cortex, while the superior medial fascicles turn upwards and head towards more dorsal cortical areas. The trajectories of the fascicles remain parallel as they turn up into the intermediate zone: they do not cross each other extensively, although individual axons and small axon bundles do appear to switch from one fascicle to another along their course.

At the border of the corpus striatum and the intermediate zone, the thalamic fibres defasciculate and form a fairly uniform array of individual fibres. Within the intermediate zone the fibres run parallel to each other in an organised fashion. Similarly to the fascicles, the ordering of the individual fibres is such that the inferior lateral fibres are destined for the more ventral cortical segment, while the more medial fibres, situated deeper in the white matter, head towards more dorsal cortical segments. This behaviour of the thalamic axons is highly conserved in all mammalian species including humans and might reflect different extracellular environments along the pathway. Although thalamic fibres from the region of the LGN reach the occipital cortical regions around E15–E16, they do not substantially invade the cortical plate for some time. Most of the labelled fibres remain restricted to the intermediate and subplate zones until E18. During this period thalamic fibres accumulate and develop local branches, but these branches are also mainly restricted to the subplate region.

5.1
The Waiting Period

When LGN fibres reach the occipital cortex (E16 in the rat and E15 in the mouse), only the lowest part of the true cortical plate is in place and the cortical plate is not yet permissive to axonal ingrowth. Moreover, most of the main target cells of the thalamic axons, the cells of layer 4, do not take up their position in the plate until after birth (Lund and Mustari 1977). Wise and Jones (1976) first used the term 'the waiting period' to describe the behaviour of commissural axons in the rat, which accumulate for a protracted period in the intermediate zone before invading the cortical plate. A similar phenomenon has been reported for thalamocortical axons in the monkey (Rakic 1976, 1977), rat (Lund and Mustari 1977) and cat (Shatz and Luskin 1986). Thalamic axons arrive below the correct regions of cortex before the cortical plate has fully formed and they accumulate and 'wait' over the subplate layer (for several weeks in cat and monkey) before invading the plate proper.

Catalano et al. (1991, 1996) and Kageyama and Robertson (1993) have questioned the existence of a waiting period in the somatosensory and visual cortex of the rat. Using DiI application in the dorsal thalamus, they demonstrated that some thalamocortical afferents invade the deep part of the cortical plate of the somatosensory and visual cortex on E18/E19. They suggest that thalamic axons simply move up through the cortical plate progressively, as the layers differentiate, always extending up to the lower edge of the dense cortical plate of newly arrived, immature neurons. However, the data of others (Erzurumlu and Jhaveri 1990; Molnár and Blakemore 1990a,b; Molnár et al. 1998a,b) indicate that, although thalamic fibres indeed invade the rodent cortex before birth, there is a waiting period of only a few days, not as long as was originally suggested by Lund and Mustari (1977). In mouse occipital cortex, thalamic fibres arrive on E15, but do not invade the cortex significantly until around E18 (Molnár et al. 1998b), similarly to the pattern observed in rat (Molnár et al. 1998a). In marsupials however, the pattern of thalamocortical development is different: the front of thalamic axons moves up through the cortical plate progressively as the layers differentiate, always extending up to the lower edge of the dense cortical plate of newly arrived, immature neurons (Molnár et al. 1995a, 1998c, Fig. 2).

5.2
Invasion of the Cortex and Establishment
of Laminar Termination Patterns

The migration of cortical plate neurons is at its peak at the beginning of the 'waiting period' for thalamic axons. The migrating cells therefore have to pass through the mass of accumulating fibres waiting in the subplate layer, as well as between the subplate cells themselves. Before E18 in mouse very few thalamic axons appear to have grown into the occipital cortical plate itself. Invasion suddenly begins around late E18, just 2–3 days before birth. The course taken by these invading thalamic axons is distinct and highly ordered: they make a 90° turn from their trajectory in the white matter and the vast majority of the fibres initially grow radially into the cortex directly above the region of the subplate in which they have been waiting (see Agmon et al. 1993). Only a small portion take more erratic routes, running somewhat obliquely upwards.

By birth a substantial number of thalamic fibres have invaded the cortical plate. Within the cortical plate the fibres follow a much less regular pattern of ascent than was apparent in the white matter. A few fibres reach the marginal zone, but the vast majority stop below the uppermost sector of the cortex, the dense cortical plate. Individual axons at P2 show arborisations typically extending over an area with a diameter of about 300 μm (Jhaveri et al. 1991; Agmon et al. 1993; Catalano et al. 1996; Molnár et al. 1998a). The extensive lateral spread of the terminal arbors of thalamic axons contrasts with the

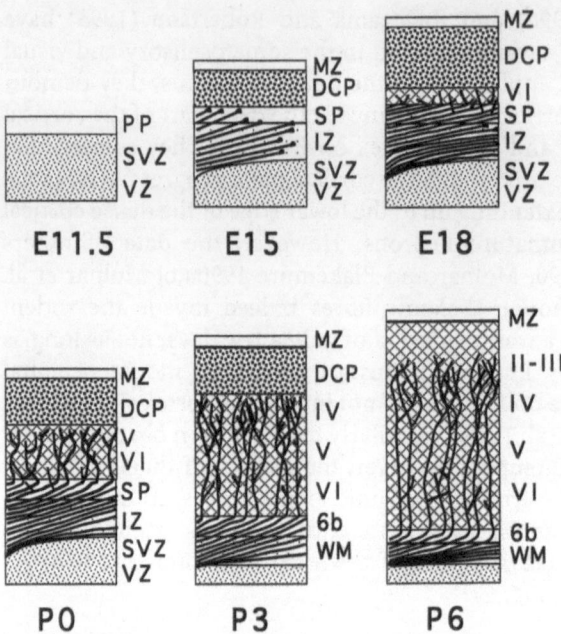

Fig. 2. Maturation of cortical lamination and ingrowth of thalamic fibres in mouse, summarised on schematic coronal sections of left somatosensory cortex [E11.5, E15, E18, at birth, P0 (i.e. E21–E22), P3 and P6]. At E11.5 the first postmitotic cells migrate to outer edge of cerebral wall to form preplate (*PP*), which is subsequently split into marginal zone (*MZ*, the future layer I) and subplate (*SP*) by arrival of neurons of the true cortical plate. When thalamocortical fibres arrive (at E15), there is a distinct subplate and cortical plate consists almost entirely of tightlypacked, immature neurons (the dense cortical plate, *DCP*). Thalamic fibres accumulate within the subplate between about E15 and E18, although some axons and side-branches penetrate the lowest part of cortical plate. At P0, there is no significant axonal invasion of DCP. Even at P3, most thalamic fibres stop below the still-migrating cells. Thalamic axons begin to cluster and start to form whisker-related patterns. Thalamic axons will occupy the inside of the barrels. *IZ* Intermediate zone; *VZ* ventricular zone; *SVZ* subventricular zone, *II–VI* layers 2–6 respectively; *6b* sub-layer of layer 6; *WM* white matter. (Modified after Molnár et al. 1998c)

remarkable order with which they are arranged at the junction between white and grey matter. In favourable preparations whole arrays of individual axons can be seen entering the cortex, and their strict parallel arrangement is most impressive. The fibres heading for ventral areas run superficial to fibres destined for more dorsal cortical segments. Since the earlier fibres enter the earlier developing ventral cortical areas first, this type of fibre ordering reflects the timing of its establishment. It appears that the topographic array of arriving fibres may be maintained by distinct molecular signals, as the fibres invade the cortex, to transform the array into a precise topographic pattern of axons entering the cortical sheet.

Some individual thalamic axons can be seen to branch into layer 6a and within the lower third of the dense cortical plate, but most of the terminal arbors are in the putative layer 4. By about P2, most axons have bifurcated at least once at around the top of layer 5, have arborised directly below the dense cortical plate, in what is presumably destined to become layer 4, and appear to have lost their growth cones. A few fibres grow all the way up through the dense cortical plate to the marginal zone, where they branch and appear to terminate within a fairly narrow laminar range. Between P2 and about P8 (when the cortex has achieved its mature lamination) migrating neurons continue to arrive above the arborisations of thalamic afferents in putative layer 4 and the supragranular layers form. These experiments demonstrate clearly that thalamic fibres arrive in the subplate layer and, at least in the occipital cortex, wait under the cortical plate in a topographic pattern. This is not to say that this topography cannot and will not change during the periods of thalamocortical fibre accumulation below the cortex and invasion of the cortical plate (see Molnár 1998; Molnár et al. 1999a).

6
The Thalamocortical Pathways are Modified in Regions where Transient Cells are Located During Development

Thalamocortical connectivity, examined in adult brains with anterograde and retrograde tracing techniques, shows substantial rearrangement in three specific regions of the pathway: (1) the thalamic reticular nucleus, (2) the internal capsule and (3) the white matter below the cortex (see Nelson and Le Vay 1985; Bernardo and Woolsey 1987; Mitrofanis and Guillery 1993). These regions coincide with zones where first-generated cells in the forebrain are distributed: in the thalamic reticular nucleus, internal capsule and subplate (Luskin and Shatz 1985a,b; Bayer and Altman 1991; Mitrofanis 1992). These cells are much more apparent during development than in the adult and it was suggested that some of them disappear by programmed cell death during development. The rearrangements observed in these regions might have occurred as a result of interactions between the developing projections and the transient neurons. A number of studies (see McConnell et al. 1989; Blakemore and Molnár 1990; Shatz et al. 1990; De Carlos and O'Leary 1992; Erzurumlu and Jhaveri 1992; Molnár and Blakemore 1995a,b) have suggested that the initial topography of the thalamocortical projections is established with the assistance of these cells.

6.1
The Handshake Hypothesis

The handshake hypothesis (Blakemore and Molnár 1990; Molnár and Blakemore 1995a,b) proposes that axons from the thalamus and from the

early-born cortical preplate cells meet and intermingle in the basal telencephalon, whereafter thalamic axons grow over the scaffold of preplate axons
and become "captured" for the waiting period in the subplate layer of the
corresponding part of the cortex. To test this idea numerous workers have
attempted to label the thalamocortical and corticothalamic axons with different dyes from corresponding regions of the thalamus and cortex at the
same time.

Bicknese et al. (1994a) and Miller et al. (1993), in the rat and hamster
respectively, tried to discover whether ascending thalamic axons grow along
the same routes as those traversed by the groups of subplate axons from
corresponding cortical areas. Their double-labelling experiments, however,
were performed well after the arrival of the thalamic fibres and after the
establishment of layer 6 and 5 connections. They concluded that the labelled
axon bundles from the cortex and from the thalamus run in separate fascicles, with the afferent thalamocortical axons travelling more superficially, in
the subplate, and the efferent corticothalamic axons deeper in the intermediate zone. This is very similar to the adult pattern seen in the rat (Woodward
and Coull 1984) and cat (Nelson and LeVay 1985). Unfortunately, the location
of the subplate axons in relation to the afferent and efferent pathway was
not determined in these studies. In earlier stages, however, similar double-
labelling experiments of Erzurumlu and Jhavery (1992) and Bicknese and
Pearlman (1992) revealed that the growing thalamocortical and earliest corticofugal axons were intermingled within the internal capsule as implied by
the single back-labelling study of Blakemore and Molnár (1990) and Molnár
et al. (1998a,b). The spatially distinct pathways of corticofugal and corticopetal pathways in the adult does not exclude the possibility that the earliest
thalamic and subplate axons do selectively fasciculate on each other during
embryonic development. Three-dimensional confocal microscopic reconstructions with identified subplate axons at early stages confirmed this suggestion (Molnár et al. 1998a,b).

7
Introduction to the Development of Barrel Cortex

Thalamic fibres arborise in layer 4, and in primary somatosensory cortex they
begin to impose a periphery-related cytoarchitectonic pattern. A mature
barrel is composed of a mass of thalamic axons in layer 4, carrying signals
from a single facial vibrissa, surrounded by a 'wall' consisting of neuronal cell
bodies whose dendrites project into the barrel lumen (Woolsey and Van der
Loos 1970; Woolsey et al. 1975; Steffen 1976). Somatosensory thalamic afferents invade the somatosensory cortex shortly before birth in rodents and
at this stage no barrel structures are evident. Thalamocortical axons begin to
assume a barrel-like pattern on the first postnatal day (P0) and the post-

synaptic features of barrels appear gradually over the following days (Woolsey and Van der Loos 1970; Wise and Jones 1978; Killackey and Belford 1979; Jhavery et al. 1991; Agmon et al. 1993; O'Leary et al. 1994). The spatial topography of this barrel development ensures that the majority of cells from each barrel respond exclusively to stimulation of their corresponding vibrissae. The barrel cortex of rodents has proved to be an extremely useful model for the investigation of activity-dependent brain development and plasticity at the molecular and cellular level.

The postnatal formation of the gross barrel pattern, assessed using markers of thalamocortical afferentation and metabolic activity, does not appear to require neural activity (Chiaia et al. 1992; Henderson et al. 1992; Agmon et al. 1993; Schlaggar et al. 1993; Schlaggar and O'Leary 1994; reviewed by O'Leary et al. 1994), although glutamate receptor blockade causes exuberant thalamic axon distribution across barrels and enlargement of the receptive fields of cells in the barrel walls (Fox et al. 1996). The pattern of barrels can be influenced by alteration in activity, and blockade of glutamate neurotransmission reduces cortical barrel plasticity in response to whisker removal (Schlagger et al. 1993; Fox et al. 1996). The importance of neural activity in the formation of cytoarchitectonic features of cortical barrels has been suggested by recent analysis of specific transgenic mice, presented below.

The plasticity exhibited by the barrel cortex during a brief sensitive period of early postnatal development presents many outstanding questions regarding mechanisms of barrel cortex development. How do the thalamic afferents change the patterning of cortical cells in the barrel field in the primary somatosensory cortex of the mouse? Do dendritic modifications in layer 4 lead to cell migration during the formation of cytoarchitectonic barrels? What are the molecular and cellular mechanisms of the changes in the early whisker-related patterns which occur shortly after a modification at the periphery?

7.1
Mutant Mice Provide New Insights into Developmental Mechanisms

Molecular and cellular mechanisms of cortical development have been explored using specific mouse mutants which have been derived from either spontaneous mutations or specific transgenic modifications. For the sake of clarity we shall divide the mutants into two groups. The first group consists of various mutants in which the thalamocortical projections (1) fail to extend along their path, (2) target the cortical plate in a different manner or (3) fail to develop the characteristic arborisation or develop arborisation, but fail to impose a whisker-related pattern. In each case, in spite of the normal sensory periphery, the characteristic cortical cytoarchitecture does not develop. In these mutants the mechanisms involved in normal thalamocortical targeting or interaction are disturbed primarily due to the lack of a receptor, ligand or other molecule. These problems can lead to abnormal thalamocortical arbor formation and/or stabilisation and defects in cortical pattern formation.

The second group is comprised of mutants in which the abnormalities have *indirect* effects on thalamocortical development. The initial thalamocortical development is not disturbed, but the changed peripheral input alters the thalamocortical projections at an early phase. The peripheral input might present the autonomously developed thalamocortical connectivity with information that is in conflict with intrinsic patterning. Depending on the relative dominance of the two phases, the interaction can lead to the reduction of the primary cortical sensory area or to the altering of the cortical topography (reversal, or compression of maps in albinism or monocular enucleation). These mutations could be extremely useful in understanding the interactions between the autonomous and the activity-dependent phases of thalamocortical development.

8
Axonal Pathfinding at the Cortico-Striatal Junction in *Tbr-1*, *Gbx-2* and *Pax-6* KO Mice

The boundary between the cortical intermediate zone and the ganglionic eminence appears to be critical for thalamocortical development. During early stages of forebrain development thalamocortical and corticofugal projections pause before they cross this boundary (Métin and Godement 1996; Molnár et al. 1998a; Molnár and Cordery 1999) and their subsequent interaction was proposed to occur exactly at this region of the primitive internal capsule, beyond which thalamic fibres cofasciculate with early corticofugal projections to reach the cortex (Molnár and Blakemore 1990a, 1995a,b; Molnár et al. 1998a). Thalamocortical projections show a distinct change in their fasciculation pattern at this border, with the fascicles separating into individual fibres as they enter the intermediate zone. Stoykova et al. (1997) and Fernandez et al. (1998) described a stripe of *Pax-6*-positive cells linking the intermediate neuroepithelium to populations of the *Pax-6*-positive neurons of the basal telencephalon in the mouse embryo at a corresponding stage of development (see Fig. 1). Semaphorin G, a member of cell-surface and secreted proteins, is expressed along this stripe at corresponding embryonic ages (Skaliora et al. 1998). Recently, Hevner and colleagues (1998) described mice with mutations of transcription factor genes expressed either in cortex (*Tbr-1*), dorsal thalamus (*Gbx-2*) or both (*Pax-6*), in which errors of corticothalamic and thalamocortical pathfinding occurred in the region of the internal capsule, away from regions of disturbed gene expression. Both thalamocortical and corticofugal projections failed to cross the middle of the internal capsule in these mutants. We do not understand the interactions between the developing reciprocal projections and the cells of the internal capsule and the transient stripe of *Pax-6-* and semaphorin-G-positive cells. It will be important to understand more of the cellular and molecular interactions in this area during development to be able to explain these interesting findings and to understand the logic of these early developmental steps in other vertebrates (Cordery and Molnár 1999).

8.1
The *reeler* Mouse

8.1.1
The reeler Phenotype

Reeler is an autosomal recessive mutation in the mouse that disorganises cortical development (Caviness and Sidman 1973; Caviness 1976, 1982; Caviness and Rakic 1978; Goffinet 1979, 1992, 1995). The protean effects of the mutation on cortical cell migration, architectonic and hodological development have been described in detail (reviewed by Caviness 1988). The gene responsible for this trait was cloned (D'Arcangelo et al. 1995; Bar et al. 1995; Hirotsune et al. 1995) and shown to encode a secreted glycoprotein named reelin, expressed in the embryonic cortex at the level of Cajal-Retzius cells. A *reeler*-associated antigen, CR-50, was also found to be expressed by embryonic Cajal-Retzius neurons and could be an epitope belonging to reelin (Ogawa et al. 1995). D'Arcangelo et al. (1997) recently reported that the CR-50 antibody recognised reelin in vitro.

8.1.2
The reeler Mutant Mouse as a Model System to Explore Mechanisms of Thalamocortical Development

The *reeler* mouse is a remarkable 'experiment of nature' (Caviness et al. 1988) allowing different aspects of cortical development to be studied. In *reeler*, as in normal mice, the earliest postmitotic cells migrate to the outer edge of the cerebral wall to form the typical polymorphic neurons of the primordial plexiform layer or preplate (Goffinet 1979, 1980; Goffinet and Lyon 1979; Pinto-Lord and Caviness 1979; Caviness 1982). However, rather than splitting the preplate and accumulating in an inside-out sequence between marginal zone and subplate, the neurons of the cortical plate gather, in a somewhat irregular but basically *outside-in* sequence, entirely *below* the preplate (Caviness 1982). Caviness and colleagues (1981) termed the abnormal unsplit preplate of *reeler* the 'superplate'.

Despite this radical disturbance of the genesis and lamination of the cortex, thalamic axons reach the appropriate cortical areas (Dräger 1976; Steindler and Colwell 1976; Caviness and Frost 1980) and terminate principally in the middle of the cortical plate, corresponding to layer 4 (Frost and Caviness 1980; Caviness et al. 1988). However, the trajectory of thalamic axons within the cortex is highly abnormal in the adult animal: instead of running radially from the white matter to layer 4, they form long loops, running up through the entire thickness of the cortex in obliquely oriented fascicles and then defasciculating and plunging down to reach layer 4 from above rather than from below (Caviness 1976; Fig. 3).

Our results suggest that the timing and the overall topography of both the corticofugal and thalamofugal projections are essentially identical in early stages in normal and *reeler* genotypes (Molnár et al. 1998b). We could find no early differences between wild-type and *reeler* mice in the appearance of cortex, thalamus or their axons (Molnár et al. 1998b). The polymorphic cells of the preplate send out their pioneering axons, in both strains, at E13 and

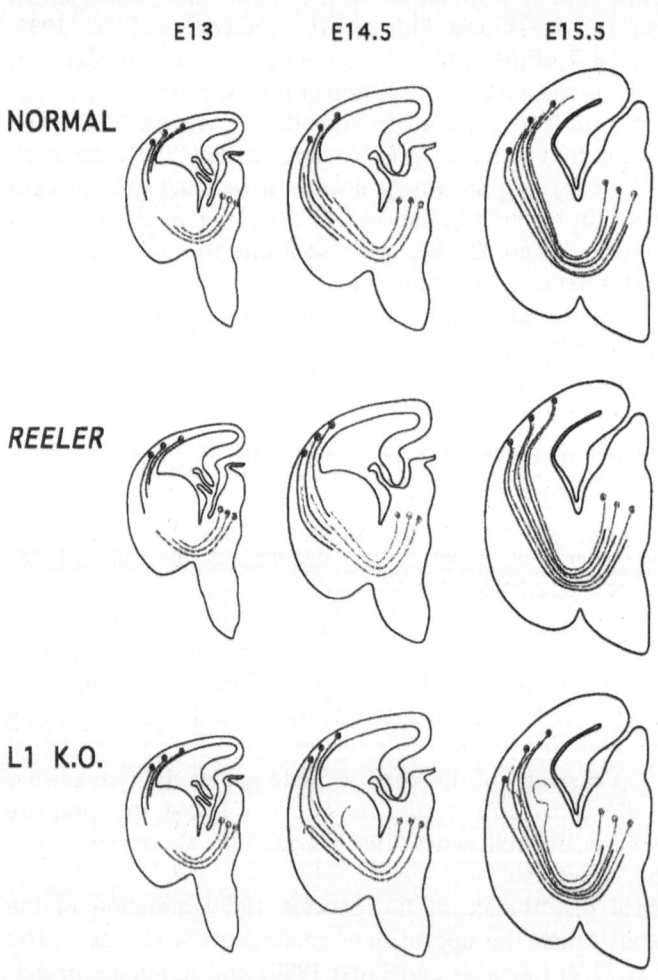

E13.5, *before* the arrival of any cells of the cortical plate proper, and before the arrival of thalamic or brainstem (Crandall et al. 1992) afferents. These descending axons approach the primitive internal capsule, with a similar degree of fibre order in the two strains, at about E13.5 (for the occipital cortex).

At roughly the same stage, axons from the dorsal thalamus grow in quite strict order through the primitive internal capsule and approach their corticofugal counterparts. In both wild-type and *reeler*, the thalamic fibres project to the appropriate cortical regions, closely intermix with, and possibly fasciculate upon, the pioneer corticofugal axons from those same cortical areas.

The differences in cortical structure between *reeler* and normal mouse start to become evident immediately after the arrival of the first wave of migrating cortical plate cells, which stop below the preplate in *reeler* rather than penetrating and splitting it (Goffinet 1979; Pinto-Lord and Caviness 1979). The abnormal placement of the gathering plate cells in *reeler* forces a change in the appearance of the pre-existing early corticofugal axons, leading them to form oblique bundles, running from the superplate, through the cortical plate below. Nevertheless, this does not change the paths of the distal parts of these axons, which have already turned through and under the developing striatum towards the primitive internal capsule, and made contact

←━━━→

Fig. 3. Early development of reciprocal thalamocortical and corticofugal projections in wild-type (*above*), *reeler* mutant (*middle*) and L1 KO (*below*) mice at different embryonic ages (E13, E14.5, E15.5). Similarly to Fig. 1, each schematic drawing represents an imaginary section through the right hemisphere, revealing the entire pathway. *Above* In normal mice, at E13 early thalamic and corticofugal fibres synchronously approach the internal capsule. At E14.5 thalamocortical and earliest corticofugal (preplate) projections meet and begin to fasciculate outside primitive internal capsule, under striatum. At E15.5 thalamocortical projections reach appropriate cortical areas by associating to early corticofugal fibre scaffold. *Middle* In *reeler* mice, at E13 formation of preplate, outgrowth of descending corticofugal axons and concomitant growth of thalamic axons through primitive internal capsule, all occur indistinguishably from normal animals. However, by E14.5, as thalamic axons are approaching their target areas, cortical plate itself has started to form and distinctive differences in phenotypes begin to emerge. In *reeler* cortical plate forms under all preplate (now superplate) cells, whose pioneer axons gather into oblique fascicles, running through the thickening cortical plate. Thalamic axons follow these fibers up to superplate layer. In wild-type, the majority of corticofugal axons derive from those preplate cells that come to lie below cortical plate in subplate. At E14.5 thalamic axons run through intermediate zone and enter subplate. By comparison, in *reeler* at the same age, thalamic fibres are growing up obliquely through cortical plate and some are already entering superplate. At E15.5, the vast majority of axons in the normal animal are still restricted to subplate layer, but in *reeler* they all appear to have passed through the plate and into superplate above. This behaviour can only be observed in normals in a small portion of axons from normals, whereas almost all axons show this behaviour in *reeler*. *Below* In L1 KO mice, initial steps are indistinguishable. Subsequently fasciculation abnormalities occur and some thalamic projections get misrouted at the striato-cortical junction. Fascicules formed by thalamocortical axons in striatum appear thicker and more disorganised. Some fibre fascicules get derailed and defasciculate or turn abruptly within striatum, and never reach the striato-cortical junction to enter the cortex intermediate zone

with the array of thalamic axons advancing towards the cortex, as in the normal animal (Molnár et al. 1998b).

In *reeler*, however, the thalamic fibres immediately penetrate the cortical plate and run up diagonally to the superplate layer at the top of the cortex, where they then wait for roughly the same period of time before turning down again into the plate itself (Caviness et al. 1988; Molnár and Blakemore 1992, 1995a,b). These results are compatible with the hypothesis that thalamic axons are specifically guided towards the preplate cells by running in close association with the axons that grew down earlier to establish the corticofugal scaffold.

The intimate relationship of thalamic axons to preplate axons as they approach the cortex may also explain why they come to rest in the subplate (or for *reeler* the superplate) layer and probably form synapses on the cells there (Chun and Shatz 1988b; Friauf et al. 1990; Friauf and Shatz 1991; Herrmann et al. 1994). In *reeler* mice, the thalamic afferents arrive in the superplate at E14.5/E15 and gather for a period of some days, at least until E18, just as they do in the subplate of normal mice. Thus, a 'waiting period', in association with the appropriate target cells, occurs in both normal and *reeler* mice, even though those target cells lie in quite different locations.

Subplate neurons are known to express immunoreactivity to the surface molecules L1 (Godfraind et al. 1988; Chung et al. 1991) and fibronectin (Stewart and Pearlman 1987; Chun and Shatz 1988a; Sheppard et al. 1991), which could provide highly attractive surfaces for the growth of thalamic axons (providing they are expressing appropriate adhesion molecules) in an otherwise relatively non-permissive environment. The selective expression of glycosaminoglycans (Derer and Nakanishi 1983; Fukuda et al. 1997), chondroitin sulphate core proteins (Bicknese et al. 1994a; Miller et al. 1995) and specific peanut agglutinin lectin binding (Götz et al. 1992) has also been demonstrated in the subplate region and along fibre fascicles crossing the cortical plate in *reeler* (Bicknese et al. 1994b; Kurukulasuriya and Salinger 1996) at corresponding ages. It is yet to be determined whether thalamic fibres require more than one of these extracellular cues (sequentially or simultaneously) or whether there is a specific requirement for only one or a subset of them. Fukuda et al. (1997) used more sensitive visualisation of the neurocan molecule in acid alcohol-fixed and paraffin-embedded sections and observed that subplate projections, but not thalamic axons, express neurocan immunoreactivity in embryonic rat brain. It was previously thought that chondroitin sulphate core proteins are only expressed by subplate cells but not their projections (Bicknese et al. 1994a). Fukuda et al. (1997) demonstrated that L1 immunoreactivity was specifically localised on growing thalamic axons and using double immunolabelling techniques they showed that L1-bearing thalamocortical axons were found to extend along neurocan-positive regions. According to their study, L1-positive and neurocan-positive axons overlapped. Friedlander et al. (1994) reported that neurocan can bind L1 in vitro. The heterophilic interaction between L1-positive thalamic projections and neurocan-positive subplate cells might be important in thalamocortical pathfinding.

The existence of these privileged pathways for axon growth could explain how thalamic axons in *reeler* are able to penetrate the cortical plate and steer up to reach the equivalent cells in the superplate, while ignoring the hostile territory of cortical plate cells around them. In the *reeler* mutant, various adhesion molecules (L1, NCAM and fibronectin) and glycosaminoglycans were found to be *similarly* expressed by early-born cortical neurons, despite their different dispositions (Derer and Nakanishi 1983; Godfraind et al. 1988; Bicknese et al. 1994a).

8.2
The L1 KO Mouse

The mechanism of heterophilic fibre–fibre interaction was proposed for contact guidance throughout the internal capsule between thalamocortical and corticothalamic projections. Recently, a mouse deficient in L1 was produced by two independent groups (Cohen et al. 1997; Dahme et al. 1997). We examined the thalamocortical pathfinding and topography in this mouse in collaboration with J. Taylor and N. Mather (Human Anatomy and Genetics, Oxford). Our carbocyanide dye tracing experiments in embryonic and early postnatal L1 KO mice brains revealed that fasciculation problems occur at the striato-cortical boundary. Some fibres do not make it through and form aberrant projections into the striatum, while some cross the region normally and enter the cortex. The majority of the thalamic fibres branch and arborise normally and assume whisker-related pattern in the barrel cortex (Molnár et al. 1999b). The normal whisker-related pattern was confirmed with cytochrome oxidase (CO), Nissl stainings and 2-deoxyglucose (2-DG) experiments (Molnár et al. 1999).

8.2.1
Possible Inhibitory Factors in and Around the Internal Capsule

Not all influences on growing axons are attractive. Some molecules inhibit or discourage axon growth (Caroni and Schwab 1988a,b; Schwab and Caroni 1988; Pini 1993) and it may be that the pioneering subplate axons are constrained within the routes that they take by the existence of hostile surfaces outside those routes. Skaliora et al. (1998) described expression of the semaphorin D gene, encoding a protein that regulates axon outgrowth, in the ganglionic eminence at E15 in the rat, which seems to correspond to the region of the perireticular nucleus (Mitrofanis 1992) and also discovered a transient stripe of semaphorin G expression which might correspond to a transient intermediate zone at the striato-cortical boundary (Fernandez et al. 1998; see Fig. 1). It is conceivable that the internal capsule is a hostile en-

vironment for axon growth (Table 1). We have proposed this based on the
fasciculation pattern of thalamic axons in the region (Molnár 1994).

9
Mutants with Disturbances in Thalamocortical Interactions

A number of mutant mice have been described that fail to develop normal
cortical barrels (Cases et al. 1996; Welker et al. 1996; Iwasato et al. 1997;
Abdel-Majid et al. 1998; Hannan et al. 1998b; Table 2; Fig. 4). These various
mice reveal multiple molecular pathways in barrel cortex development. Using
different techniques, various aspects of thalamocortical development can be
dissected at a mechanistic level (see Fig. 5). Using cytochrome oxydase (CO)
histochemistry, one can gain information on the presence or absence of the
presynaptic whisker-related pattern. Simple Nissl staining reveals the presence or absence of the patterning of layer 4 cells which signal the status of the
postsynaptic modifications of dendritic development. The functional orga-

Table 1. Summary of mutant mice with altered thalamocortical development

Gene mutated	Type of mutation	Phenotype	Molecular mechanism	Reference
Reeler (rl)	Spontaneous KO	Primary defect in cortical development	No reelin	Caviness and Rakic (1978) Rakic and Caviness (1995) D'Arcangelo et al. (1997)
Disabled (mdab1) (*Scrambler* and *Yotari*)	Spontaneous KO	Same phenotype as *reeler*	Disturbed signalling through mdab1	Howell et al. (1997) Sheldon et al. (1997) Ware et al. (1997)
L1	Targeted KO	Thalamocortical axon fasciculation defects	Selective fasciculation	Cohen et al. (1997) Dahme et al. (1997)
TBR1	Targeted KO	Thalamocortical and cortico--thalamic axon elongation defect	Not known	Hevner et al. (1998)
Pax-6	Targeted KO	Th-C and C-Th axon elongation disrupted at the IC	Not known	Stoykova et al. (1997) Hevner et al. (1998) Kawano et al. (1999)
Gbx-2	Targeted KO	Th-C and C-Th axon elongation disrupted at the IC	Not known	Hevner et al. (1998)

KO, knockout; IC, internal capsule.

Table 2. Summary of mutant mice with defects in barrel cortex development

Gene mutated	Type of mutation	Phenotype	Molecular mechanism	Reference
Adenylyl cyclase I (Adcy I)	Spontaneous KO	Th-C axon segregation deficit	Signalling through presynaptic 5-HT1R	Welker et al. (1996); Abdel-Majid et al. (1998)
Monoamine oxidase A (MAO-A)	Targeted KO	Th-C axon segregation deficit	Excess 5-HT binding to presynaptic 5-HT1R	Cases et al. (1995, 1996)
Growth-associated protein-43 (GAP-43)	Targeted KO	Th-C axon segregation deficit	Presynaptic signalling via cAMP pathway	Maier et al. (1999)
NMDA Receptor subunit N1 (NMDAR-N1)	KO with low expression of knockin	Th-C axon segregation deficit	Postsynaptic NMDAR activation by glutamate	Iwasato et al. (1997)
Phospholipase C-β1 (PLC-β1)	Targeted KO	Deficit in postsynaptic barrel formation	Postsynaptic receptor signalling and protein trafficking	Hannan et al. (1998b)

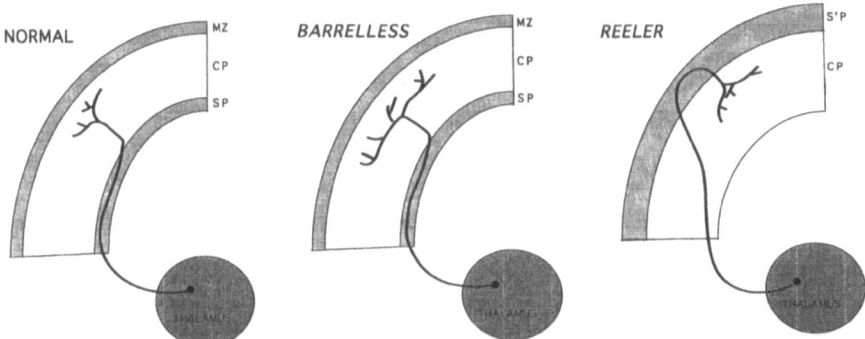

Fig. 4. Thalamocortical axons in normal, *barrelless* and *reeler* mutant mice for illustration of their trajectories and termination in cortex. In normal animals, thalamic fibres reach correct regions of cortex, enter into overlying cortical plate, then form arbors and terminate in a restricted area mostly on layer 4 neurons. In *barrelless* mutant mice, thalamocortical axons show normal extracortical trajectories and intracortical laminar targeting, but their axons form abnormal thalamocortical arbors. In normal mice, thalamocortical arbors terminate in a confined region of primary somatosensory cortex, whereas in *barrelless* they terminate in a 2–5 times larger area (Welker et al. 1996). In *reeler* mice, despite abnormality in position of early postmitotic neurons and in formation of cortical layers (Caviness 1982), thalamic fibres reach correct regions of cortex, and ultimately terminate on neurons that appear to be equivalent to layer 4 in the normal animal (Frost and Caviness 1980). Only the local pattern of innervation within grey matter seems highly abnormal, thalamic fibres looping up to top of cortex, and reaching equivalent cells in middle of cortex from above rather than from below (Caviness 1976; Frost and Caviness 1980; Caviness et al. 1988). *CP* Cortical plate; *SP* subplate; *S'P* superplate; *MZ* marginal zone

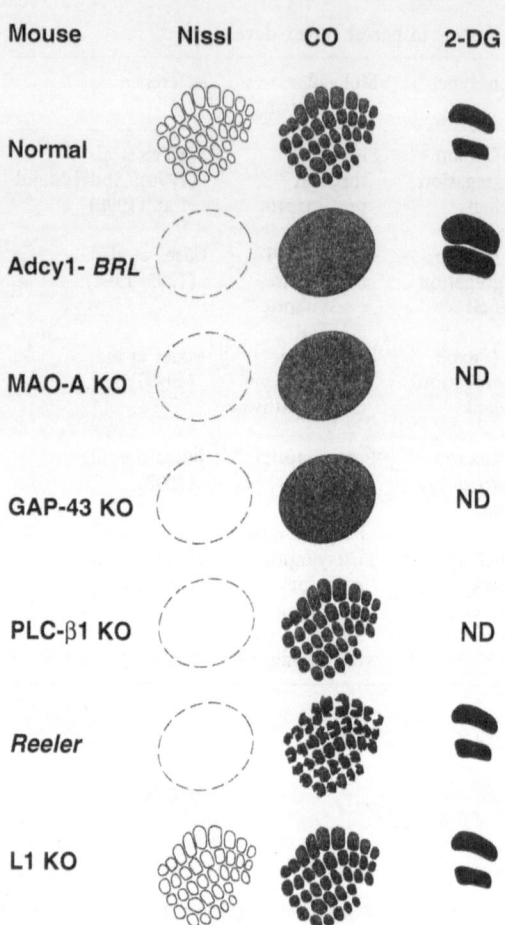

Mouse	Nissl	CO	2-DG
Normal			
Adcy1- *BRL*			
MAO-A KO			ND
GAP-43 KO			ND
PLC-β1 KO			ND
Reeler			
L1 KO			

Fig. 5. Summary of the different abnormalities in barrel development found in various mutant and KO mice. Results obtained using analysis of each type of mouse using Nissl and cytochrome oxidase (*CO*) histochemistry, and 2-deoxy-glucose (*2-DG*) autoradiography on tangential sections of somatosensory cortex are illustrated. CO staining failed to reveal periphery-related patterns in Adcy1-BRL, MAO-A KO and GAP-43 KO mice and, although it demonstrated patches with normal patterning, individual barrels were shaped irregularly in *reeler*. Nissl staining demonstrated lack of barrels in Adcy1-*BRL*, MAO-A KO, GAP-43 KO, PLC-β1 KO and *reeler*. 2-DG autoradiography, after stimulation of the first three whiskers on rows A and C, showed normal patterns in all strains of mice with the exception of Adcy1-BRL, where, inspite of segregation, 2-DG signals were more diffuse. *ND* No available data

nisation of the thalamocortical and corticocortical projections can be surveyed with 2-DG autoradiography (Welker et al. 1992).

9.1
Barrel Formation in *reeler* Mouse

2-DG mapping in combination with Nissl and CO staining has been used to examine whether or not a functional cortical representation of the mystacial whiskers can be related to a cytoarchitectonic organisation in the *reeler* mutant (Bronchti et al. 1999a). In *reeler*, examined on the coronal plane, DG autoradiography revealed a columnar activation pattern with a highest DG uptake in the intermediate layers. In the tangential plane, DG uptake showed that the cortical whisker representation in *reeler* is organised in an identical manner to normal mice. However, Nissl staining of *reeler* cortex did not show clearly defined barrel boundaries. CO staining in *reeler* revealed irregularly shaped, dense patches in a region corresponding to primary somatosensory cortex. Thus, a modified radial organisation in *reeler* does not seem to have altered an ordered, functional whisker representation, although the cytoarchitectonic pattern is less clear.

9.2
The *Barrelless* Mouse

9.2.1
The Barrelless Phenotype

Barrelless mice show a lack of segregation of thalamocortical afferents and cytoarchitectonic barrels; thus both pre- and postsynaptic whisker-related pattern formation is disturbed. In addition to the axon tracing studies which demonstrated lack of spatial segregation, 2-DG uptake studies revealed that the topology of whisker representation in the cortex was preserved (Welker et al. 1996). This indicates that the establishment of topology amongst thalamocortical axons projecting to barrel cortex occurs earlier than, and is not dependent on, their segregation into discrete barrels. Furthermore, electrophysiological analysis of these *barrelless* mice revealed defective temporal discrimination of centre from surround whisker stimulation (Welker et al. 1996). Thus, the segregation of thalamocortical axons into barrels is required for somatosensory discrimination of spatial and temporal information.

It was demonstrated on Nissl-stained sections in adult *barrelless* mice that barrels failed to appear. Labelling with dextran in primary somatosensory cortex (S1) of adult mutants revealed that thalamocortical connections terminate in layers 4 and 6, but they are continuously distributed in layer 4, rather than being confined to barrel-like structures and leaving septa between barrels relatively free of label as seen in normal mice (Welker et al. 1996). Reconstructions of *barrelless* thalamocortical axons showed that they terminate in an abnormally large area of somatosensory cortex (Gheorghita-Bächler et al. 1996; Welker et al. 1996).

The possible areal differences in the developing and adult thalamocortical topography of visual and somatosensory cortices in normal and *barrelless* animals was also examined by Molnár et al. (1996). This study demonstrated that the initial laminar and tangential targeting of the thalamocortical connections is not affected by the mutation and is initially indistinguishable from the normal. Anterograde tracing from the ventrobasal complex (VB) revealed an identical laminar pattern of thalamocortical fibre ingrowth at all stages examined. Thalamocortical fibres arrive at the cortex at E16, begin to invade the cortical plate at E18 and begin to arborise below the dense cortical plate at P0–P2 in both phenotypes. The first obvious difference is that at early postnatal ages no complete barrel patterns were observed.

9.2.2
Lack of Formation and Stabilisation of Barrel Patterns in the Mutant

In certain species (e.g. rabbit and grey squirrel), barrels appear only transiently and the pattern disappears in adulthood (Woolsey et al. 1975; Rice et al. 1985; Rice 1995). In *barrelless* animals, no whisker-related pattern appears at any time of development (Molnár et al. 1996; Welker et al. 1996). Agmon et al. (1995) and Catalano et al. (1996) demonstrated that the whisker-related pattern is not the result of axonal pruning. In normal mice, the thalamic axons enter the cortex in an organised fashion and their ingrowth into the cortical plate is precise from the very beginning. There might be substantial rearrangement within the subplate and lower cortical plate by forming transient side-branches, but this occurs somewhat earlier, before the innervation of the cortical plate (Naegele et al. 1988). Catalano and Shatz (1998) recently demonstrated that initial targeting of thalamocortical axons to visual cortex was dependent upon activity. These experiments suggest that even thalamocortical targeting can be modified and may depend on early activation patterns. Nevertheless, this initial step does not seem to be affected by the defect in the *barrelless* mouse (Molnár et al. 1996).

In spite of the substantial differences in the appearance of the thalamocortical arbors in normal and *barrelless* animals, there are parameters that remain constant for normal and *barrelless* thalamocortical axons. In *barrelless*, the total axon length was significantly longer and the tangential area spanned by the arbors was approximately three times larger than those in normals. The number of axonal boutons in *barrelless* is greater than in normal; however, the number of boutons per axonal length is 7–8 boutons/ 100 μm axon for both strains (Gheorghita-Bächler et al. 1996). This indicates that the *barrelless* mutation only affects certain aspects of thalamocortical arbor formation, while some remain unaffected. It seems that the extent of axon growth and bouton formation are regulated separately.

9.2.3
Similar Areal Differences in Thalamocortical Innervation Patterns in Normal and Barrelless Mice

In putative somatosensory and visual cortices of fixed brains of normal (NOR) and *barrelless* (brl) mice, animals showed no area-specific abnormality of the thalamocortical projections. Pairs of single crystals of carbocyanine dyes (DiI and DiAsp) placed at 250- or 500-μm distance from each other into the putative visual cortex revealed very similar patterns of back-labelling in adult and normal early postnatal animals.

At early postnatal ages, quantitative analysis demonstrated a similar low percentage of double-labelled cells in VB and LGN of both groups of mice both in somatosensory and visual thalamus. However, in LGN, the number of labelled cells were 3–5 times smaller than in VB in *both* strains of mice (Molnár et al. 1996). This suggests that the abnormality in *barrelless* is not due to spatial organization of the cell bodies of the projection neurons in VB or LGN, but merely due to a cortical defect of pattern formation after the initial correct ingrowth of thalamocortical axons.

Tangential sections through layer 4 of S1 of normal mice reveal labelling (dextran) that is confined to the inside of the barrels, leaving septa between barrels relatively free of label. In *barrelless* mice, labelling forms a continuous zone. Recent in vitro work of Bolz and colleagues demonstrated that both normal and *barrelless* thalamic fibres (from E15 thalamic explants) were growing longer and had more branches after 3 days in culture on membrane preparations from P8 primary somatosensory cortex of *barrelless* animals than from normals (Uziel et al. 1998). Since both normal and barrelless thalamic fibres exibit this different growth pattern, it was suggested that cortical membrane associated molecules contribute to the altered axonal growth in *barrelless*. This *barrelless* phenotype (Welker et al. 1996) has recently been found to be due to a mutation in the adenyl cyclase type I gene (Adcy1; Abdel-Majid et al. 1998). Adcy1 is a signalling enzyme involved in the regulation of cyclic adenosine monophosphate (cAMP) and subsequently the activity of protein kinase A (PKA). The mechanism by which this mutation may produce the *barrelless* phenotype is discussed below in relation to a different mutant mouse with a similar phenotype.

9.3
The MAO-A KO Mouse

A mutation of the monoamine oxidase A (MAO-A; Cases et al. 1995) gene has also been found to produce a similar phenotype to the Adcy I *barrelless* mice, with exuberant aborisation of thalamocortical afferents in somatosensory cortex (Cases et al. 1996). The role of excess serotonin (5-HT) in barrel formation was elegantly demonstrated in transgenic mice (Tg8) in which the integration of a transgene disrupts the gene encoding MAO-A. As a result, 5-HT concentrations in the brain increased up to nine-fold in pups while noradrenaline (NA) concentrations increased up to two-fold in these mice (Cases et al. 1996). MAO-A is an enzyme that catalyses the oxidative deamination of monoamines. This genetic mutation leads to an excess of 5-HT being released in the developing cortex. The specificity of this mechanism was verified by the application of an inhibitor of serotonin, parachlorophenylalanine, which rescued the barrels during development. CO histochemistry, a marker of metabolic activity which correlates with synaptic activity, revealed an absence of barrel patterning in the somatosensory cortex of MAO-A knockout mice. The total area of primary somatosensory cortex of these mice was found to have increased by 20%, suggesting that the increased serotonin levels had induced exuberant thalamic afferents, whose arbors failed to segregate into barrel clusters. The cells of the cortex, as examined by Nissl staining, failed to form barrels, which would be expected if the thalamacortical axons normally segregate first and cortical cells migrate into a barrel pattern later.

In this trangenic mouse, as in the *barrelless* mutant mouse (Welker et al. 1996), normal pattern formation exists in thalamus and trigeminal nuclei, but the characteristic barrel pattern in layer 4 fails to appear in the primary somatosensory cortex. Thalamocortical projections labelled with biocytin arborise in a continuous band within layer 4 with gross topographic organisation (Cases et al. 1996). Inhibition of serotonin synthesis during the first postnatal week restored barrel formation in the transgenic mouse line. Transient MAO-A inhibition, especially during thalamocortical fibre ingrowth (P0–P4), produces a phenocopy of the barrelfield alterations seen in the Tg8 mice. The role of serotonin in thalamocortical development is still not fully understood, but Lebrand et al. (1996) demonstrated that developing thalamic neurons take up exogenous 5-HT through high-affinity uptake sites located on thalamocortical axons and terminals. Thalamic neurons do not synthesise 5-HT, but the internalised 5-HT might be used for signalling.

The presynaptic release of 5-HT, which binds to the 5-HT1B receptor, links the *barrelless* defect in both MAO-A and Adcy1 mice. Adcy I is normally regulated by activation of the 5-HT1 receptor, to catalyse the formation of cAMP, which in turn inactivates PKA. This may lead to increased neurotransmitter release at thalamocortical synapses. Both mutations could therefore lead to a decrease in cAMP, via either increased activation of the presynaptic 5-HT1B receptor (Boschert et al. 1994) or defective Adcy I function, and thus affect thalamocortical neurotransmitter release. The Adcy I and MAO-A *barrelless* mice have thus revealed an important presynaptic signalling pathway, regulated by 5-HT, which is required for refinement of thalamocortical arbors into barrels in the developing somatosensory cortex.

9.4
The NMDA Receptor KO Mouse: Role of Activity in Barrel Formation

Numerous studies have demonstrated that blocking the flow of sensory information from the whiskers to the somatosensory cortex at various levels during the first postnatal week prevents the establishment of the characteristic patterning of layer 4 cells in S1 (see Schlaggar and O'Leary 1993; Schlaggar et al. 1993; Killackey et al. 1995). The cytoarchitectonic features of normal whisker barrels emerge in immature occipital cortex transplanted into the barrel area of newborn rats (Schlaggar and O'Leary 1991), suggesting that barrel formation may not require pre-established intrinsic cues in the cortex itself. Thalamic axons are capable of transmitting impulses to cortical cells during the embryonic (from E18/E19) and early prenatal period (Higashi et al. 1996; Molnár et al. 2000) and this activity could begin to influence the target structure. However, blockade of postsynaptic activity from shortly after birth, by local application of an NMDA-receptor antagonist, does not prevent barrel formation in the normal barrel cortex of the rat (Fox et al. 1996), nor in the hamster, in which the cortex and thalamic innervation are less developed at birth (Chiaia et al. 1994). NMDA receptor-dependent long-term potentiation (LTP) matches the period in which barrel structure can be modified by sensory perturbations (Crair and Malenka 1995), suggesting a role for NMDA receptors in barrel plasticity.

There is evidence from the NMDA receptor KO mice experiments (Li et al. 1994; Tonegawa et al. 1995) that the induction of barreloids requires NMDA-receptor activation and/or postsynaptic activity. In some forms of NMDA receptor KO mice, barreloids develop in the thalamus, but no barrels appear in the primary somatosensory cortex. Furthermore, the level of ectopic NMDAR1 transgene expression in NMDAR1 KO mice was critical in determining the formation of barrel patterns in somatosensory cortex (Iwasato et al. 1997).

9.5
A Specific Postsynaptic Defect in Barrel Formation Identified in PLC-β1 Knockout Mice

Receptor-mediated signal transduction plays an important role in activity-dependent modification of neuronal structure and function during cortical development. One signalling protein that has been implicated in cortical development and plasticity is PLC-β1 (Kind et al. 1997; Hannan et al. 1998a). PLC-β1 is a G-protein-coupled phosphodiesterase that hydrolyses phosphatidylinositol 4,5-biphosphate (PIP2) into the second messengers diacylglycerol (DAG) and 1,4,5-triphosphate (IP3), which in turn regulate protein kinase C and Ca^{2+} release from the endoplasmic reticulum, to control cellular processes such as protein trafficking. The PLC-β class of phosphodiesterases are activated following neurotransmission, via G proteins coupled to group I metabotropic glutamate receptors (mGluRs), muscarinic acetylcholine receptor (mAChRs) or serotonin 5-HT2 receptors.

PLC-β1 was found to be highly expressed in normal mouse somatosensory cortex during early postnatal development in a pattern correlated with barrel distribution (Hannan et al. 1998b). Localisation of PLC-β1 to the intermediate compartment-like organelle, the botrysome, in developing neurons has been observed during the sensitive period in both visual and somatosensory cortex (Kind et al. 1997; Hannan et al. 1998a). PLC-β1 immunoreactive botrysomes are present in domains resembling barrels in and above developing layer 4, corresponding to regions receiving somatosensory thalamic input. Botrysomes have been most frequently observed either at the base of, or within, cortical dendrites (Kind et al. 1997). PLC-β1 may transduce signals from thalamic afferents which have developed synapses in layer 4. There is also less intense, diffuse immunoreactivity corresponding to neuronal cell bodies and processes within the barrels. These results correlate with recent observations of PLC-β1 localization in rat somatosensory cortex (Hannan et al. 1998a) and suggest that this signalling protein may play a role in barrel development.

This proposed role of PLC-β1 was tested using KO mice (Kim et al. 1997). Examination of the somatosensory cortex of mice homozygous for this null mutation, using Nissl staining, revealed a defect in the formation of barrel structures (Hannan et al. 1999). The morphology of axons from VB which project into somatosensory cortex was investigated using carbocyanine dye (DiI) tracing. Thalamocortical axons labelled with DiI were found to segregate into a whisker-related pattern in layer 4 in both normal and PLC-β1 KO mice, demonstrating a specific postsynaptic deficit in barrel cortex development. These results suggest that PLC-β1 is required for the morphological differentiation of cortical neurons in response to synaptic activation.

PLC-β1 may be activated by group I mGluRs (mGluR1 and mGluR5), the NMDA receptor, mAChRs or 5-HT2 receptors (Dudek and Bear 1989; Carter et al. 1990; Casabona et al. 1997; Kim et al. 1997). In both visual (Dudek and Bear 1989) and somatosensory (Bevilacqua et al. 1995) cortex, IP3 turnover in response to mGluR stimulation peaks during the sensitive period and then declines. Serotonin depletion in the developing somatosensory cortex affected barrel development, but did not alter the plasticity of the barrels in response to whisker removal (Turlejski et al. 1997). Furthermore, the Adcy1 and MAO-A *barrelless* mice described above implicate 5-HT in the sensitive period of barrel formation. In support of a role for the mAChRs, carbachol-induced IP3 turnover correlates with the sensitive period in barrel cortex (Bevilacqua et al. 1995) and cholinergic input has been implicated in barrel plasticity (Zhu and Waite 1998). Importantly, PLC-β1 KO mice did show reduced phosphoinositide hydrolysis in cortical neurons (Kim et al. 1997), relative to wild-type mice, following mAChR activation. Furthermore, in the rodent somatosensory cortex, the mAChR has been shown to be transiently expressed in axons and apical dendrites of cortical pyramidal neurons, suggesting a possible role for these receptors in development (Buwalda et al. 1995).

PLC-β1 may therefore be crucial in receptor-activated signal transduction during activity-dependent cortical maturation and plasticity. It is part of a distinct postsynaptic signalling pathway, which can be genetically disrupted without affecting thalamic afferent segregation. In the absence of PLC-β1 there may be abnormal regulation of protein trafficking required for correct dendritic remodelling and cell migration during cortical development.

9.6
The GAP-43 KO Mouse

Another type of KO mouse, with a null homozygous mutation in the growth-associated protein-43 (GAP-43) gene, has recently been described. These mice appear to have a primarily presynaptic defect with aberrant segregation of thalamocortical arbors, consistent with the proposed role of GAP-43 in linking signalling with cytoskeletal rearrangements in growth cones and presynaptic terminals. Mice lacking GAP-43 expression fail to establish the whisker-related barrels in layer 4 of the primary somatosensory cortex. Thalamocortical afferents are misrouted as they enter the cortical plate and run tangentially rather than radially (Maier et al. 1999). This leads to the abnormal topography of the thalamocortical projections and the lack of barrel field. Maier et al.'s study suggests that GAP-43 is critical for the normal establishment of ordered topography in cortex. 5-HT transporter staining between hemispheres shows that, while thalamocortical axons do cluster in layer 4 of GAP-43 −/− cortex, these clusters bear no consistent relationship to the stereotyped whisker pattern or barreloid array. This irregular and unpredictable patterning of thalamocortical axon clusters is direct evidence that an ordered or consistent whisker map does not form in the absence of GAP-43.

5-HT staining confirmed that thalamocortical fibres enter the cortex and reach layer 4, but they are not segregated into barrels. Double-labelling experiments from cortex demonstrated differences in topography, with a larger proportion of double-labelled cells observed in the GAP-43 KO mice. Some crude preservation of topography, however, was suggested by similarly oriented patches of labelled cells in thalamus with respect to the cortical injection.

9.7
Overview of Mutant Mice with *Barrelless* Phenotypes

These various mutant and transgenic mice (Table 2) have provided important insights into the molecular and cellular mechanisms of cortical development. They have been used to demonstrate that the segregation of thalamocortical axons into barrels requires appropriate levels of serotonin release (which does not occur in the MOA-A KO mice) and that the binding of serotonin to the 5HT1 receptor may regulate axonal segregation via adenylyl cyclase signalling. It has also been shown that this presynaptic pathway of barrel formation can be separated from a postsynaptic pathway involving PLC-β1 signalling via neurotransmitter receptor activation. The postsynaptic pathway presumably responds to thalamocortical activity by reorientation of dendrites in layer 4 and the arrangement of cells around the barrel hollow. Segregation of thalamocortical axons can occur in the absence of such postsynaptic plasticity, thus identifying separate steps in the development of barrel cortex.

10
Mutations Indirectly Affecting Thalamocortical Development

Early gross thalamocortical topography is accomplished as early as E15.5 in the mouse, before the arrival of the peripheral afferent input to the thalamus.

For example, the first axons of the optic tract reach the level of the lateral geniculate nucleus (LGN) at around E16, but these early fibres seem to be destined for the superior colliculus (Lund and Bunt 1976). Optic tract fibres do not invade the LGN substantially until around birth in the rat (Karlsson 1966). The implication is that the initial development of the early pattern of connection between thalamus and cortex is not dependent on external afferent input from sensory pathways or other subcortical structures, but subsequently the pattern is modified as the periphery connects to this developing network of connections. Since the waiting compartment for thalamic fibres coincides with the transient cell group of the subplate and with sites of fibre rearrangement seen in the adult, it is important to understand the nature of interactions between thalamic fibres and the subplate (reviewed by Molnár et al. 1999a).

10.1
Thalamocortical Topography in Anophthalmic Mutants and After Early Binocular Enucleation

Anophthalmic mice (Godement et al. 1979; Kaiserman-Abramoff et al. 1980; Cullen and Kaiserman-Abramoff 1986), as well as bilaterally enucleated hamsters (Rhodes and Fish 1983) and ferrets (Guillery et al. 1985), develop geniculocortical and corticogeniculate interconnections with an essentially normal topography (within the precision of tracing methods). The anophthalmic mutant mouse (ZRDCT-An) provides an excellent model for the study of the role of the sensory periphery on cortical specialisation. In this mouse, the eyes and the optic nerves fail to develop and therefore the lateral geniculate nucleus (LGN) never receives any retinal afferents. Incomplete penetrance of the mutation produces microphthalmic mice, which provide interesting controls because they develop eyes with small optic nerves, which reach LGN, allowing limited vision. Combined electrophysiological recording and neuroanatomical tracing study in anophthalmic mutant mice demonstrated that throughout the occipital cortex multiunit responses to auditory clicks could be recorded (Bronchti et al. 1999b). This was never observed in normal mice. In more medial parts of the occipital pole, responses to somatosensory stimulation of the trunk were recorded in addition to the auditory response. In the microphthalmic animal, visual responses were recorded in the corresponding areas; auditory and somatosensory responses were limited to the temporal and parietal areas respectively. Injection of fluorescent tracer in the occipital cortex of anophthalmic mice revealed retrogradely labelled cells in the dorsal LGN. Microphthalmic mice had a similar labelling pattern in the LGN from corresponding occipital cortical injections. These results suggest that there is a subcortical component to the reorganisation. In anophthalmic mutant mice, somatosensory input from the dorsal column nuclei to the dorsal LGN is known to exist (Asanuma and Stanfield

1990) and Bronchti et al. (1999b) were able to trace the possible source of auditory fibres within the dorsal LGN from tracer injections into the inferior colliculus. The anophthalmic mutant mouse seems to present a compensatory development of the non-visual sensory systems comparable to the naturally blind mole rat (Bronchti et al. 1989; Heil et al. 1991; Doron and Wollberg 1994). The absence of sensory organs in mutant mice also provide excellent experimental paradigms to study cross-modal plasticity (Sur et al. 1990).

10.2
Altered Thalamocortical Topography in Albinism and After Early Monocular Enucleation

Perhaps one of the most intriguing experiments of nature for studying the interplay between the autonomous and activity-dependent processes is the albino mutant. Due to a gene defect, abnormally high levels of some intermediate metabolite of melanin synthesis occur, which might cause delay and disturbance in cell generation in the retina (Jeffery 1997). Most probably due to this delay, many developing fibres from the temporal retina are inappropriately routed at the optic chiasm into the contralateral hemisphere (Lund 1965; Guillery 1986). Altering the ratio of the crossed and uncrossed retinal input into the thalamus causes secondary changes in the thalamocortical topography. Interestingly, this alteration can manifest itself as a cortical map reversal or compression (for reviews see Guillery 1986). Early monocular enucleation can lead to similar misrouting of retinal fibres and to similar secondary changes in thalamocortical topography (Trevelyan and Thompson 1992), indicating that these changes in albino mice are secondary and not directly linked to the mutation.

We still do not know why the same abnormal routing of retinal fibres in albinos can lead to different patterns of thalamocortical connections in some animals and not in others (Guillery 1986). It is indeed conceivable that the final pattern depends on the relative maturity of the thalamocortical connections at the time when the retinal projections are beginning to exert their influence and signal the conflicting information from the sensory periphery. If this happens relatively early then there is scope for modification, which will lead to a pattern where the geniculocortical projections are the reverse of the abnormal segment of layer A1. When it is too late to modify the thalamocortical topography, the abnormal geniculate representation will reach the cortex without correction and this leads to the intracortical suppression of the entire layer A1. It would be interesting to follow the initial phases of development in both groups of animals. At present we know very little about the nature and mechanisms of these secondary changes. It would be interesting to examine whether these map reversals require activity or a specific receptor in normal and mutant animals. Further studies of these mutants could help determine the interplay between the autonomous and periphery-dependent processes of thalamocortical development and could also shed light on the pathology of strabismus.

10.3
Extranumery Vibrissae

Various inbred mouse strains produce abnormal patterns of whisker representation, including extra vibrissae (Van der Loos et al. 1986; Welker and Van der Loos 1986). These mutants are extremely useful in examining the development of the whisker-related central pattern. Welker and Van der Loos (1986) argue that early appearance of extranumery vibrissae is based on a different mechanism than the appearance of the corresponding extra barrels in the primary somatosensory cortex. The hard wiring theory would require that the mutation of the extra vibrissae would always be accompanied by the mutation of the subcortical and cortical levels. It is highly unlikely that the cortex had a related mutation which produced the extra barrels. It is much more feasible that the altered flow of sensory information changed the cortical representation and altered the cytoarchitecture at cortical and subcortical levels, as was demonstrated with early lesion experiments (Van der Loos and Woolsey 1973). Moreover, extranumery vibrissae can only change the corresponding cortical representation if their innervation reaches a threshold (Welker and Van der Loos 1986).

11
Conclusions

We have discussed some of the growing number of mutants that have interesting abnormalities in thalamocortical development. The informative nature of these mice is emphasised by studies of other mice, with mutations in genes thought to regulate thalamocortical development which do not have abnormal thalamocortical barrel formation. These include the L1 KO (Molnár et al. 1999), one of the semaphorin genes (Catalano et al. 1998) and nitric oxide synthase (Finney and Shatz 1998). While the importance of these genes may have been masked by compensatory developmental mechanisms in the knockout mice, they also highlight the utility of genetic mutations that do produce abnormalities in cortical development. It is clear that these genes act in concert with many others and our challenge is to obtain meaningful information from the analysis of mutant mice.

There are numerous recently described spontaneous neurological mouse mutants where the gene defect has been identified (e.g. *barrelless*, *tottering* and *opisthotonos*), in addition to genetically modified mice which allow mechanisms of cortical development to be dissected at molecular and cellular levels (Flecher et al. 1996; Street et al. 1997). Perhaps genetic screening will be extended soon to mammals to identify genes responsible for formation of cortical connections and cytoarchitecture in the developing cortex, as it was recently applied to the vertebrate retinotectal system by Bonhoeffer and colleagues (Karlstrom et al. 1997). There is still much to be learned about the multiple mechanisms involved in thalamocortical development and plasticity, and parallel developments expanding our knowledge of, and ability to manipulate, the mouse genome (Tonegawa et al. 1995) provide a bright future for this research.

Acknowledgements. We are indebted to Carolyn Hannan, Alla Katsnelson, Vivien Lane and Carlos Ernesto Restrepo for discussions and for their thoughtful comments on an earlier version of this manuscript, and Colin Blakemore and Egbert Welker for their support. The original work of the authors' laboratories were supported by the Medical Research Council (G9706008), Nuffield Medical Trust, UK, and the Swiss National Science Foundation (3100-56032.98).

References

Abdel-Majid RM, Leong WL, Schalkwyk LC, Smallman DS, Wong ST, Storm DR, Fine A, Dobson MJ, Guernsey DL, Neumann PE (1998) Loss of adenylyl cyclase I activity disrupts patterning of mouse somatosensory cortex. Nature Genet 19:289-291

Agmon A, Yang LT, O'Dowd DK, Jones EG (1993) Organized growth of thalamocortical axons from deep tier of terminations into layer IV of developing mouse barrel cortex. J Neurosci 13:5365-5382

Agmon A, Yang LT, Jones EG, O'Dowd DK (1995) Topological precision in the thalamic projection to neonatal mouse barrel cortex. J Neurosci 15:549-561

Altman J, Bayer SA (1979) Development of the diencephalon in the rat. V. Thymidine radiographic observations on internuclear and intranuclear gradients in the thalamus. J Comp Neurol 188:473-500

Angevine JB Jr, Sidman RL (1961) Autoradiographic study of cell migration during histogenesis of cerebral cortex in the mouse. Nature 192:766-768

Asanuma C, Stanfield BB (1990) Induction of somatic sensory inputs to the lateral geniculate nucleus in congenitally blind mice and in phenotypically normal mice. Neuroscience 39:533-545

Bar I, Lambert de Rouvroit C, Royaux I, Kritzman DB, Dernoncourt C, Ruelle D, Beckers MC, Goffinet AM (1995) A YAC contig containing the *reeler* locus with preliminary characterization of candidate gene fragments. Genomics 26:543-549

Bayer SA, Altman J (1990) Development of layer I and the subplate in the rat neocortex. Exp Neurol 107:48-62

Bayer SA, Altman J (1991) Neocortical development. Raven Press, New York

Bernardo KL, Woolsey TA (1987) Axonal trajectories between mouse somatosensory thalamus cortex. J Comp Neurol 258:542-564

Berry M, Rogers AW (1965) The migration of neuroblasts in the developing cerebral cortex. J Anat 99:691-709

Bevilacqua JA, Downes CP, Lowenstein PR (1995) Transiently selective activation of phosphoinositide turnover in layer V pyramidal neurons after specific mGluRs stimulation in rat somatosensory cortex during early postnatal development. J Neurosci 15:7916-7928

Bicknese AR, Pearlman AL (1992) Growing corticothalamic and thalamocortical axons inter-
   digitate in a restricted portion of the forming internal capsule. Soc Neurosci Abstr 18:778
Bicknese AR, Sheppard AM, O'Leary DDM, Pearlman AL (1994a) Thalamocortical axons extend
   along a chondroitin sulfate proteoglycan-enriched pathway coincident with the neocortical
   subplate and distinct from the efferent path. J Neurosci 14:3500–3510
Bicknese AR, Wang W, Sharma A (1994b) Multiple guidance cues are involved in the segregation of pioneering pathways in the cortex and internal capsule: evidence from *reeler*. Soc
   Neurosci Abstr 20:1683
Blakemore C, Molnár Z (1990) Factors involved in the establishment of specific interconnections between thalamus and cerebral cortex. Cold Spring Harbor Symp Quant Biol 55:
   491–504
Boschert U, Amara DA, Segu L, Hen R (1994) The mouse 5-hydroxytryptamine1B receptor is
   localized predominantly on axon terminals. Neuroscience 58:167–182
Boulder Committee: Angevine JB Jr, Bodian D, Coulombre AJ, Edds MV, Hamburger V,
   Jacobson M, Lyser KM, Prestige MC, Sidman RL, Varon S, Weiss P (1970) Embryonic
   vertebrate central nervous system: Revised terminology. Anat Rec 166:257–262
Bronchti G, Heil P, Scheich H, Wollberg Z (1989) Auditory pathways and auditory activation of
   primary visual targets in the blind mole rat (*Spalax ehrenbergi*). I. 2-deoxyglucose study of
   subcortical centers. J Comp Neurol 284:253–274
Bronchti G, Molnár Z, Welker E, Crocelois A, Krubitzer L (1999a) Auditory and somatosensory
   activity in the visual cortex of the anophthalmic mutant mouse. IBRO World Congr Neurosci 5:54
Bronchti G, Katznelson A, Van Dellen A, Blakemore C, Molnár Z, Welker E (1999b) Deoxyglucose mapping reveals an ordered cortical representation of whiskers in the *reeler* mutant
   mouse. Swiss Soc Neurosci Meeting 4:22
Brückner G, Mares V, Biesold D (1976) Neurogenesis in the visual system of the rat: an
   autoradiographic investigation. J Comp Neurol 166:245–256
Buwalda B, de Groote L, Van der Zee EA, Matsuyama T, Luiten PG (1995) Immunocytochemical
   demonstration of developmental distribution of muscarinic acetylcholine receptors in rat
   parietal cortex. Dev Brain Res 84:185–191
Cajal SR (1909–1911) *Histologie du système nerveux de l'homme et des vertébrés*, 2 vols. (trans.
   L. Azoulay), repr. Instituto Ramón y Cajal del CSIC, Madrid, 1952–1955
Caroni P, Schwab ME (1988a) Two membrane protein fractions from rat central myelin with
   inhibitory properties for neurite growth and fibroblast spreading. J Cell Biol 106:1281–1288
Caroni P, Schwab ME (1988b) Antibody against myelin-associated inhibitor of neurite growth
   neutralizes non-permissive substrate properties of CNS white matter. Neuron 1:85–96
Carter HR, Wallace MA, Fain JN (1990) Purification and characterization of PLC-β_m, a muscarinic cholinergic regulated phospholipase C from rabbit brain membrane. Biochem Biophys Acta 1054:119–128
Casabona G, Knopfel T, Kuhn R, Gasparini F, Baumann P, Sortino MA, Copani A, Nicoletti F
   (1997) Expression and coupling to polyphosphoinositide hydrolysis of group I metabotropic
   glutamate receptors in early postnatal and adult rat brain. Eur J Neurosci 9:12–17
Cases O, Seif I, Grimsby J, Gaspar P, Chen K, Pournin S, Müller U, Aguet M, Babinet C, Shih JC,
   De Maeyer E (1995) Aggressive behaviour and altered amounts of brain serotonin and
   norepinephrin in mice lacking MAOA. Science 268:1763–1766
Cases O, Vitalis T, Seif I, De Maeyer E, Sotelo C, Gaspar P (1996) Lack of barrels in the
   somatosensory cortex of monoamine oxidadase A-deficient mice: role of a serotonin excess
   during the critical period. Neuron 16:297–307
Catalano SM, Shatz JC (1998) Activity-dependent cortical target selection by thalamic axons.
   Science 281:559–562
Catalano SM, Messersmith EK, Goodman CS, Shatz CJ, Chedotal A (1998) Many major CNS
   axon projections develop normally in the absence of semaphorin III. Mol Cell Neurosci
   11:173–182

Catalano S, Robertson RT, Killackey HP (1991) Early ingrowth of thalamocortical afferents to the neocortex of the prenatal rat. Proc Natl Acad Sci USA 88:2999–3003

Catalano SM, Robertson RT, Killackey HP (1996) Individual axon morphology and thalamocortical topography in developing rat somatosensory cortex. J Comp Neurol 367:36–53

Caviness VS Jr (1976) Patterns of cell and fiber distribution in the neocortex of the *reeler* mutant mouse. J Comp Neurol 170:435–448

Caviness VS Jr (1982) Neocortical histogenesis in normal and *reeler* mice: A developmental study based upon [3H] thymidine autoradiography. Dev Brain Res 4:293–302

Caviness VS Jr (1988) Architecture and development of the thalamocortical projection in the mouse. In: Bentivoglio M, Spreafico R (eds) Cellular Thalamic Mechanisms. Excerpta Medica, Amsterdam, pp 489–499

Caviness VS Jr, Frost DO (1980) Tangential organization of thalamic projections of the neocortex in the mouse. J Comp Neurol 194:355–367

Caviness VS Jr, Rakic P (1978) Mechanisms of cortical development: A view from mutations in mice. Ann Rev Neurosci 1:297–326

Caviness VS Jr, Sidman RL (1973) Time of origin of corresponding cell classes in the cerebral cortex of normal and *reeler* mutant mice: An autoradiographic analysis. J Comp Neurol 170:449–460

Caviness VS Jr, Pinto-Lord MC, Evrard P (1981) The development of laminated pattern in mammalian neocortex. In: Brinkley LL, Carlson BM, Connelly TG (eds) Morphogenesis and pattern formation, Raven Press, New York, pp 103–126

Caviness VS Jr, Crandall JE, Edwards MA (1988) The *reeler* malformation, implications for neocortical histogenesis. In: Jones EG, Peters A (eds) Cerebral cortex, vol 7: Development and maturation of cerebral cortex. Plenum Press, New York, pp 59–89

Chiaia NL, Fish SE, Bauer WR, Bennett-Clarke CA, Rhoades RW (1992) Postnatal blockade of cortical activity by tetrodotoxin does not disrupt the formation of vibrissa-related patterns in the rat's somatosensory cortex. Dev Brain Res 66:244–250

Chiaia NL, Fish SE, Bauer WR, Figley BA, Eck M, Bennett-Clarke CA, Rhodes RW (1994) Effects of postnatal blockade of cortical activity with tetrodotoxin upon the development and plasticity of vibrissa-related patterns in the somatosensory cortex of hamsters. Somatosens Mot Res 11:219–228

Chun JJM, Shatz CJ (1988a) A fibronectin-like molecule is present within the developing cat cerebral cortex and is correlated with subplate neurons. J Cell Biol 106:857–872

Chun JJM, Shatz CJ (1988b) Distribution of synaptic vesicle antigens is correlated with the disappearance of a transient synaptic zone in the developing cerebral cortex. Neuron 1:297–310

Chung WW, Lagenaur CF, Yan YM, Lund JS (1991) Developmental expression of neuronal cell adhesion molecules in the mouse neocortex and olfactory bulb. J Comp Neurol 314:290–305

Coggeshall RE (1964) A study of diencephalic development in the albino rat. J Comp Neurol 122:241–269

Cordery P, Molnár Z (1999) Embryonic development of connections in turtle pallium. J Comp Neurol 413:26–54

Crair MC, Malenka RC (1995) A critical period for long-term potentiation at thalamocortical synapses. Nature 375:325–328

Crandall JE, Hassinger LC, Bonacorso J (1992) Afferents to preplate neurons in embryonic mouse cortex. Soc Neurosci Abstr 18:778

Cullen MJ, Kaiserman-Abramoff IR (1986) Cytological organization of the dorsal lateral geniculate nuclei in mutant anophtalmic and postnatally enucleated mice. J Neurocytol 5:407–424

Dahme M, Bartsch U, Martini R, Anliker B, Schachner M, Mantei N (1997) Disruption of the mouse L1 gene leads to malformations of the nervous system. Nat Genet 17:346–349

D'Arcangelo G, Miao GG, Chen SC, Soares HD, Morgan JI, Curran T (1995) A protein related to extracellular matrix proteins deleted in the mouse mutant *reeler*. Nature 374:719–723

D'Arcangelo G, Nakajima K, Miyata T, Ogawa M, Mikoshiba K, Curran T (1997) Reelin is a secreted glycoprotein recognized by the CR-50 monoclonal antibody. J Neurosci 17:23–31

De Carlos JA, O'Leary DDM (1992) Growth and targeting of subplate axons and establishment of major cortical pathways. J Neurosci 12:1194–1211

Derer P, Derer M (1990) Cajal-Retzius cell ontogenesis and death in mouse brain visualised with horseradish peroxidase and electron microscopy. Neuroscience 36:839–856

Doron N, Wollenberg Z (1994) Cross-modal neuroplasticity in the blind mole rat *Spalax Ehrenbergi*: a WGA-HRP tracing study. Neuro Report 5:2697–2701

Dräger UC (1976) *Reeler* mutant mice: physiology in primary visual cortex. Exp Brain Res Suppl 1:274–276

Dudek S, Bear MF (1989) A biochemical correlate of the critical period for synaptic modification in the visual cortex. Science 246:673–675

Erzurumlu RS, Jhaveri S (1990) Thalamic axons confer a blueprint of the sensory periphery onto the developing rat somatosensory cortex. Dev Brain Res 56:229–234

Erzurumlu RS, Jhaveri S (1992) Emergence of connectivity in the embryonic rat parietal cortex. Cerebral Cortex 2:336–352

Fernandez AS, Pieau C, Repérant J, Boncinelli E, Wassef M (1998) Expression of the *Emx-1* and *Dlx-1* homeobox genes define three molecularly distinct domains in the telencephalon of mouse, chick, turtle and frog embryos: implications for the evolution of telencephalic subdivisions in amniotes. Development 101:2099–2111

Finney EM, Shatz CJ (1998) Establishment of patterned thalamocortical connections does not require nitric oxide synthase. J Neurosci 18:8826–8838

Flecher CF, Lutz CM, O'Sullivan TN, Saughnessy FD, Hawkes R, Frankel WN, Copeland NG, Jenkins NA (1996) Absence epilepsy in *tottering* mutant mice is associated with calcium channel defects. Cell 87:607–617

Fox K (1996) The role of excitatory amino acid transmission in development and plasticity of SI barrel cortex. Prog Brain Res 108:219–234

Fox K, Schlaggar BL, Glazewski S, O'Leary DDM (1996) Glutamate receptor blockade at cortical synapses disrupts development of thalamocortical and columnar organization in somatosensory cortex. Proc Natl Acad Sci USA 93:5584–5589

Friauf E, Shatz CJ (1991) Changing patterns of synaptic input to subplate and cortical plate. J Neurophysiol 66:2059–2071

Friauf E, McConnell SK, Shatz CJ (1990) Functional synaptic circuits in the subplate during fetal and early postnatal development of cat visual cortex. J Neurosci 10:2601–2613

Friedlander DR, Milev P, Karthikeyan L, Margolis RK, Margolis RU (1994) The neuronal chondroitin sulfate proteoglycan neurocan binds to neural cell adhesion molecules Ng-CAM/L1/NILE and N-CAM, and inhibits neuronal cell adhesion and neurite outgrowth. J Cell Biol 125:669–680

Frost DO, Caviness VS Jr (1980) Radial organization of thalamic projections to the neocortex in the mouse. J Comp Neurol 194:369–393

Fukuda T, Kawano H, Ohyama K, Li HP, Takeda Y, Oohira A, Kawamura K (1997) Immunohistochemical localisation of neurocan and L1 in the formation of thalamocortical pathway of developing rats. J Comp Neurol 382:141–152

Gheorghita-Bächler F, DuBois R, Welker E (1996) Morphology of thalamo-cortical axons in the somatosensory cortex (S1) of the mouse mutant *barrelless*. Eur J Neurosci Suppl 9:93

Godement P, Saillour P, Imbert M (1979) Thalamic afferents to the visual cortex in congenitally anopthalmic mice. Neurosci Lett 13:271–278

Godement P, Vanselow J, Thanos S, Bonhoeffer F (1987) A study in developing visual systems with a new method of staining neurones and their processes in fixed tissue. Development 101:697–713

Godfraind C, Schachner M, Goffinet AM (1988) Immunohistochemical localisation of cell adhesion molecules L1, J1, N-CAM and their common carbohydrate L2 in the embryonic cortex of normal and *reeler* mice. Dev Brain Res 42:99–111

Goffinet AM (1979) An early developmental defect in the cerebral cortex of the *reeler* mouse. Anat Embryol 157:205–216

Goffinet AM (1980) The cerebral cortex of the *reeler* mouse embryo: an electron microscopic analysis. Anat Embryol 159:199–210

Goffinet AM (1992) The *reeler* gene: a clue to brain development and evolution. Int J Dev Biol 36:101–107

Goffinet AM (1995) A real gene for *reeler*. Nature 374:675–676

Goffinet AM (1999) Evolution of mammalian cortical lamination: the Reelin/DAB1 pathway and cortical evolution. In: Cardew G, Bock G (eds) Evolutionary developmental biology of the cerebral cortex. John Wiley and Sons, Chichester (in press)

Goffinet AM, Lyon G (1979) Early histogenesis in the mouse cerebral cortex: a Golgi study. Neurosci Lett 14:61–66

Götz M, Novak N, Bastmayer M, Bolz J (1992) Membrane-bound molecules in rat cerebral cortex regulate thalamic innervation. Development 116:507–519

Götz M, Williams BP, Bolz J, Price J (1995) The specification of neural fate: a common precursor for neurotransmitter subtypes in the rat cerebral cortex in vitro. Eur J Neurosci 7:889–898

Guillery RW (1986) Neural abnormalities of albinos. Trends Neurosci 9:364–367

Guillery RW, Ombrellaro M, LaMantia AL (1985) The organization of the lateral geniculate nucleus and the geniculocortical pathway that develops without retinal afferents. Dev Brain Res 20:221–233

Hannan AJ, Kind PC, Blakemore C (1998a) Phospholipase C-β1 expression correlates with neuronal differentiation and synaptic plasticity in rat somatosensory cortex. Neuropharmacology 37:593–605

Hannan AJ, Blakemore C, Shin H-S, Kind PC (1998b) Phospholopase C-β1 is required for normal development of barrels in mouse somatosensory cortex. Soc Neurosci Abstr 24:59

Hannan AJ, Blakemore C, Katsnelson A, Huber K, Bear M, Kim D, Shin H-S, Kind PC (1999) Phospholipase C-β1 mediates activity-dependent differentiation in the cerebral cortex. (submitted)

Heil P, Bronchti G, Wollberg Z, Scheich H (1991) Invasion of visual cortex by the auditory system in the naturally blind mole rat. Neuroreport 2:735–738

Henderson TA, Woolsey TA, Jacquin MF (1992) Infraorbital nerve blockade from birth does not disrupt central trigeminal pattern formation in the rat. Dev Brain Res 66:146–152

Herrmann K, Antonini A, Shatz CJ (1994) Ultrastructural evidence for synaptic interactions between thalamocortical axons and subplate neurones. Eur J Neurosci 6:1729–1742

Hevner RF, Miyashita E, Martin G, Rubenstein JLR (1998) Lack of thalamocortical connections in mutants affecting cortical (TBR-1) or thalamic (GBX-2) gene expression. Soc Neurosci Abstr 24:58

Hickey TL (1980) Development of the dorsal lateral geniculate nucleus in normal and visually deprived cats. J Comp Neurol 189:467–481

Hickey TL, Hitchcock PF (1984) Genesis of neurons in the dorsal lateral nucleus of the cat. J Comp Neurol 228:186–199

Higashi S, Molnár Z, Kurotani T, Inokawa H, Toyama K (1996) Functional thalamocortical connections develop during embryonic period in the rat: an optical recording study. Soc Neurosci Abstr 22:1976

Hirotsune S, Takahara T, Sasaji N, Hirose K, Yoshiki A, Ohashi T, Kusakabe M, Murakami Y, Muramatsu M, Watanabe S et al. (1995) The *reeler* gene encodes a protein with an EGF-like motif expressed by pioneer neurons. Nat Genet 10:77–83

Honig MG, Hume RI (1986) Fluorescent carbocyanine dyes allow living neurons of identified origin to be studied in long-term cultures. J Cell Biol 103:171–187

Howell BW, Hawkes R, Soriano P, Cooper JA (1997) Neuronal position in the developing brain is regulated by mouse *disabled-1*. Nature 389:733–736

Iwasato T, Erzurumlu RS, Huerta PT, Chen DF, Sasaoka T, Ulupinar E, Tonegawa S (1997) NMDA receptor-dependent refinement of somatotopic maps. Neuron 19:1201–1210

Jeffery G (1997) The albino retina: an abnormality that provides insight into normal retinal development. Trends Neurosci 20:165–169

Jhaveri S, Erzurumlu RS, Crossin K (1991) Barrel construction in rodent neocortex: role of thalamocortical afferents vs. extracellular matrix molecules. Proc Natl Acad Sci USA 88:4489–4493

Jones EG (1983) Summing up. In: Macchi G, Rustioni A, Spreafico R (eds) Somatosensory integration in the thalamus. Elsevier, Amsterdam, pp 385–392

Jones EG (1985) The thalamus. Plenum Press, New York

Kageyama GH, Robertson RT (1993) Development of geniculocortical projections to visual cortex in rat: evidence for early ingrowth and synaptogenesis. J Comp Neurol 335:123–148

Kaiserman-Abramoff IR, Graybiel AM, Nauta WJH (1980) The thalamic projection of cortical area 17 in a congenitally anopthalamic mouse strain. Neuroscience 5:41–52

Karlsson U (1966) Observations on the postnatal development of neuronal structures in the lateral geniculate nucleus of the rat by electron microscopy. J Ultrastruct Res 17:158–175

Karlstrom RO, Trowe T, Bonhoeffer F (1997) Genetic analysis of axon guidance and mapping in the zebrafish. Trends Neurosci 20:3–8

Kawano H, Fukuda T, Kuto K, Horie M, Uyemura K, Takenchi K, Osumi N, Eto K, Kawamura K (1999) Pax-6 is required for thalamocortical pathway formation in fetal rats. J Comp Neurol 408:147–160

Killackey HP, Belford GR (1979) The formation of afferent patterns in the somatosensory cortex of the neonatal rat. J Comp Neurol 183:285–304

Killackey HP, Rhoads RW, Bennett-Clarke CA (1995) The formation of a cortical somatotopic map. Trends Neurosci 18:402–407

Kim D, Jun KS, Lee SB, Kang N-G, Min DS, Kim Y-H, Ryu SH, Suh P-G, Shin H-S (1997) Phospholipase C isozymes selectively couple to specific neurotransmitter receptors. Nature 389:290–293

Kind PC, Kelly GM, Fryer HJL, Blakemore C, Hockfield S (1997) Phospholipase C-β1 is present in the botrysome, an intermediate compartment-like organelle, and is regulated by visual experience in cat visual cortex. J Neurosci 17:1471–1480

Kostovic I, Rakic P (1980) Cytology and time of origin of interstitial neurons in the white matter in infant and adult human and monkey telencephalon. J Neurocytol 9:219–242

Kurukulasuriya NC, Salinger WL (1996) Error correction in *reeler* brain development. Soc Neurosci Abstr 22:972

Lebrand C, Cases O, Adelbrecht C, Doye A, Alvarez C, El-Mestikawy S, Seif I, Gaspar P (1996) Transient uptake and storage of serotonin in developing thalamic neurons. Neuron 17:823–835

Levitt P, Barbe MF, Eagleson KL (1997) Patterning and specification of the cerebral cortex. Annu Rev Neurosci 18:419–439

Li Y, Erzurumlu RS, Chen C, Jhaveri S, Tonegawa S (1994) Whisker-related neuronal patterns fail to develop in the trigeminal brainstem nuclei of NMDAR1 knockout mice. Cell 76:427–437

Lund RD (1965) Uncrossed visual pathways in the hooded and albino rats. Science 149:1506–1509

Lund RD, Bunt AH (1976) Prenatal development of central optic pathways in albino rats. J Comp Neurol 165:247–264

Lund RD, Mustari MJ (1977) Development of the geniculocortical pathway in rats. J Comp Neurol 173:289–305

Luskin MB, Shatz CJ (1985a) Neurogenesis of the cat's primary visual cortex. J Comp Neurol 242:611–631

Luskin MB, Shatz CJ (1985b) Studies of the earliest-generated cells of the cat's visual cortex: Cogeneration of subplate and marginal zones. J Neurosci 5:1062–1075

Maier DL, Mani S, Donovan SL, Soppet D, Tessarollo L, McCasland JS, Meiri KF (1998) Disturbed cortical map and absence of cortical barrels in growth-associated protein (GAP)-43 knockout mice. Proc Natl Acad Sci USA 96:9397–9402

Marin-Padilla M (1971) Early prenatal ontogenesis of the cerebral cortex (neocortex) of the cat (*Felis domestica*): a Golgi study. I. The primordial neocortical organization. Z Anat Entwicklungsgesch 134:117–145

McAllister JP II, Das GD (1977) Neurogenesis in the epithalamus, dorsal thalamus and ventral thalamus of the rat: an autoradiographic and cytological study. J Comp Neurol 172:647–686

McConnell SK (1995) Strategies for the generation of neuronal diversity in the developing nervous system. J Neurosci 15:6987–6998

McConnell SK, Ghosh A, Shatz CJ (1989) Subplate neurons pioneer the first axon pathway from the cerebral cortex. Science 245:978–982

Métin C, Godement P (1996) The ganglionic eminence may be an intermediate target for corticofugal and thalamocortical axons. J Neurosci 16:3219–3235

Miller B, Chou L, Finlay BL (1993) The early development of thalamocortical and corticothalamic projections. J Comp Neurol 335:16–41

Miller B, Sheppard AM, Bicknese AR, Pearlman AM (1995) Chondroitin sulfate proteoglycans in the developing cerebral cortex: the distribution of neurocan distinguishes forming afferent and efferent axonal pathways. J Comp Neurol 355:615–628

Miller MW (1988) Development of projection and local circuit neurons in neocortex. In: Peters A, Jones EG (eds) Cerebral cortex, vol 7: development and maturation of cerebral cortex. Plenum Press, New York, pp 133–175

Mione MC, Danevic C, Boardman P, Harris B, Parnavelas JG (1994) Lineage analysis reveals neurotransmitter (GABA or glutamate) but not calcium-binding protein homogeneity in clonally related cortical neurons. J Neurosci 14:107–123

Mitrofanis J (1992) Patterns of antigenic expression in the thalamic reticular nucleus of developing rats. J Comp Neurol 320:161–181

Mitrofanis J (1994) Development of the pathway from the reticular and perireticular nuclei to the thalamus in ferrets: a DiI study. Eur J Neurosci 6:1865–1882

Mitrofanis J, Guillery RW (1993) New views of the thalamic reticular nucleus in the adult and developing brain. Trends Neurosci 16:240–245

Molnár Z (1994) Multiple mechanisms in the establishment of thalamocortical innervation. DPhil Thesis, University of Oxford, Oxford, UK

Molnár Z (1998) Development of thalamocortical connections. Springer and R.G. Landes, Berlin Heidelberg New York

Molnár Z, Cordery P (1999) Connections between cells of the internal capsule, thalamus and cerebral cortex in embryonic rat. J Comp Neurol 413:1–25

Molnár Z, Blakemore C (1990a) Relationship of corticofugal and corticopetal projections in the prenatal establishment of projections from thalamic nuclei to the specific cortical areas of the rat. J Physiol 430:104

Molnár Z, Blakemore C (1990b) Development of thalamocortical connectivity in vivo and in vitro. Soc Neurosci Abstr 16:1007

Molnár Z, Blakemore C (1992) How are thalamocortical axons guided in the *reeler* mouse? Soc Neurosci Abstr 18:778

Molnár Z, Blakemore C (1995a) How do thalamic axons find their way to the cortex? Trends Neurosci 18:389–397

Molnár Z, Blakemore C (1995b) Guidance of thalamocortical innervation. In: Bock G, Cardew G (eds) Development of the cerebral cortex. Wiley, Chichester (Ciba Foundation Symposium, 193), pp 127–149

Molnár Z, Adams R, Blakemore C (1998a) Mechanisms underlying the establishment of topographically ordered early thalamocortical connections in the rat. J Neurosci 18:5723–5745

Molnár Z, Adams R, Goffinet AM, Blakemore C (1998b) The role of the first postmitotic cells in the development of thalamocortical fibre ordering in the *reeler* mouse. J Neurosci 18:5746–5785

Molnár Z, Bronchti G, Blakemore C, Welker E (1996) Initial topological order in thalamocortical projections in the *barrelless* mutant mouse. Soc Neurosci Abstr 22:1013

Molnár Z, Knott GW, Blakemore C, Saunders NR (1998) Development of thalamocortical projections in the South American grey short-tailed opossum (*Monodelphis domestica*). J Comp Neurol 398:491–514

Molnár Z, Higashi S, Adams R, Toyama K (2000) Earliest thalamocortical interactions. In: Kossut M (ed) The Barrel Cortex. F.P. Graham Publishing Co, London (in press)

Molnár Z, Mather NK, Katznelson A, Voelker C, Bronchti G, Welker E, Taylor JSH (1999) Disturbed fasciculation, but ordered cortical termination of thalamocortical projections in L1 knock-out mice. Soc Neurosci Abstr 25:1305

Naegele JR, Jhaveri S, Schneider GE (1988) Sharpening of topographical projections and maturation of geniculocortical axon arbors in the hamster. J Comp Neurol 277:593–607

Nelson SB, LeVay S (1985) Topographic organization of the optic radiation of the cat. J Comp Neurol 240:322–330

Ogawa M, Miyata T, Nakajima K, Yagyu K, Seike M, Ikenaka K, Yamamoto H, Mikoshiba K (1995) The *reeler* gene-associated antigen on Cajal-Retzius neurons is a crucial molecule for laminar organization of cortical neurons. Neuron 14:899–912

O'Leary DDM, Ruff NL, Dyck RH (1994) Development, critical period plasticity, and adult reorganizations of mammalian somatosensory systems. Curr Opin Neurobiol 4:535–544

Pini A (1993) Chemorepulsion of axons in the developing mammalian central-nervous-system. Science 261:95–98

Pinto-Lord MC, Caviness VS Jr (1979) Determinants of cell shape and orientations: a comparative Golgi analysis of cell-axon interrelationships in the developing neocortex of normal and *reeler* mice. J Comp Neurol 187:49–70

Rakic P (1976) Prenatal genesis of connections subserving ocular dominance in the rhesus monkey. Nature 261:467–471

Rakic P (1977) Prenatal development of the visual system in the rhesus monkey. Philos Trans R Soc Lond B Biol Sci 278:245–260

Rakic P (1995) Radial versus tangential migration of neuronal clones in the developing cerebral cortex. Proc Natl Acad Sci USA 92:11 323–11 327

Rakic P, Caviness VS Jr (1995) Cortical development: View from neurological mutants two decades later. Neuron 14:1101–1104

Rhodes RW, Fish SE (1983) Bilateral enucleation alters visual callosal but not cortico-tectal or corticogeniculate projections in hamsters. Exp Brain Res 51:451–462

Rice FL (1995) Comparative aspects of barrel structure and development. In: Jones EG, Diamond IT (eds) Cerebral cortex, vol 11: the barrel cortex of rodents. Plenum Press, New York, pp 1–75

Rice FL, Gómez C, Barstow C, Burnet A, Sands P (1985) A comparative analysis of the development of the primary somatosensory cortex: interspecies similarities during barrel and laminar development. J Comp Neurol 236:477–495

Rose JE (1942a) The ontogenetic development of the rabbit's diencephalon. J Comp Neurol 77:61–129

Rose JE (1942b) The thalamus of the sheep: cellular and fibrous structure and comparison with pig, rabbit and cat. J Comp Neurol 77:469–523

Schlaggar BL, O'Leary DDM (1991) Potential of visual cortex to develop an array of functional units unique to somatosensory cortex. Science 252:1556–1560

Schlaggar BL, O'Leary DDM (1993) Patterning of the barrel field in somatosensory cortex with implications for the specification of neocortical areas. Perspect Dev Neurobiol 1:81–91

Schlaggar BL, O'Leary DDM (1994) Early development of the somatotopic map and barrel patterning in rat somatosensory cortex. J Comp Neurol 346:80–96

Schlaggar BL, Fox K, O'Leary DDM (1993) Postsynaptic control of plasticity in developing somatosensory cortex. Nature 364:623–626

Schwab ME, Caroni P (1988) Oligodendrocytes and CNS myelin are nonpermissive substrates for neurite growth and fibroblast spreading in vitro. J Neurosci 8:2381–2393

Shatz CJ (1981) Inside-out pattern of neurogenesis of the cat's lateral geniculate nucleus. Soc Neurosci Abstr 7:140

Shatz CJ (1983) The prenatal development of the cat's retinogeniculate pathway. J Neurosci 3:482–499

Shatz CJ, Luskin MB (1986) Relationship between the geniculocortical afferents and their cortical target cells during development of the cat's primary visual cortex. J Neurosci 6:3655–3668

Shatz CJ, Ghosh A, McConnell SK, Allendoerfer KL, Friauf E, Antonini A (1990) Pioneer neurons and target selection in cerebral cortical development. Cold Spring Harbor Symp Quantit Biol 55:469–480

Sheldon M, Rice DS, D'Arcangelo G, Yoneshima H, Nakajima K, Mikoshiba K, Howell BW, Cooper JA, Goldwitz D, Curran T (1997) *Scrambler* and *yotari* disrupt the disabled gene and produce a reeler-like phenotype in mice. Nature 389:730–733

Sheppard AM, Hamilton SK, Pearlman AL (1991) Changes in the distribution of extracellular matrix components accompany early morphogenetic events of mammalian cortical development. J Neurosci 11:3928–3942

Skaliora I, Singer W, Betz H, Püschel AW (1998) Differential patterns of semaphorin expression in the developing rat brain. Eur J Neurosci 10:1215–1229

Soriano E, Dumesnil N, Auladell C, Cohen-Tannoudji M, Sotelo C (1995) Molecular heterogeneity of progenitors and radial migration in the developing cerebral cortex revealed by transgene expression. Proc Natl Acad Sci USA 92:11676–11680

Steffen H (1976) Golgi-stained barrel-neurones in the somatosensory region of the mouse cerebral cortex. Neurosci Lett 2:57–59

Steindler DA, Colwell SA (1976) *Reeler* mutant mouse: maintenance of appropriate and reciprocal connections in the cerebral cortex and thalamus. Brain Res 105:386–393

Stewart GR, Pearlman AL (1987) Fibronectin-like immunoreactivity in the developing cerebral cortex. J Neurosci 7:3325–3333

Stoykova A, Götz M, Gruss P, Price J (1997) Pax-6dependent regulation of adhesive patterning, R-cadherin expression and boundary formation in developing forebrain. Development 124:3765–3777

Street VA, Bosma MM, Demas VP, Regan MR, Lin DD, Robinson LC, Agnew WS, Tempel BL (1997) The type 1 inositol 1,4,5-triphosphate receptor gene is altered in the *opistotonus* mouse. J Neurosci 17:635–645

Sur M, Pallas SL, Roe AW (1990) Cross-modal plasticity in cortical development: differentiation and specification of sensory neocortex. Trends Neurosci 13:227–233

Tan SS, Faulkner-Jones B, Breen SJ, Walsh M, Bertram JF, Reese BE (1995) Cell dispersion patterns in different cortical regions studied with an X-inactivated transgenic marker. Development 121:1029–1039

Tear G (1999) Neuronal guidance: a genetic perspective. Trends Genet 15:113–118

Tonegawa S, Li Y, Erzurumlu RS, Jhaveri S, Chen C, Goda Y, Paylor R, Silva AJ, Kim JJ, Wehner JM et al. (1995) The gene knockout technology for the analysis of learning and memory, and neural development. Prog Brain Res 105:3–14

Trevelyan AJ, Thompson ID (1992) Altered topography in the geniculo-cortical projection of the golden hamster following neonatal monocular enucleation. Eur J Neurosci 4:1104–1111

Turlejski K, Djavadian RL, Kossut M (1997) Neonatal serotonin depletion modifies development but not plasticity in rat barrel cortex. Neuroreport 8:1823–1828

Uziel D, Welker E, Bolz J (1998) Role of cortical factors in the development of mouse barrelfield. Eur J Neurosci 10(S10):277

Valverde F, Facal-Valverde MV, Santacana M, Heredia M (1989) Development and differentiation of early generated cells of sublayer VIb in the somatosensory cortex of the rat: a correlated Golgi and autoradiographic study. J Comp Neurol 290:118–140

Valverde F, De Carlos JA, Lopez-Mascaraque L (1995a) Time of origin and early fate of preplate cells in the cerebral cortex of the rat. Cereb Cortex 5:483–493

Valverde F, López-Mascaraque L, Santacana M, De Carlos JA (1995b) Persistence of early-generated neurons in the rodent subplate: assessment of cell death in neocortex during the early postnatal period. J Neurosci 15:5014–5024

Van der Loos H, Woolsey TA (1973) Somatosensory cortex: structural alterations following early injury to sense organs. Science 179:395–376

Van der Loos H, Welker E, Dörfl J, Rumo G (1986) Selective breeding for variations in patterns of mystatial vibrissae of mice. Bilaterally symmetrical strains derived from ICR stock. J Hered 77:66–82

Walsh C, Cepko CL (1993) Clonal dispersion in proliferative layers of developing cerebral cortex. Nature 362:632–635

Ware M, Fox JW, Gonzàlez JL, Davis NM, Lambert de Ruvroit C, Russo CJ, Chua SC Jr, Goffinet AM, Walsh CA (1997) Aberrant splicing of a mouse *disabled* homolog, *mdab1*, in the *scrambler* mouse. Neuron 19:239–249

Welker E, van der Loos H (1986) Quantitative correlation between barrel-field size and the sensory innervation of the whiskerpad: a comparative study in six strains of mice bred for different patterns of mystical vibrissae. J Neurosci 6:3355–3373

Welker E, Rao SB, Dörfl J, Melzer P, Van der Loos H (1992) Plasticity in the barrel cortex of the adult mouse: Effects of chronic stimulation upon deoxyglucose uptake in the behaving animal. J Neurosci 12:153–170

Welker E, Armstrong-James M, Bronchti G, Ourednik W, Gheorghita-Bächler F, DuBois R, Guernsey DL, van der Loos H, Neumann PE (1996) Altered sensory processing in the somatosensory cortex of the mouse mutant *barrelless*. Science 271:1864–1867

Wise SP, Jones EG (1976) The organization and postnatal development of the comissural projection of the somatic sensory cortex of the rat. J Comp Neurol 168:313–344

Wise SP, Jones EG (1978) Developmental studies of thalamocortical and commissural connections in the rat somatic sensory cortex. J Comp Neurol 178:187–208

Woodward WR, Coull BM (1984) Localisation and organisation of geniculocortical and corticofugal fiber tracts within the subcortical white matter. Neuroscience 12:1089–1099

Woolsey TA, Van der Loos H (1970) The structural organization of layer IV in the somatosensory region (S1) of mouse cerebral cortex. Brain Res 17:205–242

Woolsey TA, Welker C, Schwartz RH (1975) Comparative anatomical studies of the SmI face cortex with special reference to the occurrence of "barrels" in layer IV. J Comp Neurol 164:95–104

Zhu XO, Waite PME (1998) Cholinergic depletion reduces plasticity of barrel field cortex. Cereb Cortex 8:63–72

Subject Index